SOLID STATE PHYSICS

VOLUME 45

Founding Editors

FREDERICK SEITZ

DAVID TURNBULL

SOLID STATE PHYSICS

Advances in Research and Applications

Editors
HENRY EHRENREICH
DAVID TURNBULL

Division of Applied Sciences
Harvard University, Cambridge, Massachusetts

VOLUME 45

ACADEMIC PRESS, INC.
Harcourt Brace Jovanovich, Publishers

Boston   San Diego   New York
London   Sydney   Tokyo   Toronto

This book is printed on acid-free paper. ∞

COPYRIGHT © 1991 BY ACADEMIC PRESS, INC.
ALL RIGHTS RESERVED.
NO PART OF THIS PUBLICATION MAY BE REPRODUCED OR
TRANSMITTED IN ANY FORM OR BY ANY MEANS, ELECTRONIC
OR MECHANICAL, INCLUDING PHOTOCOPY, RECORDING, OR
ANY INFORMATION STORAGE AND RETRIEVAL SYSTEM, WITHOUT
PERMISSION IN WRITING FROM THE PUBLISHER.

ACADEMIC PRESS, INC.
1250 Sixth Avenue, San Diego, CA 92101

United Kingdom Edition published by
ACADEMIC PRESS LIMITED
24–28 Oval Road, London NW1 7DX

LIBRARY OF CONGRESS CATALOG CARD NUMBER: 55-12200

ISBN 0-12-607745-2
ISSN 0081-1947

PRINTED IN THE UNITED STATES OF AMERICA
91 92 93 94   9 8 7 6 5 4 3 2 1

Contents

CONTRIBUTORS TO VOLUME 45 vii
PREFACE ix

Structural Ordering in Colloidal Suspensions

AJAY K. SOOD

I.	Introduction	2
II.	Synthesis and Characterization of Colloids	8
III.	Experimental Scene	12
IV.	Interparticle Interaction Potentials	19
V.	Theoretical Aspects of Equilibrium Phases	30
VI.	Liquidlike Order, Polydispersity, Binary Mixtures, and Glass	41
VII.	Colloidal Crystals in Two Dimensions	62
VIII.	Elastic Properties and Shear Flow	66
IX.	Miscellaneous and Conclusions	70
	Acknowledgments	73
	Note Added in Proof	73

Crystal Nucleation in Liquids and Glasses

K. F. KELTON

I.	Introduction	75
II.	Classical Theory: Steady-State Nucleation	83
III.	Classical Theory: Time-Dependent Nucleation	93
IV.	Experimental Studies of Nucleation in Undercooled Liquids	103
V.	Experimental Studies of Nucleation in Glasses	123
VI.	Computer Simulations of Steady-State Nucleation	155
VII.	Beyond Classical Nucleation Theory	159
VIII.	Concluding Remarks	172
	Appendix A: Approximations to ΔG	173
	Acknowledgments	177

Glass Transition and Relaxation of Disordered Structures

FUMIKO YONEZAWA

I.	Introduction	179
II.	General Discussions	181
III.	Glass Transition	186
IV.	Computer Simulations of Glass Transition	197
V.	Simulations of Glass Transition at Constant Pressure	202
VI.	Fluctuation and Relaxation in Disordered Systems	235
VII.	Summary and Future Work	254
	Author Index	255
	Subject Index	265
	Cumulative Author Index, Volumes 1–45	289

Contributors

Numbers in parentheses indicate the pages on which the authors' contributions begin.

K. F. KELTON (75), *Department of Physics, Washington University, St. Louis, Missouri 63130*

AJAY K. SOOD (1), *Department of Physics, Indian Institute of Science, Bangalore 560 012, India*

FUMIKO YONEZAWA (179), *Department of Physics, Keio University, Yokohama, Japan*

Preface

The three articles in this volume center largely on the structural relations between liquids, crystals, and glasses and the transitions between these states.

The first article, by A. K. Sood, reviews the structures formed by dilute suspensions of monodisperse polymer particles ("polyballs"), as well as the principles of colloid dispersion that account for their stability. Recently these suspensions have attracted much attention, because interparticle spacings of the order of the wavelengths of visible light are easily produced, so that the ordering of particles in the suspension is readily studied by the diffraction of light. Depending on preparation conditions, they may exhibit crystalline or liquid-like order, and even structural and dynamic properties characteristic of a glass. Consequently, as the Sood article emphasizes, these suspensions serve as excellent models for displaying the ordering, melting, and glass forming behavior of more conventional atomic systems. Valuable insights on these behaviors have been provided by the colloidal systems, and Sood notes that they also have, in some instances, displayed phenomena such as shear melting, which have not yet been seen in conventional atomic systems.

Experience indicates that the crystallization of liquids and glasses has always occurred by the heterogeneous mode of crystal nucleation and growth. The second article, by K. F. Kelton, presents a comprehensive review of experience and theory on the nucleation component of this process. More than a century ago, J. Willard Gibbs showed that, owing to the work required to form the interface with its parent phase, the nucleus of a new forming phase can become stable and grow only after reaching a critical size, including many atoms, corresponding to a large thermodynamic barrier to nucleation. From a strict thermodynamic standpoint, the phase change could then be initiated only at a thermodynamic stability limit where the interfacial work would disappear, or at the surface of appropriately chosen heterophase impurities on which the net interfacial work of nucleus formation could, as Gibbs also showed, vanish altogether.

Gibbs' later statistical theory implied that nuclei could form, well before a stability limit was reached, by structural fluctuations. However, it remained for Volmer and Weber, almost a half century later, to initiate a theory of the kinetics of nucleus formation by, in effect, combining the Gibbs thermodynamic and statistical treatments. This theory, as elaborated and improved by later investigators, has come to be known as the "classical nucleation theory." The major assumption of the classical theory is that the interfacial work and the free energy of transition, which determine the thermodynamic barrier to nucleation, can be identified with the macroscopic values of these quantities. This assumption should be valid at small departures from phase equilibrium, but under these conditions the calculated nucleation probabilities are usually many orders of magnitude lower than any that can be measured experimentally in the times and on the specimen volumes at our disposal. Thus, in the regime where the probabilities are measurable, the calculated numbers of atoms per nucleus are, typically, only a few hundred or fewer. Under these conditions, the correspondence of the interfacial work to its macroscopic value is doubtful, and the doubts have given rise to reservations and much controversy concerning the applicability of the classical theory. Nevertheless, experience on nucleation probability is often in qualitative, or even near-quantitative, agreement with the theory, and most experimentalists continue to describe their results in terms of the theory. Kelton reviews the classical theory, especially as it applies to the nucleation of crystals in undercooled liquids and glasses, with attention to its limitations and taking account of time-dependent, as well as steady-state, nucleation. Kelton also reviews the, by now, fairly extensive experience on crystal nucleation in liquids and glasses and its interpretation in terms of the classical theory.

The experience cited by Kelton indicates that, when free of heterophase impurities, most liquids do not exhibit measurable crystal nucleation unless they are deeply undercooled. This behavior strongly suggests that considerable reconstruction of the liquid short range order must attend crystal nucleation. As a result, cooling rates often can be imposed that are sufficient in magnitude to quench, with no appreciable crystal nucleation, the liquid to a glass state. Liquids that have been quenched to glasses are distributed over every binding category—covalent, ionic, metallic, and van der Waals. As yet there is no conclusive evidence that monatomic liquids have been quenched to glasses, but computer simulations reviewed by Fumiko Yonezawa in the final article of this volume have demonstrated glass formation by monatomic model systems. This article reviews simulations of the melt-glass transition and structural relaxation in the glass carried out by Yonezawa and her associates and other investigators. A striking feature of the simulations, described by Yonezawa, is their indication that icosahedral configurations are major components of the glass structures, in accord with an early suggestion of Sir Charles Frank. Consequently, formation of the close-packed crystal structures, indeed, requires substantial reconstruction of the local order in the glass.

Henry Ehrenreich
David Turnbull

Structural Ordering in Colloidal Suspensions

AJAY K. SOOD

Department of Physics
Indian Institute of Science
Bangalore, India

I.	Introduction	2
	1. Colloids in General and the Aims of This Review	2
	2. Model Colloidal Systems	3
	3. Comparison with Conventional Atomic Systems	6
II.	Synthesis and Characterization of Colloids	8
	4. Emulsion Polymerization	8
	5. Determination of Surface Charge and Surface Potential	10
III.	Experimental Scene	12
	6. Phase Diagram	12
	7. Growth Kinetics and Time Evolution	18
IV.	Interparticle Interaction Potentials	19
	8. DLVO Theory	19
	9. Electrostatic Interaction with Both Repulsion and Attraction	27
V.	Theoretical Aspects of Equilibrium Phases	30
	10. Fluid–Crystal and bcc–fcc Phase Transitions	30
VI.	Liquidlike Order, Polydispersity, Binary Mixtures, and Glass	41
	11. The Liquidlike Short-Ranged Order	41
	12. Polydispersity Effects	52
	13. Colloidal Alloys in Binary Mixtures	55
	14. Colloidal Glasses	56
VII.	Colloidal Crystals in Two Dimensions	62
	15. Structure of 2D Crystals	63
	16. Melting of 2D Colloidal Crystals	64
VIII.	Elastic Properties and Shear Flow	66
	17. Elastic Properties	66
	18. Shear Flows	68
IX.	Miscellaneous and Conclusions	70
	19. Some Topics Not Covered	70
	20. A Few Unsolved Problems	71
	21. Summary	72
	Acknowledgments	73
	Note Added in Proof	73

I. Introduction

1. Colloids in General and the Aims of This Review

Colloids define a range of systems containing dispersed particles that are larger than atomic dimensions but small enough to exhibit Brownian motion, which prevents appreciable sedimentation in normal gravity. The particle size ranges from a few tens of angstroms to a few microns. These ubiquitous dispersions fall into two categories, namely lyophilic (solvent loving) and lyophobic (solvent fearing).[1,2] The former are thermodynamically stable, and the free energy decreases on their formation from the starting components. Some examples are gelatin sol, solutions of proteins, synthetic polymers, biopolymers, soap, and microemulsions. On the other hand, the lyophobic colloids require extra free energy in their formation due to the large interfacial free energy between the particles and solvent. These colloidal particles require a protective mechanism against their agglomeration. The mechanism can be provided either by the electric charge on the surface, resulting in electrostatic repulsion between the particles (called charge stabilization), or by the adsorption or chemical binding of large molecules to the particles (called steric stabilization). Some examples of these are aqueous suspensions of charged polystyrene spheres, aqueous solutions of gold, AgI, Fe_2O_3 and SiO_2, dispersed paints and inks, and milk.

The multicomponent nature of the dispersions in terms of size and shape of the colloidal particles make their study very complicated and sometimes intractable from a physicist's point of view. Fortunately, over the past three decades or so, the synthesis of monodisperse, spherical polymeric particles with a very narrow size distribution[3-7] has paved the way to elucidate many novel and fascinating properties of the structure and dynamics of colloids under equilibrium and nonequilibrium conditions. Under suitable experimental conditions, the arrangement of colloidal particles can mimic the structural behaviour found in conventional atomic systems—gas, liquid,

---

[1] E. J. W. Verwey and J. Th. G. Overbeek, "Theory of the Stability of Lyophobic Colloids." Elsevier, New York, 1948.

[2] J. Th. G. Overbeek, in "Physics of Complex and Supermolecular Fluids" (S. A. Safran and N. A. Clark, eds.), p. 3. Wiley, New York, 1987.

[3] Polymer latexes of highly uniform particle size were accidently discovered in 1947 at the Dow Chemical Company, USA.

[4] J. W. Vanderhoff, H. J. van den Hul, R. J. Tausk, and J. Th. G. Overbeek, in "Clean Surfaces" (G. Goldfinger, ed.). Marcel Dekker, New York, 1970.

[5] H. J. van den Hul and J. W. Vanderhoff, in "Polymer Colloids" (R. M. Fitch, ed.), p. 1. Plenum, New York, 1971.

[6] I. M. Krieger and P. A. Hiltner, in "Polymer Colloids" (R. M. Fitch, ed.), p. 63. Plenum, New York, 1971.

[7] J. Hearn, M. C. Wilkinson, and A. R. Goodall, *Adv. Colloid Interface Sci.* **14**, 173 (1981).

crystal[6,8-11] and even glasses.[12-16] The objective of this review is to bring out how the model colloidal systems have opened up new vistas in condensed matter physics. The attention will be focussed primarily on structural ordering in aqueous suspensions of charged polystyrene spheres. No attempt will be made to present a complete historical survey of various developments in the ordering phenomena that are at the interface of statistical mechanics and colloidal science.

2. Model Colloidal Systems

a. Soft Colloidal Spheres

Most of the studies have been done on aqueous suspensions of charge-stabilised polystyrene spheres. Each of these colloidal particles (also termed macroions and nicknamed "polyballs") consists of a large number of styrene polymeric chains entangled in a coil. Each of these chains starts and ends with an acidic group, like $—KSO_4$. In a solvent like water with high dielectric constant ε (~ 80), the end surface groups dissociate and provide a large electrostatic negative charge per particle ($\sim 1000e$ for a particle diameter of 1000 Å, where e is the electron charge). The counterions (cations like K^+) liberated from the polyballs and additional ions present in the solvent form a cloud around each polyball and hence screen the Coulomb interaction between them. Figure 1 shows a schematic of different types of ions present in the suspension. In the simplest picture, the interaction between the polyballs over the relevant range of particle separation is predominantly a screened Coulomb repulsion of the form[1]

$$U(r) \sim \frac{Z_P^2 e^2}{\varepsilon} \left(\frac{\exp(\kappa a_P)}{1+\kappa a_P} \right)^2 \frac{e^{-\kappa r}}{r}, \qquad (2.1)$$

where κ^{-1} is the screening length given by

$$\kappa^2 = \frac{4\pi}{\varepsilon k_B T} \left[(n_P Z_P) e^2 + \sum_\alpha n_\alpha (z_\alpha e)^2 \right]. \qquad (2.2)$$

Here n_P is the number density of the polyballs, each having a surface charge

---

[8] P. A. Hiltner and I. M. Krieger, *J. Phys. Chem.* **73**, 2686 (1969).
[9] I. F. Efremov, in "Surface and Colloidal Science" (E. Matijevic, ed.), Vol. 8, Chap. 2. Wiley-Interscience, New York, 1976.
[10] P. Pieranski, *Contemp. Phys.* **24**, 25 (1983).
[11] C. A. Castillo, R. Rajagopalan, and C. S. Hirtzel, *Rev. Chem. Enging.* **2**, 237 (1984).
[12] H. M. Lindsay and P. Chaikin, *J. Chem. Phys.* **76**, 3774 (1982).
[13] R. Kesavamoorthy, A. K. Sood, B. V. R. Tata, and A. K. Arora, *J. Phys. C* **21**, 4737 (1988).
[14] A. K. Sood, *Rev. Solid State Sci.* **3**, 523 (1989).
[15] P. N. Pusey and W. van Megen, *Nature* **320**, 340 (1986).
[16] P. N. Pusey and W. van Megen, *Phys. Rev. Lett.* **59**, 2083 (1987).

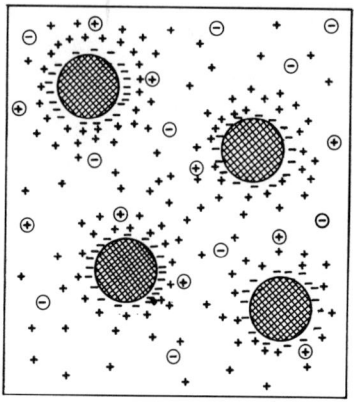

FIG. 1. Schematic illustration of the polyball colloidal suspensions. The large particles are the polyballs, each of diameter σ_P and carrying a charge $Z_P e$. The counterions, equal in number to $n_P Z_P$, are shown by $+$. The additional ions are shown by $\oplus$ and $\ominus$. The polyballs and the small ions are moving in a solvent with dielectric constant ε.

$Z_P e$ and radius a_P. The effect of the solvent is taken into account through its dielectric constant ε; T is the temperature, and k_B is the Boltzmann constant. The additional ions of type α with number density n_α and charge $z_\alpha e$ contribute to the screening in addition to the monovalent counterions whose number density is $n_P Z_P$. The potential given by Eq. (2.1) is called the Derjaguin–Landau–Verwey–Overbeek (DLVO) potential.[1] It is clear from Eqs. (2.1) and (2.2) that the strength and range of the repulsive interaction can be easily controlled by many experimental parameters, the most convenient of them being the polyball number density n_P and excess ion concentration n_α. The former is typically measured in terms of the volume fraction ϕ defined as $\phi = \pi \sigma_P^3 n_P / 6$, where the polyball diameter is $\sigma_P = 2 a_P$. When the screening length $\kappa^{-1} \ll a_S$ (the average interparticle distance), the interactions between the polyballs are negligible and particles perform free Brownian motion as in the gas phase of atomic systems. As interactions develop, the particles exhibit spatial correlations over two or three interparticle spacings (liquidlike short-range order), and for sufficiently strong interactions the colloidal particles freeze into a body-centered cubic (bcc) or face-centered cubic (fcc) structure with lattice constants of a few particle diameters (order of optical wavelengths).[10] When the suspension crystallizes, beautiful iridescence is observed due to Bragg diffraction of visible light.

A novel use of the colloidal crystalline arrays as Bragg diffraction devices to reject Rayleigh scattered light in Raman instrumentation has been developed.[17] Since the interaction can be fairly long ranged, the ordered phase can be formed in very dilute suspensions ($\phi \sim 0.005$).

In addition to the gas, liquid, and crystalline states, the glassy state can also be formed in the colloids.[14] This has been demonstrated recently in

[17] P. L. Flaugh, S. E. O'Donnell, and S. A. Asher, *Appl. Spectrosc.* **38**, 847 (1984).

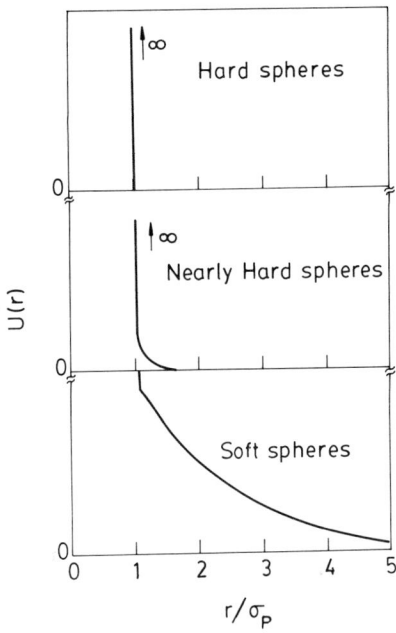

FIG. 2. A sketch of different types of interaction potentials $U(r)$: Hard spheres (as in billiard balls); nearly hard spheres (as in colloidal suspensions of sterically stabilised PMMA particles); soft spheres (as in aqueous suspensions of charged polystyrene spheres).

monodisperse colloids at high particle density[18] ($\phi > 0.2$) and in dilute binary mixtures containing two different particle diameters.[12,13] As in atomic systems, the glassy state is characterized by absence of Bragg diffraction, distortion in the static structure factor, finite rigidity to low-frequency shear, and much slower diffusion than in the liquid state.

Charge-stabilised monodisperse colloids can also be formed from silica spheres prepared by hydrolysis and subsequent polymerization of tetraethoxysilane (TES), and order–disorder transitions have been studied in these systems as well.[19–21]

b. *Nearly Hard Colloidal Spheres*

In addition to the charged colloids that have long-range interaction (soft spheres), there are two other model colloidal systems in which a number of interesting studies have been carried out. These systems interact through a

[18] E. B. Sirota, H. D. Ou-Yang, S. K. Sinha, P. M. Chaikin, J. D. Axe, and Y. Fujii, *Phys. Rev. Lett.* **62**, 1524 (1989).
[19] W. Strober, A. Fink, and E. Bohn, *J. Colloid Interface Sci.* **26**, 62 (1968).
[20] A. K. van Helden and A. Vrij, *J. Colloid Interface Sci.* **78**, 312 (1980); A. K. van Helden, J. W. Jansen, and A. Vrij, *J. Colloid Interface Sci.* **81**, 354 (1981).
[21] A. P. Philpse and A. Vrij, *J. Chem. Phys.* **88**, 6459 (1988).

steep repulsive potential and are good examples of nearly hard sphere systems. Figure 2 shows a schematic of interparticle potentials for hard sphere, nearly hard sphere, and soft sphere systems. Some examples of nearly hard sphere systems are (i) polymethylmethacrylate particles stabilised sterically by poly-12-hydroxystearic acid[14,15,22] and (ii) colloidal silica spheres sterically stabilised by stearyl chains grafted onto the surface.[23] The control parameter as a function of which one observes changes in structural ordering in nearly hard colloidal sphere systems is the volume fraction ϕ. The disorder–order transition occurs close to $\phi \sim 0.5$, in agreement with computer simulations.[24]

Ordering is seen not only in synthetic model colloidal systems but also in natural colloids. For example, bcc order had been observed for bushy stunt virus (almost spherically shaped particles) as far back as 1941.[25] The fcc closed-packed structure resulting in brilliant iridescence in suspensions of Tipula (iridescent) virus (having icosahedral structure) was reported by Williams and Smith.[26] There are a large number of other natural colloids like opals which show a fascinating variety of ordered structures.[10]

3. Comparison with Conventional Atomic Systems

1. In suspensions the colloidal particles are in equilibrium with a solvent with fixed chemical potential. The statistical mechanics of the colloidal suspensions can be carried out in the same way as in the case of the atomic liquids by treating the colloidal particles as "supramolecules" dispersed in a fluctuating background. The analogue of the "bare" potential between molecules is the potential of the average force between the macroions.[27,28]

2. An important difference between the conventional and colloidal systems is the length scale that makes the particle density differ by a factor $\sim 10^{10}$ (recall in atomic systems, $n_p \sim 10^{22} \, \text{cm}^{-3}$, whereas in the colloids $n_p \sim 10^{12} \, \text{cm}^{-3}$). The small particle density makes the direct measurement of thermodynamic properties like the equation of state and specific heat difficult (recall osmotic pressure $P \sim n_p k_B T$). Further the large difference in n_p between atomic and colloidal systems results in the same magnitude of

[22] L. Antl, J. W. Goodwin, R. D. Hill, R. H. Ottewill, S. W. Owens, and S. Papworth, *Colloids Surf.* **17**, 67 (1986).
[23] C. G. de Kruif, J. W. Jansen, and A. Vrij, in "Physics of Complex and Supermolecular Fluids" (S. A. Safran and N. A. Clark, eds.), p. 315. Wiley, New York, 1987.
[24] W. G. Hoover and F. H. Ree, *J. Chem. Phys.* **19**, 3609 (1968).
[25] J. D. Bernal and I. Fankuchen, *J. Gen. Physiol.* **25**, 111 (1941).
[26] R. C. Williams and K. Smith, *Nature* **179**, 119 (1957).
[27] W. G. McMillan and J. E. Mayer, *J. Chem. Phys.* **13**, 276 (1945).
[28] A. Vrij, E. A. Nieuwenhuis, H. M. Fijnaut, and W. G. M. Agterof, *Faraday Discuss. Chem. Soc.* **65**, 7 (1978).

difference between the elastic constants, colloidal crystals being $\sim 10^{10}$ weaker. This can be qualitatively understood by noting that the elastic constant $G \sim U n_p$, where U is the interaction energy between the particles.[29] The energy scale U is of the same order of magnitude (~ 10 eV) in both types of systems, as can be guessed from the fact that colloidal crystals are as stable at room temperature as the atomic solids. The typical range of elastic constants of the colloidal solids is 0.5 to 1000 dynes/cm^2. This value is so small that a simple jerk of the sample is sufficient to shear-melt the colloidal solids (crystals as well as glasses). Further an extremely low value of G results in many exotic nonlinear flow behaviours as a function of applied shear.[30-34]

3. Another difference between the colloids and conventional solids is the presence of intervening solvent resulting in Brownian dynamics rather than ballistic dynamics in the colloids. This does not influence the thermodynamic properties of the system but has consequences for the dynamics. In colloidal crystals the longitudinal sound modes are overdamped by backflow of the solvent, and only transverse phonons of small wavevector are underdamped.[35] This is in contrast to the usual crystals, where both longitudinal and transverse sound modes can propagate.

4. In spite of the progress made in the synthesis of colloidal particles, there is still a finite distribution, usually very narrow, in the size of the particles. The polydispersity in size as well as charge of the particles does not have an analogue in atomic systems. The polydispersity can have significant influence on the structure and dynamics of the colloids.[36-39]

5. It is instructive to compare the experiments on colloids with the molecular dynamics computer simulation "experiments," which are done to understand cooperative behaviour in condensed matter. Experiments on colloids have many of the advantages of computer experiments such as the ability to follow the motion in real space and time without disadvantages such as periodic boundary conditions, small numbers of particles, and the very long runs required for equilibration. The last of these becomes important in the study of the amorphous state and close to phase transitions.

[29] R. S. Crandall and R. Williams, *Science* **198**, 293 (1977).
[30] D. A. Weitz, W. D. Dozier, and P. M. Chaikin, *J. Phys. (Paris) Colloq.* **46**, C3-257 (1985).
[31] Jean-Marc di Meglio, D. A. Weitz, and P. M. Chaikin, *Phys. Rev. Lett.* **58**, 136 (1987).
[32] N. A. Clark and B. J. Ackerson, *Phys. Rev. Lett.* **44**, 1005 (1980).
[33] B. J. Ackerson and N. A. Clark, *Phys. Rev. Lett.* **46**, 123 (1981); *Physica A* **118**, 221 (1983).
[34] B. J. Ackerson and N. A. Clark, *Phys. Rev. A* **30**, 906 (1984).
[35] A. J. Hurd, N. A. Clark, R. C. Mockler, and W. J. O'Sullivan, *Phys. Rev. A* **26**, 2869 (1982).
[36] E. Dickinson, R. Parker, and M. Lal, *Chem. Phys. Lett.* **79**, 578 (1981).
[37] E. Dickinson and R. Parker, *J. Phys. Lett. (Paris)* **46**, L229 (1985).
[38] J. L. Barrat and J. P. Hansen, *J. Phys. (Paris)* **47**, 1547 (1986).
[39] P. N. Pusey, *J. Phys. (Paris)* **48**, 709 (1987).

II. Synthesis and Characterization of Colloids

4. EMULSION POLYMERIZATION

Polymeric monodisperse particles with diameter $<3\,\mu$m are synthesised by the emulsion polymerization technique.[7,40] The standard deviation in diameter defined as

$$\delta = [\langle \sigma_P^2 \rangle - \langle \sigma_P \rangle^2]^{1/2}/\langle \sigma_P \rangle, \tag{4.1}$$

where

$$\langle \sigma_P^n \rangle = \int \mathscr{P}(\sigma_P)\sigma_P^n \, d\sigma_P \tag{4.2}$$

is typically 1 to 2%. In Eq. (4.2), $\mathscr{P}(\sigma_P)$ is the distribution function for the diameters of the particles. Larger particles ($\sigma_P > 3\,\mu$m) are usually produced by suspension polymerization or by a newer and better process called swollen emulsion polymerization.[41] Following Bangs,[42] we shall briefly describe how the polyball colloids are synthesised by emulsion polymerization.

Surfactants like sodium dodecyl sulphate (SDS) are first dissolved in water in concentration above the critical micelle concentration. A surfactant molecule (shown in Fig. 3a) has a hydrophilic end (polar group like the sulphonic acid group in SDS) and a hydrophobic (hydrocarbon) tail. Above the critical micelle concentration, these surfactant molecules associate with one another to form spherical objects called micelles, wherein the hydrophobic tails are together in the center and the hydrophilic heads are on the surface of the sphere surrounded by the external water phase (Fig. 3b). The monodispersity arises naturally due to the tendency of the micelles to have the same number of surfactant molecules. After the micelles are formed, styrene in the monomer form is added, which enters and swells the micelles (Fig. 3c). The next step involves addition of a water-soluble polymerization initiator, such as potassium persulphate ($K_2S_2O_8$), that forms —SO_4K or sulphate-ion-free radicals (denoted by a dot . in Fig. 3d). The reaction starts by formation of oligomers with —KSO_4 as end groups (Fig. 3d). The reaction between the free radicals and styrene molecules occurs in micelles to scavenge the unreacted styrene to form polystyrene chains. The polymerization is terminated by interaction of two free radicals (Fig. 3d). Thus, each polymer chain has a —KSO_4 acidic group on both ends. The polymeric particle thus formed (Fig. 3e) has —KSO_4 groups and an adsorbed surfactant polar head

---

[40] H. J. van den Hul and J. W. Vanderhoff, *Brit. Polym. J.* **2**, 121 (1970).

[41] J. Ugelstad, P. C. Mork, A. Berge, T. Ellingsen, and A. A. Khan, in "Emulsion Polymerization" (I. Piirma, ed.), Chap. 11. Academic Press, New York, 1982.

[42] L. B. Bangs, "Uniform Latex Particles." Seradyn Inc., Indianapolis, 1987.

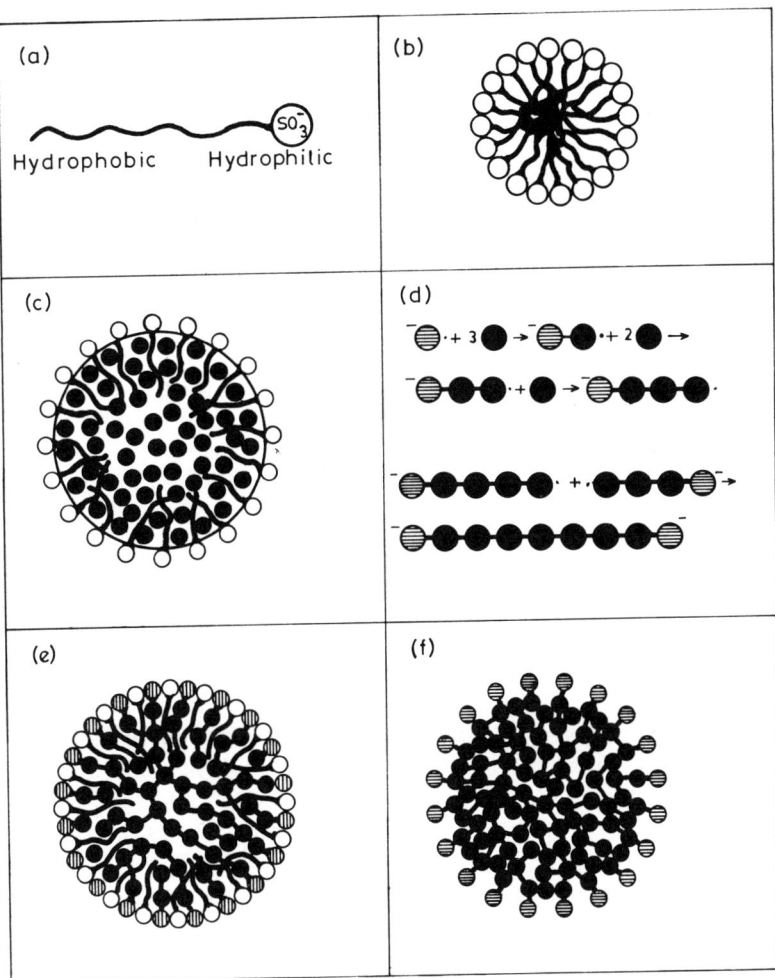

FIG. 3. Various steps in the synthesis of monodisperse polystyrene spheres. (a) A sulfonate surfactant molecule. The hydrophilic head and hydrophobic tail are marked. (b) Micelle. (c) Micelle swollen with styrene. (d) The shaded circle with a dot represents the free radical $-KSO_4$. If it is ionised, it carries a negative charge as shown here. The free radicals react with styrene molecules to form an oligomer, and the polymerization is terminated by interaction of two free radicals. (e) Uniform latex particle with surfactant still on its surface. (f) Latex particle after ion exchange clear-up. The shaded circles are either undissociated $-KSO_4$ or SO_4^- if the end groups are ionised. Taken from L. B. Bangs, "Uniform Latex Particles." Seradyn Inc., Publisher, 1987.

(sulphonate group) on the surface. The latter are easily removed from the surface by the action of ion exchange resins. The end groups when dissociated liberate counterions K^+ in the solvent and leave $-SO_4^-$ on the surface of the particle, resulting in net negative charge on the polyball.

All the sites on the surface of the particle are not dissociated. The degree of dissociation (f) varies from 0.05 to 0.4, generally decreasing with increasing surface charge.[5,43-45] The factor f depends on experimental parameters like the additional ion concentration n_α, temperature, and so on. Another way of saying the same thing is that a majority of counterions remain electrostatically bound to the macroion within a few atomic diameters of the polyball surface (the so-called Stern layer).[1,43,45] What enters in the interaction potential is the surface charge $Z_P e$ determined by the dissociated end groups. If Z is the total number of ionizable sites on the surface of the particle, then $Z_P = fZ$. It is essential to determine Z_P accurately in order to compare the theoretical predictions of the phase diagram, pair distribution function of the colloidal liquid, and so on, with the experiments.

The particle size, shape, and polydispersity can be directly measured by using a transmission electron microscope. The hydrodynamic radius a_H of the spherical polyballs can also be estimated with dynamic light scattering, which measures the diffusion constant D_0 of the polyballs ($D_0 = k_B T/6\pi\eta a_H$, where η is the viscosity of the solvent).[46]

5. Determination of Surface Charge and Surface Potential

a. Surface Charge

The surface charge $Z_P e$ can be determined by measuring the conductivity of the suspension as a function of the particle density.[43,47] The measured conductivity K has the form

$$K = (n_P Z_P |e| \Lambda(H^+))/N_{AV} + K_b, \qquad (5.1)$$

where $\Lambda(H^+)$ is the specific conductivity of protons, N_{AV} is Avogadro's number, and K_b is the background conductivity given by the initial ionic strength. A plot of K versus $n_P \Lambda(H^+)/N_{AV}$ yields $Z_P|e|$ from the slope and K_b from the intercept. In Eq. (5.1) $\Lambda(H^+)$ is used on the assumption that all the counterions have been converted into H^+ when the suspension is treated with a mixed bed of ion exchange resins. The error in Z_P is typically $\pm 25\%$. In another method, the pH of the purified suspension is measured[43] to estimate

---

[43] D. W. Schaefer, *J. Chem. Phys.* **66**, 3980 (1977).
[44] K. Ito, N. Ise, and T. Okubo, *J. Chem. Phys.* **82**, 5732 (1985).
[45] R. J. Vold and M. J. Vold, "Colloid and Interface Chemistry." Addison-Wesley, Reading, MA, 1983.
[46] B. J. Berne and R. Pecora, "Dynamic Light Scattering." Wiley, New York, 1976.
[47] Y. Monovoukas and A. P. Gast, *J. Colloid Interface Sci.* **128**, 533 (1989).

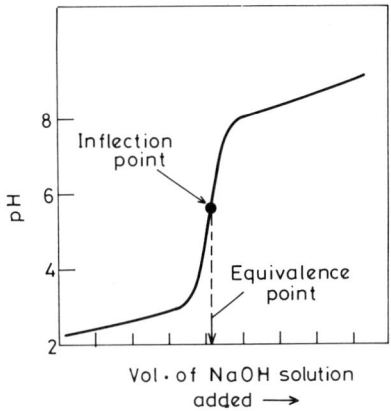

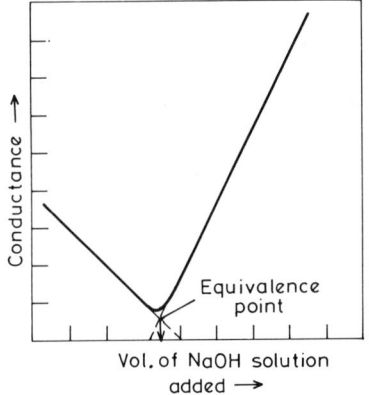

FIG. 4. (a) Typical conductometric titration and (b) potentiometric titration curves of polystyrene suspensions. The equivalent point gives the total number of sites on the surface of the polyball.

the H^+ concentration, which is again assumed to be equal to the counterion concentration. The error in Z_P is of the same order as in the conductivity measurement. The surface charge $Z_P e$ has also been determined by transference measurements.[44]

The total number of dissociable sites, Z, on the polyball surface can be estimated[5] from conductometric or potentiometric titration curves shown in Fig. 4. Knowing the equivalence point (marked in Figs. 4a and b), Z is given by[42]

$$Z = (1.004\pi) V_b C_b \rho_s \sigma_P^3 / W,$$

where V_b is the volume of base at the equivalence point (ml), C_b is the concentration of base (milliequivalent/ml), ρ_s is the polymer bulk density (g/ml), and W is the weight of particles titrated (g).

Recently Kesavamoorthy et al.[48] have demonstrated the use of Raman scattering to determine the surface charge Z_P, similar to its use in estimating the degree of dissociation in some polyelectrolytes. The degree of dissociation is estimated from the relative ratios of the intensities of the Raman modes at 1001 and 1032 cm^{-1} arising from dissociated and undissociated —SO$_4$K end groups. The charge on the polyball has been shown to decrease as ionic strength (n_α) increases. This can be qualitatively rationalised by arguing that the balance of the dissociation reaction —SO$_4$K $\leftrightarrows$ SO$_4^-$ + K$^+$ will shift to the left side due to the presence of excess ions. It was also shown[48] that the charge on the particle does not follow any particular dependence on its diameter. This is in contrast to the Pincus hypothesis that the charge on the polyball is proportional to its diameter.[49]

b. *Surface Potential ψ_0*

The potential on the surface ψ_0 of the particle is as important a parameter as the surface charge. There is no rigorous theory relating the surface charge to surface potential. The total charge measured by conductometric titration determines the surface potential ψ_0. The surface charge, which takes into account undissociated endgroups and counterions in the Stern layer, determines what is known as the outer Helmholtz layer ψ_δ.[5]

In a linearised Debye–Huckel theory, ψ_0 is related to Z_P in the simplest case by

$$\psi_0 = \frac{Z_P e}{a_P \varepsilon (1 + \kappa a_P)}. \qquad (5.2)$$

The surface potential that can be measured by electrophoresis is the zeta potential ψ_ζ. This is the potential at the surface called the shear plane that separates the colloidal particle, plus those ions that move with it, from the medium. The two potentials ψ_δ and ψ_ζ seem to be experimentally indistinguishable and are taken to be equal.[45]

III. Experimental Scene

6. PHASE DIAGRAM

a. *Polyball Suspensions: Soft Sphere Systems*

As can be seen from Eqs. (2.1) and (2.2), the experimental parameters that can be tuned to control the strength and range of the interparticle repulsive interaction are Z_P, ε, n_P, n_α, and, to some extent, temperature. The temper-

---

[48]R. Kesavamoorthy, T. Sakuntala, and A. K. Arora. To appear in *J. Phys. E.*
[49]P. Pincus, *Bull. Am. Phys. Soc.* (1983), as quoted in W. H. Shih and D. Stroud, *J. Chem. Phys.* **79**, 6254 (1983).

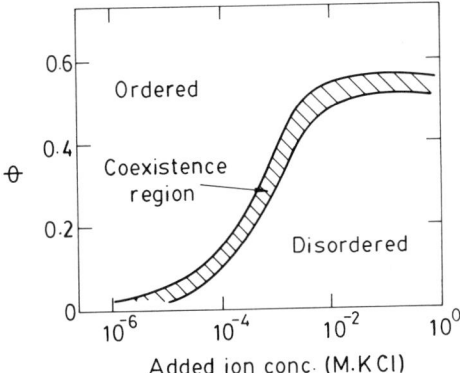

FIG. 5. Phase diagram of a monodisperse latex. The added ion concentration is expressed in mol/litre. Taken from S. Hachisu and Y. Kobayashi, *J. Colloid Interface Sci.* **46**, 470 (1974).

ature enters as the product $\varepsilon k_B T$ in determining κ. If the solvent is a noninteracting dipolar liquid, ε would follow a Curie–Weiss law and $\varepsilon k_B T$ would be independent of temperature. However, water molecules interact and $\varepsilon \sim T^{-1.5}$. Though this can have interesting consequences,[50] temperature is a weak variable. The dielectric constant ε and effective surface charge can be varied somewhat by using different solvents, but continuous variation over wide ranges is difficult. Therefore, as mentioned in Section 2, the most convenient experimental parameters are the particle density n_P or volume fraction ϕ and the excess ion concentration n_α. The latter can be easily altered by adding HCl or any salt, such as KCl, that dissociates into singly charged ions. The effect of divalent ions is different and has not been studied systematically.[51]

Many investigations have been made to observe the liquid (disordered) to crystal (ordered) transition and the structural transition from the bcc lattice to the fcc lattice. So far, structures other than bcc and fcc have not been seen in monodisperse colloidal suspensions. Some of the earlier observations of colloidal crystal formation were by Luck et al.,[52] Hiltner and Krieger,[6,8] and van den Hul and Vanderhoff.[5] A phase separation of aqueous suspensions of polyballs into ordered and disordered regions was observed by Hachisu et al.[53,54] The ordered phases always settled at the bottom of the sample cells and were recognised by their iridescence. Figure 5 shows their results.[53,54]

[50] P. M. Chaikin, P. Pincus, S. Alexander, and D. Hone, *J. Colloid Interface Sci.* **89**, 555 (1982).
[51] L. Gulbrand, B. Jonssons, H. Wennerström, and P. Linse, *J. Chem. Phys.* **80**, 2221 (1984); J. O. Bockris and A. K. N. Reddy, "Modern Electrochemistry". Plenum, New York, 1970.
[52] W. Luck, M. Klier, and H. Wesslau, *Ber. Bunseges, Phys. Chem.* **67**, 75 (1963).
[53] S. Hachisu, Y. Kobayashi, and A. Kose, *J. Colloid Interface Sci.* **42**, 342 (1973).
[54] S. Hachisu and Y. Kobayashi, *J. Colloid Interface Sci.* **46**, 470 (1974).

No structural information on the crystalline phase was obtained, and therefore the structural transformations occurring at low ionic strength within the ordered phase were not probed. Another deficiency in their study was the absence of data on surface charge of the polyballs, which prevents quantitative analysis of the data. Later, using reflection measurements, Takano and Hachisu[55] found that the coexistence region is narrower than that shown in Fig. 5.

The colloidal crystal structure determination, using Bragg scattering of visible laser light, showed that a bcc structure is formed at low volume fractions ($\phi < 0.0053$), and at $\phi = 0.0105$, bcc and fcc structures coexist.[56] The experiments were done on a polyball monodisperse suspension with particle diameter 1000 Å and charge of the order of $10^3 e$. Recently a comprehensive study of the crystal–liquid and bcc–fcc transitions has been presented for the well-characterized polyball suspension of particle diameter $\sigma_P = 1334 \pm 10$ Å and charge $1200 \pm 40e$. The crystal structures were identified by analyzing the Kossel lines, which were first observed in the colloidal crystals by Clark et al.[57] and emphasised by Pieranski et al.[10,58] in the accurate determination of the crystal structures. The Kossel diffraction images are formed by a divergent beam that can be produced either by placing a thin sheet of scatterer outside the sample container (pseudo-Kossel lines)[57,58] or by imperfections in the crystallites[59] (intrinsic Kossel lines). The latter method has been used to identify the phase transformation between fcc and bcc structures and fcc–bcc coexistence in the polyball colloidal crystals.[59] Returning to the results of Monovoukas and Gast, Fig. 6a shows their order–disorder phase diagram in the parameter space of ϕ and added ion concentration of KCl.[47] The full lines have been drawn as guides to indicate the approximate melting and freezing boundaries. The phase diagram in Fig. 6a, qualitatively similar to that of Hachisu et al. (Fig. 5),[53,54] shows that as the ionic strength (n_α) increases the ordered phase is formed at higher volume fractions. This can be understood qualitatively as follows. The increase in the ionic strength decreases the Debye screening length κ^{-1}, and hence the particle density or ϕ has to be increased to make the interparticle separation ($\sim n_P^{-1/3}$) comparable to κ^{-1}. The results of a detailed investigation of bcc–fcc transition occurring at very low volume fraction and ionic strength, marked by the shaded portion in Fig. 6a, is presented in Fig. 6b.[47] The bcc structure is stable for $\phi < 0.008$ and ionic strength lower than 2.76×10^{-6} moles/litre of KCl.

---

[55] K. Takano and S. Hachisu, *J. Colloid Interface Sci.* **66**, 124 (1978); 66, 130 (1978).
[56] R. Williams and R. S. Crandall, *Phys. Lett.* **48A**, 225 (1974).
[57] N. A. Clark, A. Hurd, and B. J. Ackerson, *Nature (London)* **281**, 57 (1979); B. J. Ackerson and N. A. Clark, *Phys. Rev. Lett.* **46**, 123 (1981).
[58] P. Pieranski, E. Dubois-Violette, F. Rothen, and L. Strzelecki, *J. Phys. (Paris)* **42**, 53 (1981).
[59] T. Yoshiyama, I. Sogami, and N. Ise, *Phys. Rev. Lett.* **53**, 2153 (1984).

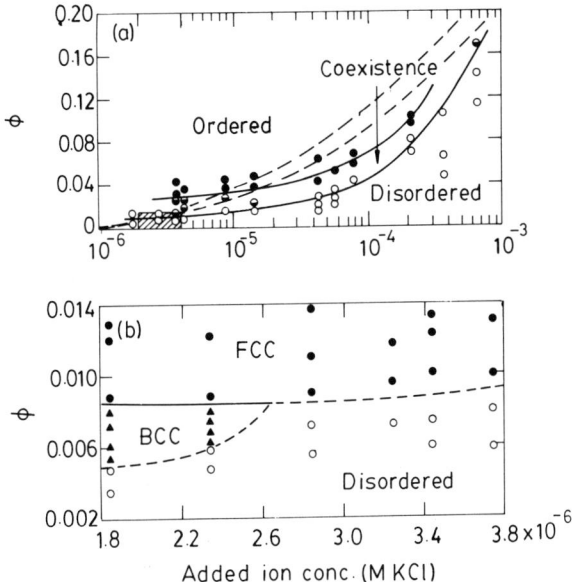

FIG. 6. (a) Experimental phase diagram showing ordered (crystal) and disordered (fluid) phases. Filled (open) circles correspond to crystals (disordered phase) occupying the whole volume of the suspension. The solid lines are drawn as guides to mark melting and freezing boundaries. The dotted lines are the calculated phase boundaries of freezing and melting, using the effective hard sphere model [see Eqs. (10.1) and (10.2)]. The hatched portion is shown in part (b). (b) Filled circles (triangles) correspond to the fcc (bcc) phase, and open circles indicate disordered samples. The broken line approximately marks the melting–freezing curves, and the solid line marks the fcc–bcc phase boundary. Taken from Y. Monovoukas and A. P. Gast, *J. Colloid Interface Sci.* **128**, 533 (1989).

Another recent determination of the phase diagram is the high-resolution, small-angle synchrotron x-ray study of a suspension of charged polystrene spheres of diameter 910 Å suspended in a solvent consisting of 90% methanol + 10% water (by volume).[18] The dielectric constant of the solvent is 38. The effective charge on the particles was estimated from the measured shear modulus of some of the crystalline samples. The shear modulus was 170 dynes/cm^2 for a sample of $\phi = 0.12$, which gives an effective charge of $\sim 135e$ per polyball. The different phases have been identified by the measured structure factors $S(Q)$, where Q is the wave vector covering the range 4×10^5 cm^{-1} to 3.4×10^6 cm^{-1}. Figure 7 shows the measured phase diagram, wherein liquid (open circles), bcc (solid squares), fcc (open triangles), coexistence of bcc and fcc (open squares), and glass (filled circles) phases are identified.[18] The solid lines, drawn as guides, mark the phase boundaries. We shall comment on the glass "phase" later. For very low ionic concentration of

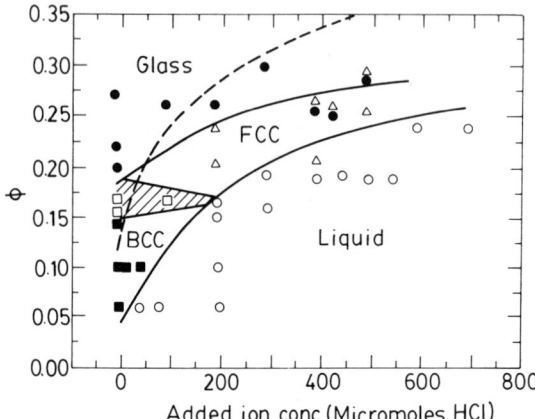

FIG. 7. Phase diagram of the polyballs suspension. Solid square: bcc crystal; open triangles: fcc crystal; open circles: liquid; filled circles: glass. The open squares in the hatched area mark the coexistence of bcc and fcc crystals. The solid lines are guides to mark phase boundaries. The dashed line is the fcc–liquid theoretical phase boundary from molecular dynamics simulations of Robbins et al. (Ref. 106 and see discussion in Section 10c). Taken from E. B. Sirota, H. D. Ou-Yang, S. K. Sinha, P. M. Chaikin, J. D. Axe, and Y. Fuji, *Phys. Rev. Lett.* **62**, 1524 (1989).

added HCl (less than 200 micromoles), the successive phases observed as ϕ is increased are liquid, bcc, fcc, and glass. The differences in the ranges of ϕ and ionic strength in Figs. 6 and 7 are due to the differences in the effective charges of the particles and the dielectric constants of the solvents.

b. *Nearly Hard Sphere Systems*

Pusey and van Megen[15,60] carried out a detailed study of the phase diagram of suspensions of sterically stabilised PMMA particles that interact through a steep repulsive potential. The solvent was a mixture of decalin and carbon disulphide with a composition chosen so as to match closely the refractive index of the particles. This index matching provides nearly transparent samples ideally suited for visual observation and light scattering measurements. Also the interparticle attractions due to van der Waals forces are minimized, leading to enhanced stability of the suspensions. As mentioned, the control parameter is the volume fraction.

Figure 8 shows the phase diagram as a function of the core volume fraction ϕ_c (related to the physical size of the PMMA particles). The ordinate is the fraction of the total sample volume occupied by the crystalline phase. The numbers on the arrows at the top of the figure indicate the samples in Ref. 15,

[60]P. N. Pusey and W. van Megen, in "Physics of Complex and Supermolecular Fluids" (S. A. Safran and N. A. Clark, eds.), p. 673. Wiley, New York, 1987.

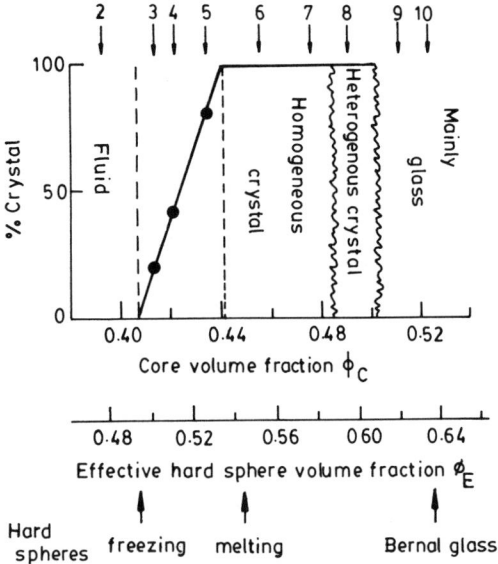

FIG. 8. Phase diagram of colloidal suspensions of sterically stabilised PMMA particles. The ordinate marks the fraction of the sample volume occupied by the crystalline phase. The core volume fraction ϕ_c is related to the actual size of the PMMA particles, and ϕ_E is the effective volume fraction. The arrows at the bottom mark the volume fractions obtained from computer simulations at freezing and melting transitions and for random close packing of hard spheres (i.e., Bernal glass). The arrows with numbers at the top of the figure refer to the sample number. Reprinted by permission from Nature vol. 320, pp. 340. Copyright © 1986, Macmillan Magazines Ltd.

wherein beautiful colored photographs of the iridescent colloidal crystals are presented. We refer the reader to Refs. 15, 60, and 61 for details. The freezing starts at $\phi_c = 0.407$. By equating this value to 0.494, the freezing concentration for hard spheres in computer simulations,[24] the effective hard sphere volume fraction ϕ_E is indicated in Fig. 8 ($\phi_E = 0.494\phi_c/0.407$).[15] The observed phase behaviour, fluid→fluid+crystal→crystal→random close-packed Bernal glass at the indicated ϕ_E values is in general agreement with the hard sphere model. Presumably the glass "phase" here is metastable, and the equilibrium phase is still crystalline.

An important difference between the conditions of the usual experiments on colloidal suspensions and those on atomic systems is that the former are done at constant volume (or volume fraction) and temperature, whereas the latter are studied at constant pressure and temperature. Therefore, in colloids, if the ϕ is between ϕ_{freezing} and ϕ_{melting}, the equilibrium state will consist of coexisting fluid and crystal phases in such a way that the osmotic pressure of the suspension will adjust to the value dictated by the equation of state. On the other hand, at constant pressure, it is difficult to set the temperature exactly at the melting temperature, and hence the equilibrium state of the system will not be in a coexistence state but in a single phase crystal or fluid.[61]

7. Growth Kinetics and Time Evolution

Slow dynamics in the colloids allows us to follow in real time the growth of the colloidal crystallites from the liquid phase. These studies throw light on the growth mechanisms of the ordered phase. In one such experiment, bcc colloidal crystals were shear-melted, and the subsequent recrystallization behaviour of the metastable liquid phase was studied by recording on video the size of the growing crystallites in real time.[62] It was found that the liquid–crystal interface was rough and that crystallites grew with a time-independent interface velocity. The limiting growth velocity, v_g ($\sim 2 \times 10^{-3}$ cm/s), is found to be the mean velocity for a particle freely diffusing on the surface of the crystallite over some distance d and is given by $v_g = 4D_0/d$, where $D_0 = k_B T/\sigma \pi \eta a_p$ is the single-particle diffusion coefficient. Note that it is the free-diffusion coefficient rather than the self-diffusion coefficient controlled by interparticle interactions that governs the growth velocity.[62]

The time evolution of the ordering proceeds by way of many intermediate steps, as has been recently demonstrated by analysis of the (intrinsic) Kossel diffraction pattern.[63] The successive steps are as follows[63]: (i) Two-dimensional hexagonal closed-packed (hcp) layers parallel to the cuvette (sample container) wall, (ii) random layered structure, (iii) layered structure with one sliding degree of freedom [e.g., sliding the two-dimensional hcp layers freely along one direction, [($1\bar{1}0$) or ($11\bar{2}$)] keeping the interplanar distance along (111) constant], (iv) stacking disorder structure in which two-dimensional hcp layers (usually termed A, B, and C) are stacked together in a random sequence, (v) ordered stacking structures with different periods, (vi) fcc structure with (111) twin, (vii) normal fcc structure, which was the terminal structure for $\phi > 0.03$. For dilute suspensions ($\phi < 0.02$), the next step in the order formation was bcc twin structure with twin plane ($11\bar{2}$) or ($\bar{1}12$), both of which are normal to the cuvette surface. After this step the normal bcc structure appeared. (It took several hours for dilute suspensions and minutes for concentrated suspensions after the samples were placed into the cuvettes, to observe the iridescence from the crystallites.)

The effect of the steepness of the repulsive part of the potential on the process of homogeneous nucleation has been shown to be significant in

---

[61] W. van Megen, P. N. Pusey, and P. Bartlett. *Phase Transitions.* **21**, 207 (1990).
[62] D. J. W. Aastuen, N. A. Clark, L. K. Cotter, and B. J. Ackerson, *Phys. Rev. Lett.* **57**, 1733 (1986).
[63] I. S. Sogami and T. Yoshiyama. *Phase Transitions.* **21**, 171 (1990).

computer simulations of crystal growth.[64] The incubation time for the onset of nucleation as well as the time required for the growth are smaller for softer potentials. These predictions are not easy to verify in real atomic systems since the atomic potentials cannot be altered much. However, the ease of tuning the interparticle potential in the colloids offers such a possibility. Recently Smits *et al.* have studied crystallization in sterically stabilised suspensions of colloidal silica particles with different adsorbed molecular chains (e.g., stearyl chains for steeper potential and polyisobutene for a less steep potential).[65] Their results confirm the computer simulation results.

IV. Interparticle Interaction Potentials

Theoretical analyses of the equilibrium structure and properties of the interacting colloidal suspensions require a reliable input of the interparticle interaction potential. The total interaction potential $U_T(r)$ is generally written[1] as a sum of the London–van der Waals attraction $U_A(r)$ and the electrostatic repulsion $U_R(r)$: $U_T(r) = U_A(r) + U_R(r)$, where r is the distance between the centres of the polyballs. Here the repulsive interaction arises due to the Coulomb interaction between the negatively charged polyballs mediated by small ions (counterions and additional ions, if present). This is the essence of the DLVO theory,[1] which has been used most extensively in the colloid literature. Later in this section, we shall discuss some recent calculations,[66-68] pointing out that the Coulomb interaction mediated by small ions does lead to both repulsive and attractive interparticle potentials. First we shall present results of the DLVO theory.[1,45]

8. DLVO Theory

a. *Repulsive Interaction*

The surface charge on the polyballs affects the distribution of small ions in the medium leading to the formation of an electric double layer around the former. The effect of electrostatic forces between the polyballs and ions (small ions will be referred to as ions) is counteracted by the thermal motion of the ions. Locally at some distance r from the surface of the polyball, the average concentration of the ions is governed by the Boltzmann distribution

$$n_i(r) = n_{i0} \exp[-z_i e\psi(r)/k_B T], \tag{8.1}$$

[64] R. D. Mountain and A. C. Brown, *J. Chem. Phys.* **80**, 2730 (1984).
[65] C. Smits, J. S. Van Duijneveldt, J. K. G. Dhont, H. N. W. Lekkerkerker, and W. J. Briels, *Phase Transitions* **21**, 157 (1990).
[66] I. Sogami, *Phys. Lett.* **A 96**, 199 (1983).
[67] I. Sogami and N. Ise, *J. Chem. Phys.* **81**, 6320 (1984).
[68] I. Sogami, in "Ordering and Organization of Ionic Solutions" (N. Ise and I. Sogami, eds.), p. 624. World Scientific, Singapore, 1988.

where $n_i(r)$ is the number density of ions of type i of valence z_i, and n_{i0} is the mean density of ions of type i. Here $\psi(r)$ is the electrostatic potential at r governed by the Poisson equation

$$\varepsilon \nabla^2 \psi(r) = -4\pi e \sum_i z_i n_i(r). \tag{8.2}$$

Combining Eqs. (8.1) and (8.2) gives the Poisson–Boltzmann (PB) equation. Assuming the valency of positive and negative ions to be equal to 1, the PB equation is

$$\varepsilon \nabla^2 \psi(r) = -4\pi e \sum_i \left[n_{i0} \exp\left(\frac{-e\psi}{k_B T}\right) - n_{i0} \exp\left(\frac{e\psi}{k_B T}\right) \right]. \tag{8.3}$$

The counterions released from the polyballs are positively charged with a number density $n_P Z_P$. There must be equal numbers of additional positive and negative ions to maintain charge neutrality. Therefore, in general, the concentration of positive ions will always be larger than that of the negative ions. Equation (8.3) is highly nonlinear and cannot be solved analytically in two and three dimensions. A great simplification is achieved when all electrostatic potential differences are much smaller than the thermal energy, $e\psi/k_B T \ll 1$. Then the PB equation can be approximated by its linearised form, the Debye–Huckel (DH) equation. Apart from getting a simple, analytical, and integrable potential, the DH equation has the basic feature of superposition inherent in linear equations. Then the interaction between the polyballs is an effective two-body potential.

The DH equation for the simple case of a symmetrical electrolyte is

$$\nabla^2 \psi = \kappa^2 \psi, \tag{8.4}$$

where

$$\kappa^2 = 4\pi e^2 (2n_0)/\varepsilon k_B T \tag{8.5}$$

and $2n_0$ is the average number density of small ions. The solution of Eq. (8.4) for a spherical particle of radius a_P, surface potential ψ_0, and surface charge $Z_P e$ along with the conditions $\psi = \psi_0$ at $r = a_P$ and $d\psi/dr \to 0$ as $r \to \infty$ is

$$\psi = \psi_0 (a_P/r) \exp\{-\kappa(r - a_P)\}. \tag{8.6}$$

Demanding that

$$\left(\frac{d\psi}{dr}\right)_{r=a_P} = -\frac{Z_P e}{\varepsilon a_P^2} \tag{8.7}$$

gives the relation between ψ_0 and Z_P expressed in Eq. (5.2).

When two polyballs are considered, their fields will be superimposed, causing an increase of the potential. So the relation between ψ_0 and Z_P will be different from Eq. (5.2). Further, one has to know what happens to ψ_0 and Z_P when two particles are brought close together. The expressions for the interaction potential are derived, assuming that either the potential ψ_0 or the charge $Z_P e$ remains constant.[1] If charge is assumed to remain constant during the approach of the polyballs, the potential energy of interaction is

$$U_R(r) = Z_P e [\psi(r) - \psi_\infty]. \qquad (8.8)$$

Verwey and Overbeek[1] have shown that for the case of small polyballs with extended double layers such that $\kappa a_P \lesssim 2.5$, $U_R(r)$ is given by

$$U_R(r) = \varepsilon a_P^2 \psi_0^2 \frac{\exp\{-\kappa(r - 2a_P)\}}{r} \varphi, \qquad (8.9)$$

where φ is a complicated function of κa_P and r/a_P that depends on whether the surface potential or charge is assumed to be constant [see Eqs. (81) and (83) of Ref. 1]. Because the parameter φ depends weakly on r and is always found to be between 0.6 and 1.0, it is generally taken to be equal to 1, giving the following simple approximate expression for U_R:

$$U_R(r) = \varepsilon a_P^2 \psi_0^2 \frac{\exp\{-\kappa(r - 2a_P)\}}{r}. \qquad (8.10)$$

The expression for κ is given by Eq. (2.2), wherein the explicit contributions of the monovalent counterions and additional ions of valency (usually 1) are displayed. For $\kappa a_P > 2.5$, the potential, with the assumption that ψ_0 remains constant, can be approximated by[1]

$$U_R^\psi(r) = \frac{\varepsilon a_P \psi_0^2}{2} \ln[1 + \exp\{-\kappa(r - 2a_P)\}]. \qquad (8.11a)$$

The corresponding interaction for the assumption of constant charge is[69]

$$U_R^Z(r) = U_R^\psi - \frac{\varepsilon a_P \psi_0^2}{2} \ln[1 - \exp\{-2\kappa(r - 2a_P)\}]. \qquad (8.11b)$$

The two potentials given by Eqs. (8.11a) and (8.11b) are numerically indistinguishable except at very short separations.

We shall discuss the case of small κa_P, which is relevant for dilute charged suspensions with low ionic strengths. In order to express $U_R(r)$ in terms of the surface charge eZ_P, we use the simplest approximate relation between ψ_0 and Z_P, given by Eq. (5.2), to get

$$U_R(r) = \frac{Z_P^2 e^2}{\varepsilon} \left(\frac{\exp(\kappa a_P)}{1 + \kappa a_P} \right)^2 \frac{e^{-\kappa r}}{r}. \qquad (8.12)$$

[69] G. R. Wiese and T. W. Healy, *Trans. Faraday Soc.* **66**, 490 (1970).

Very often in the literature, the repulsive potential $U_R(r)$ is taken to be as in Eq. (8.12) but without the geometrical factor (GF) $[\exp(\kappa a_P)/(1+\kappa a_P)]^2$; that is,

$$U_R^Y(r) = \frac{Z_P^2 e^2}{\varepsilon} \frac{e^{-\kappa r}}{r}, \qquad (8.13)$$

where the superscript Y reminds us that the potential is referred to as a Yukawa potential, whereas Eq. (8.12) is referred to as the DLVO potential.[70] The omission of the GF does not make any significant difference for very dilute suspensions having $\kappa a_P \ll 1$. The GF takes into account that the part of the volume of the suspension occupied by the polyballs is not available to the screening small ions. Further, its inclusion is like incorporating the hard core repulsion since the magnitude of $U_R(r = a_P) \gg k_B T$. Some theoretical predictions using the Yukawa potential can be artifacts of the neglect of the GF. Two such artifacts are as follows: (i) Shih and Stroud have shown that the use of the Yukawa potential predicts a reentrant melting behaviour; that is, for fixed additional ion concentration, the liquid first crystallizes, then remelts as ϕ is increased. No such reentrance is seen when the GF is incorporated. (ii) Molecular dynamics computer simulations have calculated the equation of state (i.e., compressibility $PV/Nk_B T$ as a function of ϕ) for both the DLVO potential and the Yukawa potential.[70] Here P is the osmotic pressure and $n_P = N/V$. Figure 9 shows the results for the liquid and solid branches for the DLVO potential (top curves) and of the liquid branch for the Yukawa potential (bottom curve).[70] The compressibility shows a minimum at $\phi \sim 0.02$ in the DLVO case, and for $\phi > 0.02$, the compressibility behaves as in a conventional atomic liquid since the GF essentially mimics the hard core part of the repulsive potential. However, the use of the Yukawa potential yields a monotonic decrease of the compressibility, implying an unphysical result that the liquid is infinitely compressible.

b. London–van der Waals Attraction

The attractive potential due to the London–van der Waals interaction is given by[71]

$$U_A(r) = -\frac{A}{6}\left[\frac{2a_P^2}{r^2 - 4a_P^2} + \frac{2a_P^2}{r^2} + \ln\left(\frac{r^2 - 4a_P^2}{r^2}\right)\right], \qquad (8.14)$$

where A is the Hamaker constant, which depends on the difference between the polarizabilities of the polyballs and the solvent. It can be made smaller by matching the refractive indices of the particles and the solvent. For more

---

[70]R. O. Rosenberg and D. Thirumalai, *Phys. Rev. A* **36**, 5690 (1987).
[71]H. C. Hamaker, *Physica* **4**, 1058 (1937).

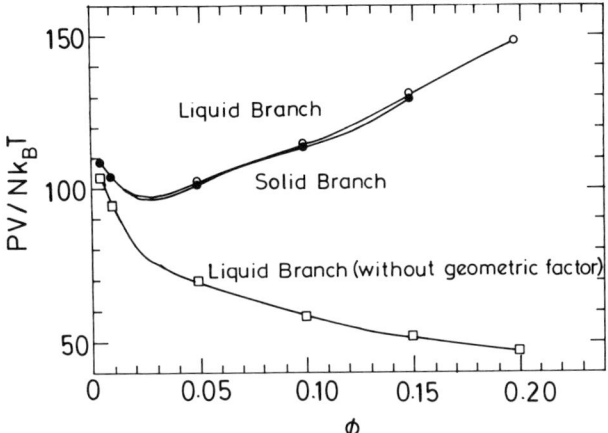

FIG. 9. Equation of state for the polyballs calculated from MD simulations for the DLVO potential (top two curves) and the Yukawa potential (lower curve). Taken from R. O. Rosenberg and D. Thirumalai, *Phys. Rev. A* **36**, 5690 (1987).

recent and rigorous formulations of the attractive interaction, we refer to many papers cited by Pailthorpe and Russel.[72] For completeness, we may add that the van der Waals interaction can lead to a repulsive potential if the Hamaker constant A is different for the two interacting particles, one lower and one higher than that of the medium. Equation (8.14) ignores the effects of retardation, which have been incorporated later.[73]

Figure 10 shows a plot of $U_T(r)/k_B T$ versus r/σ_P for a suspension with the parameters $\sigma_P = 1000$ Å, $Z_P = 600$, $n_P = 5 \times 10^{12}$ cm^{-3}; that is, $\phi = 0.0026$, $n_\alpha = 10 n_P Z_P$, $\varepsilon = 78$, and $A = 10^{-13}$ erg. The value of κa_P is 0.86, and $U_R(r)$ is given by Eq. (8.12). The radius of average interparticle spacing $a_s = n_P^{-1/3}$ is $5.8\sigma_P$. When the particle surfaces are less than 6 Å apart, attraction dominates. The primary minimum near $r = \sigma_P$ has a depth of more than $1000 k_B T$, and the repulsive barrier of height $716 k_B T$ occurs at $r = 1.007\sigma_P$. Thus the particles cannot overcome the repulsive barrier and are prevented from seeing the primary minimum resulting in stable dispersions. Since the attractive interaction falls slower ($\sim r^{-6}$) than the repulsive interaction ($\sim e^{-\kappa r}$), the total potential energy shows a negative portion, called the secondary minimum. It is shown in the inset of Fig. 11. The secondary minimum occurs at $\sim 15\sigma_P$ and has a depth of $\sim 4 \times 10^{-9} k_B T$, too small to bind. When the ionic strength of the suspension increases (κa increases)

[72] B. A. Pailthorpe and W. B. Russel, *J. Colloid Interface Sci.* **89**, 566 (1982).
[73] J. H. Schenkel and J. A. Kitchener, *Trans. Faraday Soc.* **56**, 161 (1960).

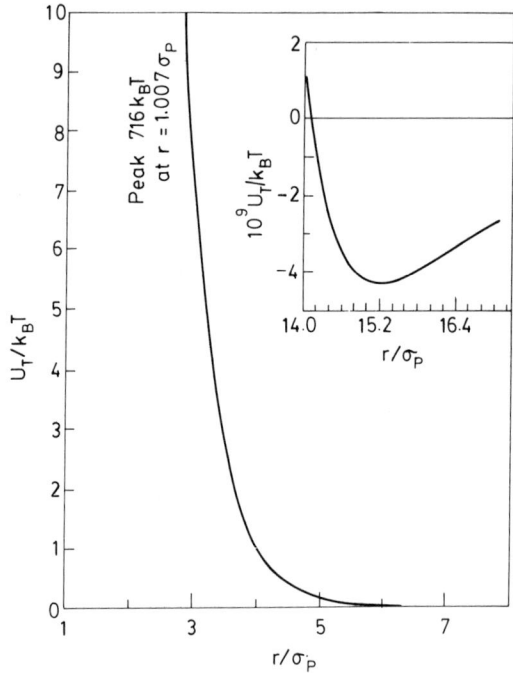

FIG. 10. The calculated total potential energy $U_T(r) = U_A(r) + U_R(r)$, where $U_A(r)$ and $U_R(r)$ are given by Eqs. (8.14) and (8.12), respectively. The parameters of the suspension are $\sigma_P = 1000$ Å, $Z_P = 600$, $n_P = 5 \times 10^{12}$ cm^{-3}, $n_\alpha = 10 n_P Z_P$, $\varepsilon = 78$, and $A = 10^{-13}$ erg. The inset shows the secondary minimum.

sufficiently, the repulsive barrier height is reduced to less than $k_B T$. The particles then feel the strong attraction of the primary minimum, and the suspension coagulates. In stable suspensions, particles never sample the primary minimum, and hence $U_R(r)$ is the dominant interaction.

In the foregoing discussion, the solvent is approximated by a medium of dielectric constant ε, and the small ions are taken to be point charges. At very high volume fractions, the discreteness of the solvent and counterions cannot be neglected.[74]

c. *Applicability of DLVO Theory and Effective Charge*

Linearization of the Poisson–Boltzmann equation, which yields the DLVO potential, is valid if $e\psi < k_B T$. The largest value of ψ is ψ_0, and

---

[74] M. Lozada-Cassou and D. Henderson, *Chem. Phys. Lett.* **127**, 392 (1986).

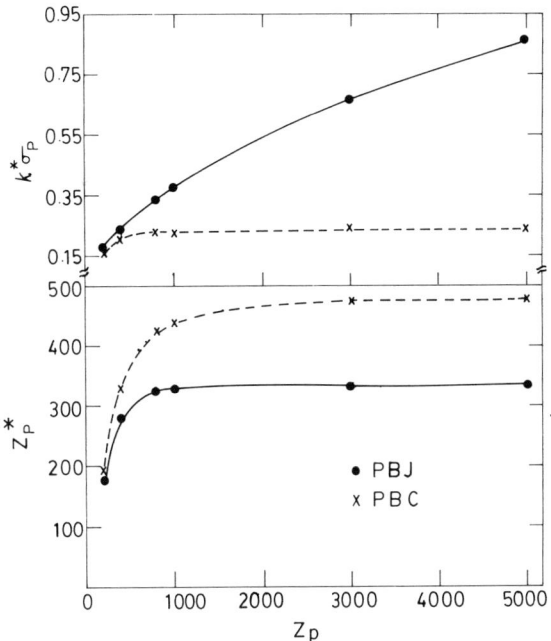

FIG. 11. Effective charge Z_P^* and effective screening parameter $K^*\sigma_P$ versus the basic charge Z_P, calculated in PBJ and PBC models. Data taken from W. Hartl and H. Versmold, *J. Chem. Phys.* **88**, 7157 (1988).

therefore the condition $e\psi_0 < k_B T$, or, using Eq. (5.2),

$$\frac{Z_P e^2}{a_P \varepsilon (1 + \kappa a_P) k_B T} < 1 \qquad (8.15)$$

must be satisfied for the DLVO potential in Eq. (8.12) to hold good. This condition [Eq. (8.15)] is often violated in polyball systems. However, away from the polyball surfaces, condition $e\psi < k_B T$ is met and the PB equation can be linearised. It has been argued[75–77] that in dilute colloidal suspensions the interaction U_R can still be described by the DLVO potential [Eq. (8.12)] even though the condition given by Eq. (8.15) is not satisfied. This is achieved by replacing Z_P and κ by the effective charge Z_P^* and a rescaled Debye screening parameter κ^*. The qualitative reasoning is that near the

[75] B. Beresford-Smith, D. Y. C. Chan, and D. J. Mitchell, *J. Colloid Interface Sci.* **105**, 216 (1985).
[76] S. Alexander, P. M. Chaikin, P. Grant, G. J. Morales, P. Pincus, and D. Hone, *J. Chem. Phys.* **80**, 5776 (1984); L. Belloni, M. Drifford, and P. Turq, *Chem. Phys.* **83**, 147 (1984).
[77] W. Härtl and H. Versmold, *J. Chem. Phys.* **88**, 7157 (1988).

particle surface where the condition in Eq. (8.15) does not hold, the counterions can be strongly bound (binding energy $> k_B T$) to the surface and therefore the charge is renormalised downward. Two approaches have been taken to calculate the effective charge Z_P^* (or the effective surface potential ψ_0^*):

1. Poisson–Boltzmann–cell (PBC) model[76,77]
2. Poisson–Boltzmann–jellium (PBJ) approximation.[75,77]

In the PBC model the polyballs are considered to be uniformly distributed in the suspension in the form of a regular lattice. Each polyball is confined in a spherical Wigner–Seitz cell that is charge neutral to have overall charge neutrality. Since the electric field on the spherical cell surface with no net charge inside is zero (Gauss's law), the electrostatic potential ψ can be taken to be zero on the cell surface. The linearization approximation is thus valid near the cell surface as long as the potential obtained by integrating the electric field remains smaller than the thermal energy. This ensures that the form of the potential $e^{-\kappa r}/r$ given by the DH equation is correct. The coefficients in the renormalised DH equation are obtained by matching the solution for ψ and its first three derivatives with the numerical solution of the PB equation in the cell. Figure 11 shows the Z_P^* and $\kappa^* \sigma_P$ as a function of the bare charge Z_P for a suspension of particle diameter $\sigma_P = 750$ Å and $n_P = 2.94 \times 10^{12}$ cm^{-3} ($\phi = 0.00065$) in both models.[77] It can be seen from Fig. 11 that Z_P^* approaches a saturation value $(Z_P^*)_{max}$ as Z_P increases. This maximum value was found by Alexander et al.[76] to be $\sim 15 a_P/l_B$, where $l_B = e^2/\varepsilon k_B T$ is the Bjerrum length (~ 7 Å for water at room temperature). The maximum value Z_P^* in Fig. 11 in the PBC model is, however, $\sim 9 a_P/l_B$.[77] The charge renormalization is small for $Z_P/(Z_P^*)_{max} \lesssim 0.4$. The effect of additional ions is not significant; Z_P^* changes by approximately 10% for added ions up to five times the concentration of the counterions.[76]

In the jellium model it is assumed that apart from the polyball under consideration, the remaining polyballs are replaced by a continuous medium (jellium) having the same average charge per unit volume. The PB equation with the jellium approximation becomes[75,77]

$$\varepsilon \nabla^2 \psi = -4\pi e n_P Z_P, \quad r < a_P,$$

$$= -4\pi e \left[n_P Z_P + \sum_i n_i z_i \exp\left(-\frac{z_i e \psi(r)}{k_B T}\right) \right], \quad r > a_P, \quad (8.16)$$

where i refers to the small ions. Equation (8.16) is solved numerically with the boundary conditions that $\psi(r) \to 0$ as $r \to \infty$ and $(d\psi/dr)_{r=a_P}$ is determined by Z_P [cf. Eq. (8.7)]. The solution of Eq. (8.16) has the asymptotic form

$$\psi(r) \to \frac{e\bar{\psi}(r)}{\varepsilon k_B T} \frac{e^{-\kappa r}}{r}. \quad (8.17)$$

The quantity $e\bar{\psi}(r)$ has the meaning of the Z_P^*. The results obtained by Hartl and Versmold[77] in the PBJ approximation are plotted in Fig. 11. The saturation value $(Z_P^*)_{max}$ is lower, and $K^*\sigma_P$ increases faster than that in the PBC model. Comparison of the particle density dependence of the measured structure factor $S(Q)$ of the colloidal liquids with the theoretical calculations suggests that the PBC model is better than the PBJ approximation.

The exact relation between the experimentally measured surface charge by conductivity measurements [Eq. (5.1)] and Z_P^* is not clear. It has been suggested that the shear modulus of the colloidal crystal can be used to estimate Z_P^* [see Eq. (17.1)].[18]

9. ELECTROSTATIC INTERACTION WITH BOTH REPULSION AND ATTRACTION

In recent times there has been considerable controversy on the important issue of whether the electrostatic interaction between the polyballs is purely repulsive of the DLVO form (as given in the previous section) or has an attractive part as well. In order to avoid confusion, it may be restated that we are not discussing the London–van der Waals attractive interaction that results in primary and secondary minima in the total potential. The experimental observations on the polyball suspensions that have been thought to support the existence of a long-range weak attraction in addition to the short-range repulsion are the following: (i) Ise and his group[59,78,79] have observed a coexistence of localised ordered structures and disordered regions. The interparticle spacing in the ordered structure is found to be less than the average value expected from the particle number density. (ii) Arora et al.[80] have observed a novel reentrant phase behaviour. A homogeneous suspension with complete disorder (gaslike) exhibits phase separation into a rare (low polyball concentration) and a dense (high polyball concentration) phase as the concentration of addition ions n_α is reduced. The suspension once again becomes homogeneous (with liquidlike or crystal-like order) on further reduction in n_α. (iii) Using optical microscopy, Kesavamoorthy et al.[81] have reported the presence of stable voids in the crystalline ordered phase. All these observations cannot be explained without invoking the presence of an attractive electrostatic interaction.

Sogami[66] and Sogami and Ise[67] have developed a theory of the electrostatic interaction in a macroionic solution. It is based on a calculation of the total electrostatic energy E of the macroionic solution by solving the linearised PB equation. The Helmholtz free energy F obtained from E gives a

---

[78]N. Ise, T. Okubo, M. Sugiura, K. Ito, and H. J. Nolte, *J. Chem. Phys.* **78**, 536 (1983).

[79]For a recent review, see N. Ise, H. Matsuoka and K. Ito, *Macromolecules* **22**, 1 (1989).

[80]A. K. Arora, B. V. R. Tata, A. K. Sood, and R. Kesavamoorthy, *Phys. Rev. Lett.* **60**, 2438 (1988).

[81]R. Kesavamoorthy, M. Rajalakshmi, and C. Babu Rao, *J. Phys. Condens. Matter* **1**, 7149 (1989).

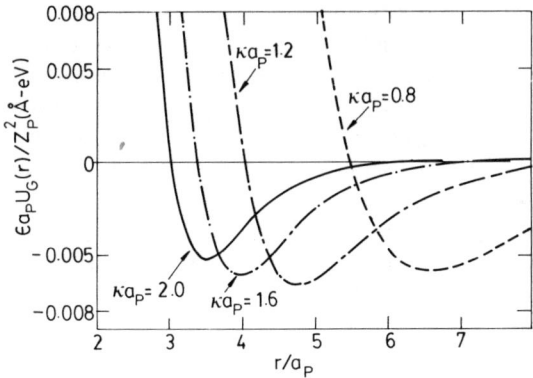

FIG. 12. Plot of $\varepsilon a_P U_G(r)/Z_P^2$ versus r/a_P [Eq. (9.2)] for various values of κa_P. Taken from I. Sogami and N. Ise, *J. Chem. Phys.* **81**, 6320 (1984).

purely repulsive potential $U_F(r)$ of the form $e^{-\kappa r}/r$ in complete agreement with the DLVO theory. However, the Gibbs free energy G, as calculated from F and which is argued to be appropriate for describing the interaction of the polyballs under isobaric conditions, yields an interparticle potential $U_G(r)$. The potential $U_G(r)$ is different from $U_F(r)$ and has both a repulsive part at short distances and attraction at large distances. The interaction potentials $U_F(r)$ and $U_G(r)$ are given by[66,67]

$$U_F(r) = \frac{Z_P^2 e^2}{\varepsilon} \left[\frac{\sinh(\kappa a_P)}{\kappa a_P}\right]^2 \frac{e^{-\kappa r}}{r} \tag{9.1}$$

and

$$U_G(r) = U_F(r)[1 + \kappa a_P \coth(\kappa a_P) - \tfrac{1}{2}\kappa r], \tag{9.2}$$

where κ is the same as in Eq. (2.2).

Figure 12 shows a plot of $U_G(r)$ as a function of r/a_P for different values of κa_P.[67] The pair potential has a minimum at $r = R_{min}$ that depends on κa_P, as sketched in Fig. 13a. When $\kappa a_P \ll 1$, $\kappa R_{min} \sim 4.6$. The depth of the minimum of the potential $U_G(r = R_{min})$ has a nonmonotonic dependence on κa_P (Fig. 13b).[67] Typical values of the parameters are $Z_p \sim 600$, $a_P \sim 1000$ Å, $\varepsilon = 78$, and $U_G(r = R_{min}) \sim k_B T$ at $\kappa a_P \sim 1$.

Later, Overbeek[82] criticised the theory of Sogami and Ise and argued that the contribution of the solvent molecules to the electrostatic free energy had

---

[82] J. Theodoor, G. Overbeek, *J. Chem. Phys.* **87**, 4406 (1987).

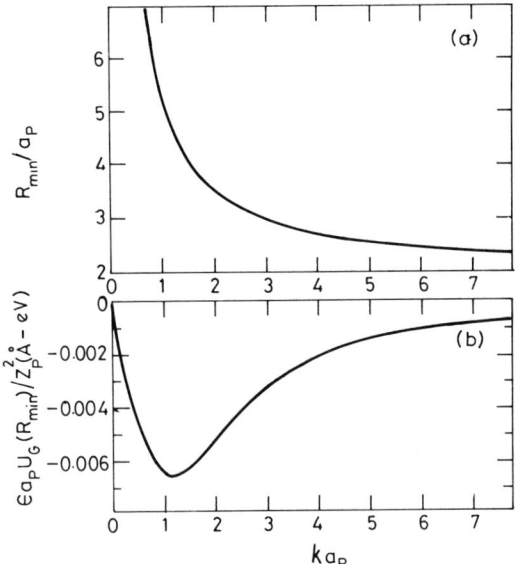

FIG. 13. (a) Position R_{min} of the minimum of the pair potential in Eq. (9.2) versus κa_P. (b) The value of U_G $(r = R_{min})$ versus κa_P. Taken from I. Sogami and N. Ise, *J. Chem. Phys.* **81**, 6320 (1984).

not been taken into account. The attractive term in $U_G(r)$ is apparently cancelled by a term derived from the contribution of the solvent. Recently, Smalley[83] has shown that Overbeek's criticism of the Sogami and Ise theory violates the Gibbs–Duhem equation, a fundamental thermodynamic relation given by

$$N_P \, d\mu_P + \sum_i N_i \, d\mu_i + N_{solvent} \, d\mu_{solvent} = 0. \tag{9.3}$$

Here N_P, N_i, and $N_{solvent}$ are the total numbers of macroions, small ions, and solvent molecules with chemical potentials μ_P, μ_i, and $\mu_{solvent}$, respectively. The Gibbs–Duhem relation states that the solvent is not an independent component when considering the thermodynamic properties of the suspension. As it stands, Smalley's calculations agree with the theory of Sogami and Ise and point out that one should reexamine the validity of the DLVO potential more carefully. One should also evaluate carefully the validity of Smalley's arguments in favour of the Sogami potential for charged colloids.

[83] M. V. Smalley. To appear in *Mol. Phys.*

V. Theoretical Aspects of Equilibrium Phases

Having seen that the charged polyball colloidal suspension can mimic the structural behaviour of conventional liquids and solids, one would like to understand theoretically the disorder–order, i.e., fluid–crystal and bcc–fcc transitions in the colloidal polyball suspensions. Theoretical analyses, analytical as well as computer simulation, have been done so far based on the DLVO or Yukawa potential, and some of these will be highlighted here.

10. Fluid–Crystal and bcc–fcc Phase Transitions

a. *Effective Hard Sphere Model for Fluid–Crystal Transition*

Early speculations regarding a phase transition from a fluid to a crystalline state in a system with a purely repulsive potential are due to Kirkwood.[84] The first evidence for such a phase transition came from the computer simulation results of Alder and Wainwright[85] on a finite system of hard spheres that showed two distinct branches in the equation of state, the lower-density branch corresponding to the hard sphere liquid and the higher-density branch to the fcc crystal. The subsequent detailed Monte-Carlo computer simulations of Hoover and Ree[24] determined that the hard sphere liquid begins to crystallize at the volume fraction $\phi_F = 0.494$ (freezing transition) and the transition is complete at $\phi_M = 0.545$ (melting transition). The transition is first order. Figure 14 shows the equation of state of the hard sphere system from the molecular dynamics computer simulation by Alder *et al.*[86] A corresponding curve for an ideal gas is also shown in Fig. 14 for comparison purposes.[87] Here V_0 is the volume necessary to confine N particles in the state of the closest packing (i.e., $\phi = 0.74 V_0/V$). The transition pressure P_m is found to be $P_m V_0/Nk_B T = 8.6$. If a conventional reduced pressure $PV/Nk_B T$ is used, the reduced transition pressure is 11.6 $[P_m V/Nk_B T = (P_m V_0/Nk_B T)(V/V_0) = 8.6 \times 1.35 = 11.6]$.[87] Hoover *et al.*[88] have also observed the fluid–crystal transition for soft spheres (the inverse power-law potentials). The fluid–solid transition due to purely repulsive forces is often referred to as the Kirkwood–Alder transition.

The nature of the hard sphere phase transition has been reasonably well explained in recent years by the application of density functional theories of freezing (for a recent review see Haymet[89]). These theories bring out the

---

[84] J. G. Kirkwood, *J. Chem. Phys.* **7**, 919 (1939).
[85] B. J. Alder and T. E. Wainwright, *J. Chem. Phys.* **27**, 1208 (1957).
[86] B. J. Alder, W. G. Hoover, and D. A. Young, *J. Chem. Phys.* **49**, 3688 (1968).
[87] K. Takano and S. Hachisu, *J. Chem. Phys.* **67**, 2604 (1977).
[88] W. G. Hoover, S. C. Gray and K. W. Johnson, *J. Chem. Phys.* **55**, 1128 (1971).
[89] A. D. J. Haymet, *Ann. Rev. Phys. Chem.* **38**, 89 (1987).

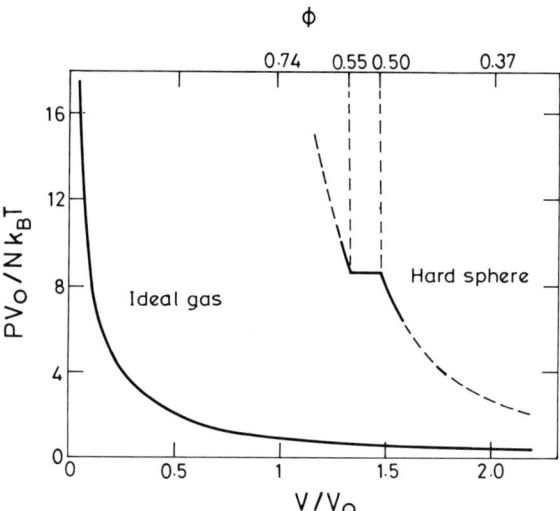

FIG. 14. Equation of state for hard sphere system and for ideal gas. Taken from K. Takano and S. Hachisu, *J. Chem. Phys.* **67**, 2604 (1977).

occurrence of the freezing transition due to a competition between two forms of entropy: S_C (configuration) and S_E (excluded volume). At low volume fractions, $S_C > S_E$, and hence the hard sphere system is disordered. At high ϕ, $S_E > S_C$ due to particle localization, which results in ordered packing. This mechanism is sometimes referred to as the excluded volume effect: the available volume of space for a particle to move freely is reduced by the repulsion of other particles.

The interaction between the polyballs is not of the hard sphere type (cf. Fig. 2); however, the polyball suspension can be approximated to the lowest order by a collection of effective hard spheres whose effective hard sphere diameter σ_{HS} is larger than the actual physical diameter. The order–disorder transition in the charged latex particles suspension can then be cast in terms of the Kirkwood–Alder transition[53,87,90] with appropriate empirical procedures for defining σ_{HS}. For example, Brenner[91] chooses σ_{HS} as that center-to-center distance between a pair of spheres at which the electrostatic interaction energy is of the order of $k_B T$. Barnes et al.[92] define σ_{HS} in terms of the thickness of the double layer and take $\sigma_{HS} = \sigma_P + y/\kappa$, where the parameter y was chosen empirically to be 3.8 to give the best fit to the

---

[90]M. Wadati and M. Toda, *J. Phys. Soc. Jpn.* **32**, 1147 (1972).
[91]S. L. Brenner, *J. Phys. Chem.* **80**, 1473 (1976).
[92]C. J. Barnes, D. Y. C. Chan, D. H. Everett, and D. E. Yates, *J. Chem. Soc. Faraday Trans. II* **74**, 136 (1978).

experimental results of Hachisu et al.[53] The effective hard sphere diameter in the Barker–Henderson perturbation theory is[93]

$$\sigma_{HS} = \sigma_P + \int_{\sigma_P}^{\infty} \left[1 - \exp\left\{-\frac{U_R(r)}{k_B T}\right\}\right] dr. \quad (10.1)$$

For known σ_{HS}, the effective volume fraction is

$$\phi_{eff} = \phi(\sigma_{HS}/\sigma_P)^3. \quad (10.2)$$

The phase boundary separating the liquid-like disordered phase and coexistence region is given by $\phi_{eff} = 0.494$, whereas the boundary separating the coexistence and the crystalline region satisfies $\phi_{eff} = 0.545$. Monovoukas and Gast[47] have calculated the phase diagram by taking Eq. (10.1) along with the DLVO potential $U_R(r)$ under the constant charge approximation. The dotted lines in Fig. 6 show their calculations. It can be seen that the trend is reproduced correctly, but the overall agreement is rather poor.

The effective hard sphere model can serve as a good reference system in perturbation theories for colloidal liquids, similar to what has been done extensively for atomic liquids (see, for example, McQuarrie[94]). In these theories the radial distribution function $g(r)$ and configurational Gibbs free energy can be calculated for the colloidal liquids.[11] These calculations along with the Gibbs free energy of the colloidal crystals based on the cell models have been used to construct the order–disorder phase diagram.[11] The perturbation theories do not work as well for the colloidal liquids in the weak screening limit (κa_P small) as for the strong screening limit. This is because the choice of an appropriate reference potential is not easy when the repulsive interaction is long range.

The colloidal crystals are analogous to the Wigner crystals. The latter are formed in a system of interacting electrons in a neutralising background of positive charges when the electron density is above the critical value.[95] The Wigner crystal is known to be bcc. This analogy has been exploited to predict the phase boundary in the colloids.[96]

b. *Analytical Theories*

The equilibrium crystalline phase at $T = 0$ can be found by performing a generalised Madelung lattice sum for the different lattice configurations (bcc, fcc, and hcp). The lattice sum, for the DLVO potential, is[70]

$$E = \frac{Z_P^2 e^2}{\varepsilon} \left(\frac{\exp(\kappa a_P)}{1 + \kappa a_P}\right)^2 \frac{N}{2} \sum_l \frac{\exp(-\kappa |\mathbf{r}|)}{|\mathbf{r}(l)|}, \quad (10.3)$$

[93] J. A. Barker and D. Henderson, *J. Chem. Phys.* **47**, 2856 (1967).
[94] D. A. McQuarrie, "Statistical Mechanics." Harper and Row, New York, 1976.
[95] C. M. Care and N. H. March, *Adv. Phys.* **24**, 101 (1975).
[96] S. Marcelja, D. J. Mitchell, and B. W. Ninham, *Chem. Phys. Lett.* **43**, 353 (1976).

where $|\mathbf{r}(l)|$ is the modulus of the lattice vector for the lattice under consideration. It is found that the differences in internal energies between any two structures (bcc and fcc or bcc and hcp) are extremely small (~ 1 part in 10^5),[70,97–99] and hence it is the entropic contribution at finite temperatures that decides the relative stability of different lattice structures (recall that the Helmholtz free energy $F = E - TS$). Nevertheless, the lattice sums do provide a zeroth-order clue to the occurrence of stable crystalline phases. The results are as follows: (i) For $\kappa a_s < 1.72$ the bcc lattice is more stable, and for $\kappa a_s > 1.72$ the fcc is a more stable structure. This can be easily appreciated by looking at the low and high κ limits. $\kappa = 0$ corresponds to the one-component plasma (OCP) or Wigner crystal, which is bcc. At the other extreme, $\kappa \to \infty$ is the hard sphere system that crystallizes into fcc, and therefore for some intermediate value of κa_s the transition from bcc to fcc should occur. (ii) $E_{hcp} > E_{fcc}$,[99] showing that the fcc is preferred over the hcp structure, in agreement with the observations.

The fcc to bcc and melting transitions for finite temperatures and κ have been considered in the self-consistent harmonic approximation by Hone et al.,[100] Rosenberg and Thirumalai,[70] and Shih et al.[101] In these theories, the behaviour of the liquid is either not taken into account[70,100] or is accounted for very crudely.[101] This is in contrast to the density functional theory (DFT) of freezing,[102] wherein the freezing is related to the instability of the liquid phase against some density waves characterizing the crystalline state. Alexander et al.[103] have applied the DFT to predict the freezing curve in the polyball suspensions.

(1) *Theories Based on the Self-Consistent Harmonic Approximation* Hone et al.[100] calculated the free energies of the bcc and fcc crystalline phases within a modified self-consistent harmonic approximation. The melting phase boundary is estimated from the semiempirical Lindemann rule, which states that the root-mean-squared displacement $\langle u^2 \rangle^{1/2}$ attains a specific fraction (W) of the interparticle spacing at the melting instability:

$$[\langle u^2 \rangle / a_s]^{1/2} = W. \tag{10.4}$$

---

[97] J. M. Silva and B. J. Mokrass, *Phys. Rev. B* **21**, 2972 (1980); M. Inoue and M. Wadati, *J. Phys. Soc. Jpn.* **50**, 1027 (1981).

[98] P. M. Chaikin, P. Pincus, S. Alexander, and D. Hone, *J. Colloid Interface Sci.* **89**, 555 (1982).

[99] P. M. Chaikin, J. M. di Meglio, W. D. Dozier, H. M. Lindsay, and D. A. Weitz, in "Physics of Complex and Supermolecular Fluids" (S. A. Safran and N. A. Clark, eds.), p. 65. Wiley, New York 1987.

[100] D. Hone, S. Alexander, P. M. Chaikin, and P. Pincus, *J. Chem. Phys.* **79**, 1474 (1983).

[101] W. Y. Shih, I. A. Aksay, and R. Kikuchi, *J. Chem. Phys.* **86**, 5127 (1987).

[102] T. V. Ramakrishnan and M. Yussoff, *Phys. Rev. B* **19**, 2775 (1979).

[103] S. Alexander, P. M. Chaikin, D. Hone, P. A. Pincus, and D. W. Schaefer, *Phys. Chem. Liq.* **18**, 207 (1988).

The $\langle u^2 \rangle$ is again calculated self-consistently. Before we present the calculated phase diagrams, we should mention the convenient parameters used to plot them. The DLVO potential, $U_R(r)$, of Eq. (8.12) can be rewritten in dimensionless form as

$$\beta U_R(r) = \Gamma \left[\frac{\exp(\lambda a_P/a_s)}{1 + \lambda a_P/a_s}\right]^2 \frac{\exp(-\lambda r/a_s)}{r/a_s}, \tag{10.5}$$

where

$$\Gamma = Z_P^2 e^2 / \varepsilon a_s k_B T, \tag{10.6}$$

$\lambda = \kappa a_s$, $a_s = n_P^{-1/3}$, and $\beta = (k_B T)^{-1}$ Thus the DLVO potential is characterized by three parameters, Γ, λ, and a_P/a_S, compared with the two parameters Γ and λ in the Yukawa potential.[104]. It is clear that the parameter Γ is related to the relative magnitude of interparticle Coulomb repulsion compared with $k_B T$. This is analogous to the coupling constant Γ_{OCP} defined for the OCP,

$$\Gamma_{OCP} = U_R(r_0)/k_B T,$$

where r_0 is the Wigner–Seitz radius defined by

$$4\pi r_0^3/3 = 1/n_P. \tag{10.7}$$

Thus the appropriate parameters used to plot the equilibrium phase diagram of the charged colloids have been taken to be Γ^{-1} and λ.[70,100,101]

The calculated bcc–fcc phase boundary (using the Yukawa potential) obtained by Hone et al. is shown by the solid curve in Fig. 15b. The melting curve using the Lindemann criterion of $W = 0.158$ is shown by the dotted line. Two features in the phase diagram may be observed: (i) There is a reentrant behaviour bcc → fcc → bcc as λ increases at constant $\Gamma^{-1} < 0.008$. When the phase diagram is translated in the parameter space of n_P and n_α, the reentrant behaviour is seen for a particle charge larger than some value depending on ε and a_P. (ii) The fcc phase is not observed for $\Gamma^{-1} > 0.008$.

Rosenberg and Thirumalai[70] have calculated the phase diagram, using the self-consistent phonon (SCP) theory and taking the interaction potential to be the DLVO form [Eq. (8.12)]. The main idea of the SCP theory is to approximate the hamiltonian of the crystal by a trial hamiltonian of a reference system consisting of a collection of linear harmonic oscillators. The spring constants and the thermally averaged displacement–displacement correlation functions are determined variationally. The latter is used in Eq. (10.4) to determine the melting curve from the Lindeman criterion. The Gibbs–Bogolyubov inequality states that

$$F \leq F_{\text{Trial}} \equiv F_0 + \langle V - V_0 \rangle_0, \tag{10.8}$$

[104] D. Thirumalai, *J. Phys. Chem.* **93**, 5637 (1989).

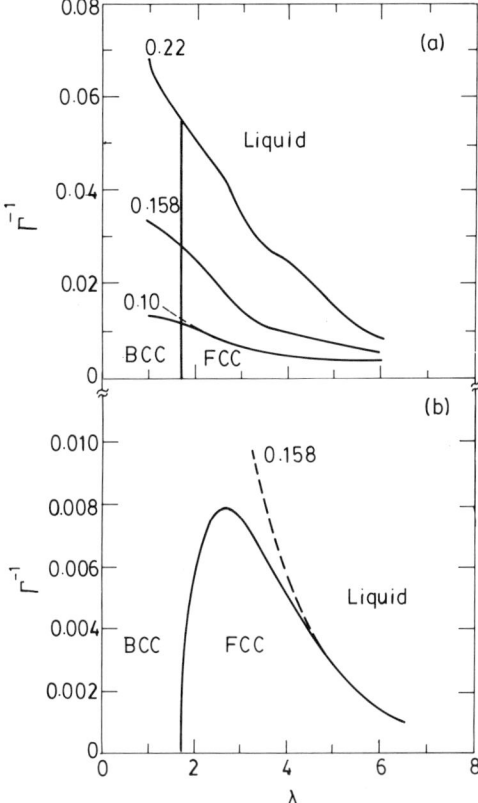

FIG. 15. (a) Calculated phase diagram based on self-consistent phonon theory. The melting curves are for the Lindermann parameter $W = 0.10$, 0.158, and 0.22 [Eq. (10.4)]. Taken from R. O. Rosenberg and D. Thirumalai, *Phys. Rev. A* **36**, 5690 (1987). The dotted line is from the theory of W. Y. Shih, I. A. Aksay, and R. Kikuchi, *J. Chem. Phys.* **86**, 5127 (1987). (b) Calculated phase diagram. The dotted line is the melting curve calculated from the Lindermann parameter $W = 0.158$. Taken from D. Hone, S. Alexander, P. M. Chaikin, and P. Pincus, *J. Chem. Phys.* **79**, 1474 (1983).

where F_0 is the free energy of the reference system, and $\langle V - V_0 \rangle_0$ is the mean potential energy difference between the actual system and the reference system, evaluated in the reference system. The inequality means that F can be approximated to be the minimum of F_{Trial} with respect to the variational parameters. The calculation was done in two ways, one of which took the reference system to be Einstein oscillators; the other, Debye oscillators. Figure 15a shows the phase diagram calculated by Rosenberg and Thirumalai,[70] wherein the vertical line at $\lambda = 1.72$ is the phase boundary between the bcc and fcc phases. The melting curves with W equal to 0.10, 0.158, and 0.22 are also plotted as solid lines in Fig. 15a. The phase diagram calculated by using the Debye spectrum for the oscillators is identical to that obtained by using the Einstein approximation.

The method of Shih et al.[101] is similar to that of Rosenberg and Thirumalai, with the difference that the melting curve is obtained by calculating the free-energy difference between the liquid and crystalline states. The reference system for estimating variationally the liquid free energy is taken to be the hard sphere system. The dotted line in Fig. 15a shows the melting phase boundary.[101] The bcc–fcc boundary is exactly the same as that given by Rosenberg and Thirumalai. The following features of the phase diagram in Fig. 15a are noteworthy.

1. The bcc–fcc phase boundary is independent of Γ^{-1}, and the bcc–fcc transition occurs at the same value of λ ($=1.72$) as that in lattice sum calculations of the internal energies at $T = 0$.

2. The phase diagram is completely different from that of Hone et al. (Fig. 15b). There is no reentrant phase transition as a function of κa_S.

We shall see later that none of the theories discussed so far agree with the experiments on the colloids and the molecular dynamics simulations. This is in contrast with the successful application of the SCP theory in predicting melting of Ar. The clue for this difference lies in the fact that the Lindemann parameter W in Eq. (10.4) is ~ 0.10 in argon, compared with $W \sim 0.2$ for the colloids. The latter is obtained from the molecular dynamics simulations.[70,105,106] A large value of W means that large-amplitude motions are significant near the melting of the colloidal crystals, and, hence, theories based on harmonic approximations are inadequate. Note from Fig. 15a that the melting curve of Shih et al.[101] coincides with that of the SCP theory obtained by using the Lindemann criterion $W = 0.10$. Since the parameter W^2 is inversely proportional to the sum of the squares of the frequencies,[70] it appears that the calculations of Shih et al.[101] based on the self-consistent phonon (Einstein) approximation neglect low-frequency modes that are known to be prevalent near melting.

c. *Computer Simulations*

Numerous studies based on Monte-Carlo computer simulations with the aim of computing osmotic pressures, order–disorder phase diagram, and radial distribution function of the colloidal fluid have been reported. We refer the reader to the reviews by van Megen and Snook[107] and by Castillo et al.[11]

---

[105] K. Kremer, M. O. Robbins, and G. S. Grest, *Phys. Rev. Lett.* **57**, 2694 (1986).
[106] M. O. Robbins, K. Kremer, and G. S. Grest, *J. Chem. Phys.* **88**, 3286 (1988).
[107] W. van Megen and I. Snook, *Adv. Colloid Interface Sci.* **21**, 119 (1984).

Here we shall summarize only the recent results of molecular dynamics (MD) simulations.[70,105,106] A comprehensive study of the phase diagram of a system of particles interacting through a Yukawa potential [Eq. (8.13)] has been carried out by Robbins et al.[106] Two sets of parameters have been chosen to represent the phase diagram: (i) $k_B T/U_a$ and λ, where $U_a = (Z_P^2 e^2/\varepsilon a_S)\exp(-\lambda)$; that is,

$$k_B T/U_a = \Gamma^{-1} e^\lambda \tag{10.9}$$

Here U_a is the Yukawa interaction energy between two polyballs separated by a distance a_S. (ii) $\tilde{T}$ and λ, where $\tilde{T} = k_B T/(m\omega_E^2 a_S^2)$. Here the characteristic energy scale has been taken to be a typical phonon (in the Einstein approximation) energy; ω_E is the phonon frequency, which can be easily shown to be related to U_t, the total energy per particle with all particles occupying lattice sites. The parameter m is the mass of a polyball.

The MD simulations were done at constant volume using periodic boundary conditions, and most of the results have been presented for the total number of particles $N = 500$. Figure 16a shows the phase diagram as a function of $\tilde{T}$ and λ The same data have been replotted (Fig. 16b) in the parameter space of $k_B T/U_a$ and λ. The open circles indicate points where the initial states were crystalline and then melted (i.e., the liquid phase is stable). The full black dot on the ordinate in Figs. 16a is the result of Monte-Carlo simulation of the melting transition of OCP ($\kappa = 0$), taken from the work of Pollock and Hansen.[108] The stable bcc and fcc phases are shown by filled triangles and circles, respectively. The full line is the bcc–fcc phase boundary between the two crystalline phases, obtained by using the energy-distribution function method.[106] In this method the free-energy difference is evaluated between two systems with potentials U_1 and U_2. To use the method for fcc and bcc phases with the same potential, it is necessary to realise that the two structures are related by a simple geometrical transformation. A bcc lattice can be transformed into a fcc lattice with the same density by stretching one axis by a factor of $2^{1/3}$ and contracting each of the other two by $2^{-1/6}$. The calculation can be interpreted as the free-energy difference between two systems with interaction potentials $U_1 = U[r = (x^2 + y^2 + z^2)^{1/2}]$ for bcc and $U_2 = U(s)$ for fcc, where $s^2 = 2^{-1/3}(x^2 + y^2) + 2^{2/3}z^2$. The fcc–bcc phase boundary could not be studied by the Parinello and Rahman algorithm of molecular dynamics at constant pressure with a deformable box.[106] In this algorithm the transformation from bcc to fcc is expected to occur through the simple deformation connecting the two structures. That it could not be observed may suggest that the free-energy barrier between the bcc and fcc phases is not small. This conclusion should be reexamined after doing the simulations using the DLVO potential.

[108] J. P. Hansen, *Phys. Rev. A* **8**, 3096 (1973); E. L. Pollock and J. P. Hansen, *Phys. Rev. A* **8**, 3110 (1973).

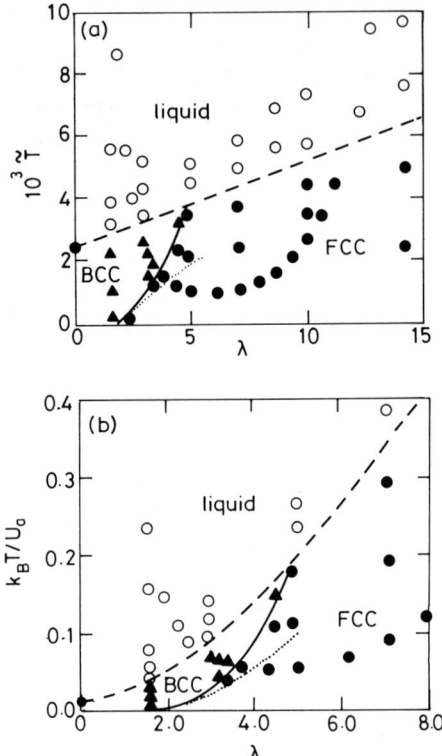

FIG. 16. (a) Phase diagram of Yukawa system from MD simulations. The dashed line is calculated from the Lindermann criterion with $W = 0.19$. The dotted line is the result of lattice dynamics calculation of the fcc–bcc phase boundary. (b) Data shown in part (a) are replotted for the parameters $k_B T/U_a$ and λ. Taken from M. O. Robbins, K. Kremer, and G. S. Grest, *J. Chem. Phys.* **88**, 3286 (1988).

The dashed lines in Fig. 16a and b have been determined[106] by taking the starting phase to be crystalline and using the Lindemann criterion of Eq. (10.4) with $W = 0.19$. It can be seen that the correlation between the open circles showing a stable liquid phase and the Lindemann criterion of $W = 0.19$ is very good. The dotted line in Fig. 16a (and in Fig. 16b) comes from lattice dynamical calculations in the harmonic approximation for the bcc to fcc phase transition. The following important observations have been brought out by Robbins *et al.*[106]

1. At $\lambda = 0$ (OCP limit), the stable phase is bcc crystal at low temperatures, which melts at high temperature without going through any solid–solid phase transition. When λ lies between 1.72 and 4.9, a transition from fcc

to bcc occurs with increasing $\tilde{T}$ or $k_B T/U_a$. For $\lambda > 4.9$, the stable phase remains fcc before melting. The bcc phase is not even metastable for $\lambda > 7.67$. It was found that even for those values of λ where the stable crystalline phase near melting is fcc, the bcc-like order was seen in the liquid phase. The bcc character of the liquid phase near melting seems to be in line with the ideas of Alexander and McTague.[109] They proposed that for a weakly first-order melting transition, the bcc structure is always the stable phase near the melting temperature.

2. The higher stability of the bcc phase with respect to the fcc phase as $\tilde{T}$ increases has been attributed to Zener's suggestion that the entropy of the bcc phase is higher due to lower-frequency shear modes (recall entropy $\propto -\ln \omega_E$) in the more open bcc structure than in the close-packed fcc structure. This was confirmed by the lattice dynamics calculations.[106] For $2 < \lambda < 5$, the sound velocities of shear modes were about 50% smaller in the bcc phase.

MD simulations using the DLVO potential, Eq. (8.12), have been carried out by Rosenberg and Thirumalai.[70] The complete phase diagram has not been determined. The effect of the number of particles N on the results of the simulations was examined, and it was noticed that reliable results for $\lambda < 4$ are obtained for $N \sim 1000$. One noteworthy difference between the two MD simulations is the following: Robbins et al.[106] found that starting from the liquid state, spontaneous crystallization was not found except in two MD runs. In contrast to this, Rosenberg and Thirumalai observed that the liquid spontaneously freezes into a bcc phase at a critical density. This different behaviour may be due to the effect of the geometrical factor in the potential or to finite size effects.

d. *Comparisons between Analytic Theories, Simulations, and Experiments*

In order to compare the results of the analytic theories (Fig. 15a and b) with the computer simulations of Robbins et al. (Fig. 16a and b), it is helpful to replot the bcc–fcc phase boundary and the melting curve of Fig. 16b in the parameter space of Γ^{-1} and λ. This can be easily done, and the results are shown by lines in Fig. 17. Also shown in Fig. 17 are the results of DFT calculations[103] (filled circles) for freezing the colloidal liquid into the bcc crystal. By comparing Figs. 17 and 15a and b, we see clearly that none of the theories agree with the MD simulations. Our conclusion regarding the poor agreement between the DFT calculations (filled circles) and the MD simulations of the freezing curve (dotted line in Fig. 17)[106] is in contrast to the claim of Alexander et al.[103] They plot [Fig. 3 of Ref. 103] $\ln(1/T^*)$ versus λ ($T^* = \varepsilon k_B T/Z_P^2 e^2 \kappa$), where the differences between the MD simulations and their calculations do not appear to be significant. We feel that our com-

---

[109] S. Alexander and J. P. McTague, *Phys. Rev. Lett.* **41**, 702 (1978).

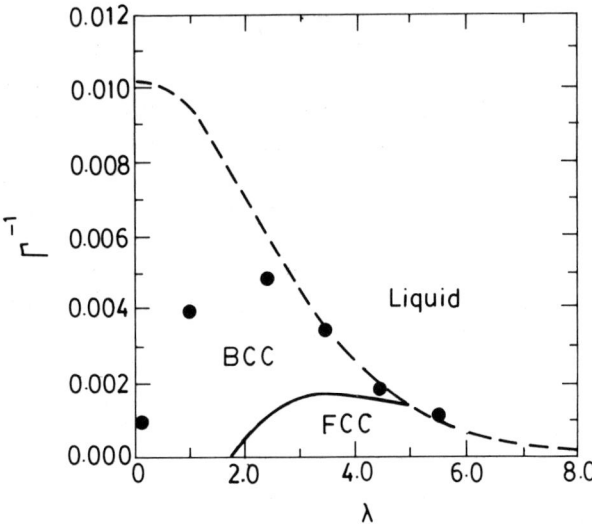

FIG. 17. The dashed and solid lines are the MD simulation results (already shown in Fig. 16b) replotted as a function of Γ^{-1} and λ. The filled dots are the results of density functional theory. Taken from S. Alexander, P. M. Chaikin, D. Hone, P. A. Pincus, and D. W. Schaefer, *Phys. Chem. Liq.* **18**, 207 (1988).

parison is more meaningful, and the discrepancy is therefore real. The natural length with which to rescale the temperature is a_s, not κ^{-1}, which itself is the control parameter along the abscissa. This point deserves further attention.

A comparison of the experimental phase diagram of the charged polyball suspensions has been recently done with the MD simulation results of Robbins *et al.*[47] The data shown in Fig. 6b have been converted to $k_B T/U_a$ versus λ by using $Z_P = 1200$ and are shown along with the MD results in Fig. 18a. The agreement is very poor. Monovoukas and Gast[47] found that the agreement becomes exceedingly good by renormalising the surface charge to 880, as shown in Fig. 18b. However, such is not the case with the experimental results of Sirota *et al.*[18] shown in Fig. 7. The dashed line in Fig. 7 is the fcc–liquid phase boundary of Robbins *et al.*[106] The poor agreement between the experimental data and MD simulations did not improve by using different renormalised charges on the particles. The inappropriateness of the MD simulations has been attributed to the use of the Yukawa potential—that is, neglect of the geometric factor in the DLVO potential, which becomes important at high volume fractions.

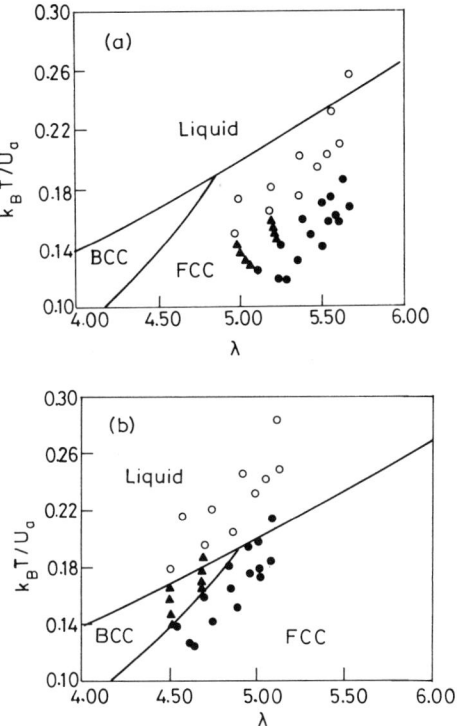

FIG. 18. (a) Comparison of experimental results with MD simulations of Robbins et al. Filled triangles: bcc crystal; filled circles: fcc; open circles: liquid. The solid lines are the results of MD simulations. (b) The same experimental data replotted after renormalising the effective charge from 1200 to 880. Taken from Y. Monovoukas and A. P. Gast, J. Colloid Interface Sci. **128**, 533 (1989).

VI. Liquidlike Order, Polydispersity, Binary Mixtures, and Glass

11. The Liquidlike Short-Ranged Order

The colloidal suspensions can be regarded as classical supermolecular fluids, and these systems offer an opportunity to understand the much investigated complexities of the liquid state of matter.[110] Quantum effects are absolutely negligible because the deBroglie thermal wavelength Λ, defined as

$$\Lambda = (2\pi\hbar^2/mk_\text{B}T)^{1/2},$$

[110] J. P. Hansen and I. R. McDonald, "Theory of Simple Liquids," second edition. Academic Press, New York, 1986.

where m is the mass of the colloidal particle ($\sim 10^{-15}$ g), is much less than the mean interparticle separation, a_s ($\sim 10^{-4}$ cm). One of the central issues in the study of atomic liquids or liquidlike short-range order in the colloids is the experimental measurement and the theoretical description of the pair distribution function $g(\mathbf{r})$ or its Fourier transform—the structure factor for density fluctuations $S(\mathbf{Q}) = \langle \rho_\mathbf{Q} \rho_{-\mathbf{Q}} \rangle / N$, where $\rho_\mathbf{Q}$ is the $\mathbf{Q}$th Fourier component of the density fluctuations. Ideally we would wish to extract direct information about the interparticle interactions from the measured $S(Q)$. But unfortunately this has not been done unambiguously, and widely different results have been extracted from the same data.[111] Levesque et al.[112] have recently used an iterative predictor–corrector method based on the modified hypernetted chain equation to extract the pair interaction potential for dense classical liquids. No such attempt has been made for the colloidal systems. Instead, one usually takes an indirect route; namely, given the interaction potential $U(r)$, the $S(Q)$ or $g(r)$ is calculated within a theoretical model.[43,113] A comparison of the calculated and the measured structure factors shows how good the interaction potential and the assumptions (generally called closure approximations) in the particular theoretical model are. Further, for pairwise additive total potential energy, all the thermodynamic properties (internal energy, osmotic pressure, etc.) of the system can be expressed in terms of $g(r)$ and the pair potential. For example, osmotic pressure is given by[110]

$$\frac{PV}{Nk_BT} = 1 - \frac{2\pi n_P}{3k_BT} \int_0^\infty g(r) \left(\frac{dU(r)}{dr}\right) r^3 \, dr. \quad (11.1)$$

For an isotropic suspension, the structure factor $S(Q)$ can be expressed in terms of the radial distribution function $g(r)$ as

$$S(Q) = 1 + \frac{4\pi n_P}{Q} \int_0^\infty [g(r) - 1] r \sin(Qr) \, dr. \quad (11.2)$$

Conversely,

$$g(r) = 1 + \frac{1}{2\pi^2 n_P r} \int_0^\infty [S(Q) - 1)] Q \sin(Qr) \, dQ. \quad (11.3)$$

We shall first discuss experimental methods for measuring $S(Q)$.

a. *Experimental Methods*

The most direct methods of probing the structure of macromolecular suspensions are (i) scattering of electromagnetic radiation (laser light scatter-

---

[111] W. S. Howells and J. F. Enderby, *J. Phys. C* **5**, 1277 (1972).
[112] D. Levesque, J. J. Weiss, and L. Reatto, *Phys. Rev. Lett.* **54**, 451 (1985).
[113] N. K. Ailawadi, *Phys. Rep.* **57**, 241 (1980).

ing[114-116] and small-angle x-ray scattering[18]) and small-angle neutron scattering,[117] and (ii) direct optical microscopy.[79,118]

(1) *Scattering Experiments*

The scattering wave vector **Q** is given by $\mathbf{Q} = \mathbf{k}_i - \mathbf{k}_s$, where $\mathbf{k}_i$ and $\mathbf{k}_s$ are the wave vectors of the incident and scattered radiations. The magnitude of **Q** is

$$Q = [4\pi\mu_m \sin(\theta/2)]/\lambda_i, \qquad (11.4)$$

where θ is the scattering angle ($0 < \theta < \pi$), μ_m is the refractive index of the medium, and λ_i is the wavelength of the incident radiation in vacuum. The required range of Q in an experiment depends on the particle density n_P of the liquid. The position of the first peak Q_1 in the $S(Q)$ of the liquid is related to n_P by $n_P = A(Q_1/2\pi)^3$, where $A = (1/\sqrt{2}$ or $\frac{4}{3}\sqrt{3})$, depending on whether the liquid freezes into the bcc or fcc structure, respectively. The range of Q typically lies between 2×10^4 cm^{-1} and 3×10^6 cm^{-1}, which can be covered by using different laser radiations[113-115] or small-angle x-ray[18] or neutron scattering.[117]

Consider light scattering from a scattering volume V_s containing N_s spherical colloidal particles. The intensity of light scattered by the solvent is negligible compared with that scattered quasielastically by the particles. The instantaneous value of the scattered electric field at the scattering angle θ (in the far-field limit) to the incident laser beam is given by

$$E_S(\mathbf{Q}, t) \propto \sum_{i=1}^{N_s} \exp[i\mathbf{Q} \cdot \mathbf{r}_i(t)], \qquad (11.5)$$

where $\mathbf{r}_i(t)$ is the position of the centre of mass of the ith particle. It can easily be shown that the time-averaged scattered intensity in the Rayleigh–Gans region is given by[46,116]

$$I_S(Q) = AP(Q)S(Q), \qquad (11.6)$$

where $P(Q)$ is the particle scattering form factor, which for a spherical particle of radius a_P is given by

$$P(Q) = [3\{\sin(Qa_P) - Qa_P \cos(Qa_P)\}/(Qa_P)^3]^2, \qquad (11.7)$$

---

[114] J. C. Brown, P. N. Pusey, J. W. Goodwin, and R. H. Ottewill, *J. Phys. A* **8**, 664 (1975).
[115] A. K. Sood, *Hyperfine Interactions* **37**, 365 (1987).
[116] M. Kerker, "The Scattering of Light and Other Electromagnetic Radiation." Academic Press, New York, 1969.
[117] B. J. Ackerson, J. B. Hayter, N. A. Clark, and L. Cotter, *J. Chem. Phys.* **84**, 2344 (1986).
[118] K. Ito, H. Okumura, H. Yoshida, Y. Ueno, and N. Ise, *Phys. Rev. B* **38**, 10852 (1988); H. Yoshida, K. Ito, and N. Ise. To appear in *J. Am. Chem. Soc.*

and $S(Q)$ is the interparticle structure factor given by

$$S(Q) = \frac{1}{N_s} \sum_{i=1}^{N_s} \sum_{j=1}^{N_s} \langle \exp\{i\mathbf{Q} \cdot (\mathbf{r}_i(t) - \mathbf{r}_j(t))\} \rangle. \tag{11.8}$$

In Eq. (11.6) A is a constant, which for vertically polarised scattered light is given by[116]

$$A = \frac{9\pi^2 \mu_m (m^2 - 1)^2 V_s n_P v_P^2}{\lambda_i^4 (m^2 + 2)^2} \frac{I_0}{R^2}. \tag{11.9}$$

Here $m = \mu_P/\mu_m$, μ_P is the refractive index of the particle ($=1.611$ for polystyrene spheres), $v_P = 4\pi a_P^3/3$, I_0 is the intensity of incident radiation, and R is the distance between the scattering volume and the detector. The quantity $AP(Q)$ can be obtained experimentally by measuring the scattered intensity $I_S^0(Q)$ of the suspension after increasing the salt concentration (n_α) such that the particles are spatially uncorrelated; that is, $S(Q) = 1$, $I_S^0(Q) = AP(Q)$, and

$$S(Q) = I_S(Q)/I_S^0(Q). \tag{11.10}$$

In most experimental setups, the detector is rotated to vary the scattering angle θ and, hence, the scattering vector $\mathbf{Q}$. In the setup we have used,[119] the incident laser beam is rotated by means of mirrors mounted on a rotating arm together with $\theta/2$ rotation of a half-wave quartz retardation plate to achieve a fixed incident polarization. A photometer employing 18 fixed scattering angles and optical fibers to transmit the scattered light to a single photomultiplier tube has been described.[120]

The foregoing discussion assumes that there is no multiple scattering. This assumption is valid if n_P is small or if the refractive indices of the solvent and the particle are matched (i.e., $m \simeq 1$). Multiple scattering results in depolarization of light. A simple procedure to correct for multiple scattering has been suggested by Grüner and Lehmann.[121] Figure 19a shows the measured structure factor (solid dots) of a typical colloidal suspension, with $\sigma_P = 1090$ Å and $n_P = 1.33 \times 10^{12}$ cm^{-3}.[122] In these experiments, the concentration of the stray ions (n_α) is controlled by having a mixed bed of ion exchange resins at the bottom of the sample cell.

$S(Q)$ can also be measured indirectly from dynamic light scattering (DLS),

---

[119] A. K. Arora, *J. Phys. E* **17**, 1119 (1984).
[120] H. R. Haller, C. Destor, and D. S. Cannell, *Rev. Sci. Instrum.* **54**, 973 (1983).
[121] F. Grüner and W. Lehmann, *J. Phys. A* **13**, 2155 (1980).
[122] B. V. R. Tata, R. Kesavamoorthy, and A. K. Sood, *Molec. Phys.* **61**, 943 (1987).

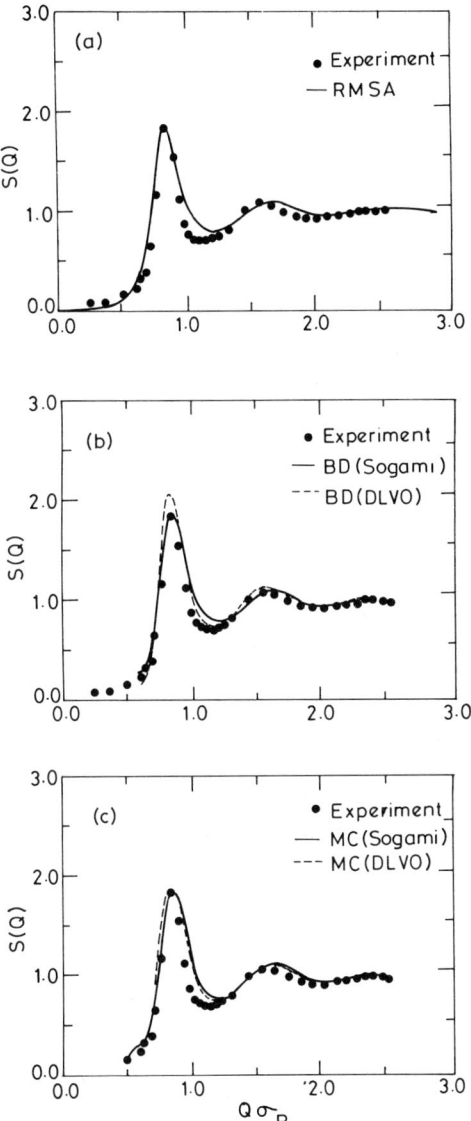

FIG. 19. (a) Comparison of experimental $S(Q)$ (filled dots) with RMSA calculations (solid line). (b) Comparison of calculated $S(Q)$ based on Sogami potential [Eq. (9.2)] and the DLVO potential [Eq. (8.12)], using Brownian dynamics computer simulations, with the experiments (filled dots). (c) Comparison of calculated $S(Q)$ based on MC simulations for the two potentials with the experimental data. Taken from B. V. R. Tata, A. K. Sood, and R. Kesavamoorthy, *Pramana J. Phys.* **34**, 23 (1990).

which is a real-time analysis of scattered intensity fluctuations.[114,123] The normalised autocorrelation function of the scattered field amplitude is

$$g^{(1)}(Q, \tau) = \frac{\langle E_s(Q, t)E_s^*(Q, t+\tau)\rangle}{\langle |E_s(Q, t)|^2\rangle} \quad (11.11)$$

$$= F(Q, \tau)/S(Q), \quad (11.12)$$

where the dynamic structure factor[46]

$$F(Q, \tau) = \frac{1}{N_s} \sum_{i=1}^{N_s} \sum_{j=1}^{N_s} \langle \exp\{i\mathbf{Q}\cdot[\mathbf{r}_i(0) - \mathbf{r}_j(\tau)]\}\rangle. \quad (11.13)$$

It has been shown (for a review of DLS from charged colloids, see Pusey and Tough[124]) that the initial decay rate of $g^{(1)}(Q, \tau)$ for liquidlike suspensions is

$$\frac{d}{dt}g^{(1)}(Q, t)|_{t=0} = K_1(Q) = -D_{\text{eff}}Q^2, \quad (11.14)$$

where

$$D_{\text{eff}} = D_0/S(Q) \quad (11.15)$$

and D_0 is the free-diffusion constant of a polyball. $K_1(Q)$ is called the first cumulant. Knowing D_0 and the measured $K_1(Q)$, one can infer $S(Q)$.

b. *Liquid State Theories, Simulations, and Comparison with Measured S(Q)*

As mentioned before, a major objective of liquid theory is to calculate the pair distribution function $g(r)$ or its Fourier transform $S(Q)$ for a given interparticle potential $U(r)$. Numerous approaches to this end, which were developed originally for conventional liquids,[110] have been applied to the colloidal liquids. We shall mention only four well-known approximations: Kirkwood superposition (KS), Percus–Yevick (PY), hypernetted chain (HNC), and the mean spherical approximation (MSA) [as well as the rescaled mean spherical approximation (RMSA)].

(1) *KS Approximation*

Starting from the definition of the distribution function in terms of a pair-additive interaction potential, Born, Green, and Yuon (BGY) derived an exact integral equation hierarchy.[110] It is possible to solve the BGY equation only if one uses a closure equation to break the hierarchy. The simplest closure equation is the KS approximation, wherein the three-body distribution function is approximated by a product of two-body distribution

---

[123] D. W. Schaefer and B. J. Berne, *Phys. Rev. Lett.* **32**, 1110 (1974).
[124] P. N. Pusey and R. J. A. Tough, in "Dynamic Light Scattering: Applications of Photon Correlation Spectroscopy" (R. Pecora, ed.), p. 85. Plenum, New York, 1985.

functions. Schaefer[43] has shown that as in atomic fluids, the KS approximation does not work well in colloidal liquids.

(2) HNC, PY, and MSA Approximations

These approximations are not meant to be used with the BGY equation. Instead, they are analysed through the Ornstein–Zernike (OZ) equation

$$h(r) = C(r) + n_P \int h(|\mathbf{r} - \mathbf{r}'|) C(\mathbf{r}') \, d\mathbf{r}', \qquad (11.16)$$

where $h(r) = g(r) - 1$, and $C(r)$ is the direct correlation function. The OZ equation has two unknown quantities $h(r)$ and $C(r)$, and hence one more equation is needed to get the solutions. This additional equation, the closure condition, is different for the three approximations:[110]

$$C(r) = [1 - \exp(\beta U(r))] g(r), \qquad \text{(PY)}$$

$$C(r) = -\beta U(r) + g(r) - 1 - \log[g(r)], \qquad \text{(HNC)}$$

$$C(r) = -\beta U(r), \quad \text{for } r > \sigma_P,$$
$$h(r) = -1, \quad \text{for } r < \sigma_P. \qquad \text{(MSA)}$$

In contrast to the PY and HNC approximations where the solution of the OZ equation can be obtained only numerically, MSA leads to closed-form analytical solutions for the DLVO potential [Eq. (8.12)].[125] It has been found that the MSA gives reliable structure factors only for sufficiently concentrated suspensions ($\phi \geqslant 0.2$). For the dilute polyball suspensions, MSA yields nonphysical negative contact values of the radial distribution function, $g(r = \sigma_P^+)$. This difficulty has been overcome by Hansen and Hayter[126] in the RMSA, wherein the particle diameter σ_P is rescaled while maintaining constant the Coulomb coupling, defined as

$$\Gamma_c = 2\beta U_R(r = 2r_0).$$

The physical argument rests on the observation that for strong Coulomb coupling the contact configurations are extremely rare because the value of the potential at contact $U_R(r = \sigma_P^+)$ is much larger than $k_B T$. This implies that the Coulomb interaction alone results in $g(r) = 0$ for $r \leqslant r_0$, without the hard core playing any physical role.[126] Thus the characteristic length of physical interest should be $2r_0$ rather than σ_P.

Schaefer[43] has compared the measured $S(Q)$ with the numerical solutions of the OZ equation in HNC and PY approximations for the DLVO potential. The HNC closure approximation gave better agreement than the PY

[125] J. B. Hayter and J. Penfold, *Molec. Phys.* **42**, 109 (1981).
[126] J. P. Hansen and J. B. Hayter, *Molec. Phys.* **46**, 651 (1982).

approximation, as for the OCP.[110] The RMSA calculations match very well with the HNC results[126] and the experimental data. The latter is demonstrated by the solid curve in Fig. 19a. The concentration of additional ions n_α was taken as an adjustable parameter and was found to be $2.1 \times 10^{15}\,\text{cm}^{-3}$. The charge Z_P was taken to be 600. It is fair to say that the RMSA is the most useful method for calculating the $S(Q)$ of the colloidal suspensions in the liquid phase over a wide range of particle densities and ionic strengths, provided the potential is of the form $\exp(-\kappa r)/r$. Figure 20a shows the calculated $S(Q)$ in RMSA for a suspension having $\sigma_p = 0.091\,\mu\text{m}$, $Z_P = 500$ with different values of n_α.[128,129] The DLVO potential is used. The corresponding radial distribution functions $g(r)$ obtained from the Fourier transform [Eq. (11.3)] are shown in Fig. 20b. It is clear from these figures that the structure weakens, and Q_1, the position of the first peak position in $S(Q)$, shifts to higher values as n_α increases. A serious limitation of the MSA or RMSA is that closed-form analytic solutions cannot be obtained for other forms of the interaction potentials (e.g., the logarithmic form of $U_R(r)$ valid when $\kappa a_P > 2.5$ [Eq. (8.11b)] or the combined repulsive and attractive potentials, such as Sogami's potential [Eq. (9.2)]). In such cases computer simulations are the most useful.

A good agreement between the measured and the calculated $S(Q)$, as shown in Fig. 19, tells us more about the validity of the closure approximation used in the theoretical model than about the best choice of the interaction potential. The inability to extract a unique $U(r)$ from the $S(Q)$ is shown in Fig. 19b and c, wherein $S(Q)$ has been calculated[127] by Brownian dynamics[130] (Fig. 19b) and Monte-Carlo simulations (Fig. 19c), using both the DLVO potential [Eq. (8.12)] and the Sogami potential [Eq. (9.2)]. The parameters of the DLVO and the Sogami potentials are the same as those taken in the RMSA calculations (Fig. 19a) except that n_α was slightly reduced to $1.75 \times 10^{15}\,\text{cm}^{-3}$ in the Sogami potential in order to get a good fit of $S(Q)$ with the data. Recall that $n_\alpha = 2.1 \times 10^{15}\,\text{cm}^{-3}$ in the DLVO potential. The two potentials $U(r)$ are plotted in Fig. 21. The arrow marks the average interparticle separation $2r_0$. The interaction at $r = 2r_0$ is repulsive in one case and attractive in the other. It is immediately clear from Figs. 19b and 19c that the $S(Q)$ obtained using the Sogami potential agrees as well with the experimental $S(Q)$ as that obtained by using the DLVO potential.

[127] B. V. R. Tata, A. K. Sood, and R. Kesavamoorthy. *Pramana J. Phys.* **34**, 23 (1990).
[128] R. Kesavamoorty, "Structure and Elastic Properties of Dilute Charged Colloids," Ph.d. Thesis, University of Madras, 1987 (unpublished).
[129] B. V. R. Tata, R. Kesavamoorthy, and A. K. Arora, *Molec. Phys.* **57**, 369 (1986).
[130] D. L. Ermak and Y. Yeh, *Chem. Phys. Lett.* **24**, 243 (1974); K. Gaylor, I. Snook, and W. van Megen, *J. Chem. Phys.* **75**, 1682 (1981).

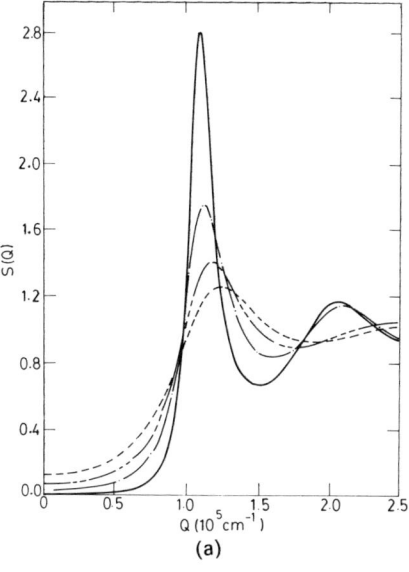

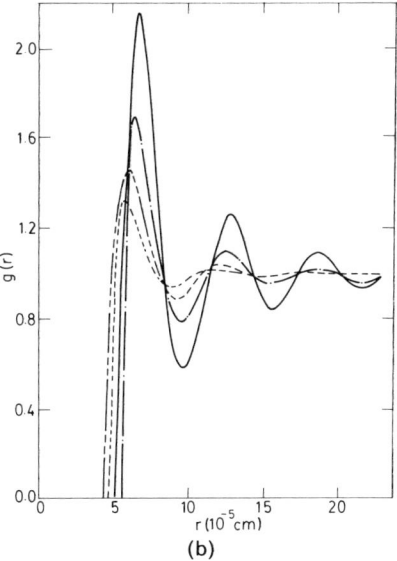

FIG. 20. (a) Calculated $S(Q)$ under RMSA. The parameters of the suspensions are $n_P = 4 \times 10^{12}$ cm^{-3}, $\sigma_P = 0.091$ μm, $Z_P = 500$, and $n_\alpha = 0$ (———), $n_\alpha = 2.5 n_P Z_P$ (—·—), $n_\alpha = 5 n_P Z_P$ (— — —), and $n_P = 7.5 n_P Z_P$ (-----). (b) $g(r)$ obtained from the Fourier transform of $S(Q)$ shown in part (a). The legends are the same as in part (a). Taken from R. Kesavamoorthy, Ph.D. Thesis, unpublished (1987).

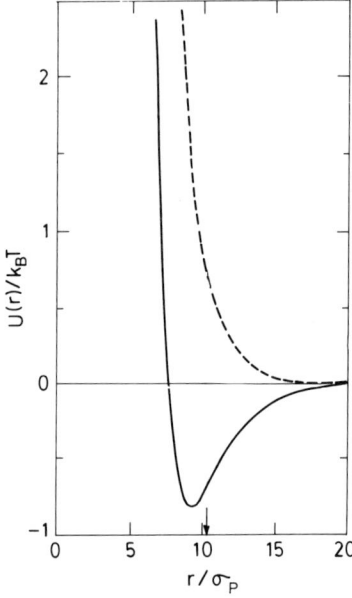

FIG. 21. The interaction potential versus r/σ_P. The solid line is the Sogami potential, and the dotted line is the DLVO potential for $\sigma_P = 0.109\,\mu m$, $Z_P = 600$, $n_P = 1.33 \times 10^{12}\,cm^{-3}$, $n_\alpha = 2.1 \times 10^{15}\,cm^{-3}$ (for the DLVO potential), and $n_\alpha = 1.75 \times 10^{15}\,cm^{-3}$ (for the Sogami potential). The arrow marks the average interparticle separation $2r_0$. Taken from B. V. R. Tata, A. K. Sood, and R. Kesavamoorthy, Pramana J. Phys. 34, 23 (1990).

c. Measured S(Q) near Freezing

It is important to measure the height S_{max} of the first peak in $S(Q)$, for the liquidlike suspensions near freezing, to examine the validity of the Hansen–Verlet freezing criterion.[131] The criterion, which originated from computer simulations of simple atomic fluids, says that $S_{max} \sim 2.85$ in a liquid near freezing. The height and width of the first peak in $S(Q)$ reflect the degree and range of spatial correlations in liquids, and hence the Hansen–Verlet criterion implies that the freezing transition occurs when the correlations in the liquid phase build up to a certain level.

Kesavamoorthy et al.[132] measured the $S(Q)$ very close to the colloidal crystal–liquid interface ($\sim 200\,\mu m$ above the interface). The measured $S(Q)$ can be taken to correspond to the liquid in equilibrium with the crystal (i.e., at the threshold of freezing). In these experiments, the crystallites were formed at the bottom of the sample cell and the crystal–liquid interface moved toward the top as a function of time. Figure 22 shows $S(Q)$ for three suspensions with different particle diameters $0.109\,\mu m$, $0.11\,\mu m$, and $0.12\,\mu m$, at the crystal (bcc)–liquid interface. In all cases, it is found that $S_{max} = 2.1 \pm 0.1$. These experiments may be compared with those of Härtl and Versmold,[77] shown in

[131] J. P. Hansen and L. Verlet, Phys. Rev. 184, 150 (1969).
[132] R. Kesavamoorthy, B. V. R. Tata, A. K. Arora, and A. K. Sood, Phys. Lett. A 138, 208 (1989).

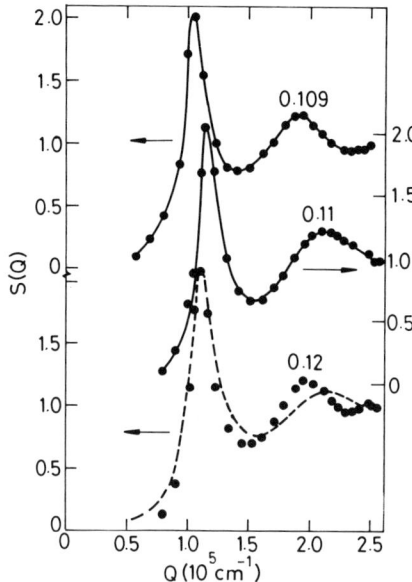

FIG. 22. Measured $S(Q)$ for three suspensions with particle diameters 0.109 μm, 0.11 μm, and 0.12 μm. The full curves are guides. The dotted line is the calculated $S(Q)$ using RMSA. Taken from R. Kesavamoorthy, B. V. R. Tata, A. K. Arora, and A. K. Sood, *Phys. Lett.* A **138**, 208 (1989).

Fig. 23, wherein the height of the first peak lies between 2.5 and 2.9. There is an important difference between the two experiments, namely the $S(Q)$ in Fig. 23 are measured for the shear-melted crystalline samples. The S_{max} in Fig. 23 will correspond to a highly metastable liquid, very far from equilibrium, and not to the equilibrium liquid at the threshold of freezing. Now if the measured S_{max} at the crystal–liquid interface is taken to be that of the liquid near freezing[132] (which we believe it is), why is it much smaller than 2.85? The reason is not clear at present. One obvious possibility is the polydispersity in the size or charge. The standard deviation of 2.5% in particle diameter (as quoted by the suppliers of the samples) decreases S_{max} by only 1%. Preliminary computer simulations[133] have shown that the charge polydispersity has to be ~34% to decrease S_{max} from 2.85 to 2.1. Such a large polydispersity in charge is very unlikely in the suspensions studied. Moreover, the polydispersity will reduce both S_{max} and S_n, where S_n is the value of $S(Q)$ at Q corresponding to [211] ([311]) diffraction if the liquid freezes into bcc (fcc). The two-order parameter density functional theory

[133] B. V. R. Tata, 1989 (unpublished).

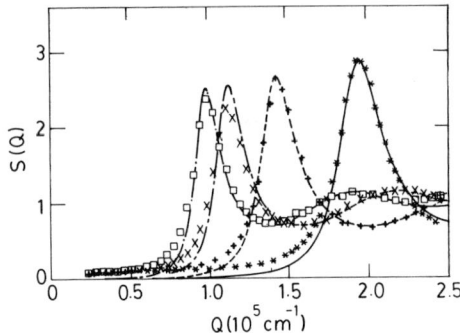

FIG. 23. Measured $S(Q)$ for different n_P. (i) $n_P = 2.94 \times 10^{12}$ cm^{-3} (□□□). (ii) $n_P = 4.4 \times 10^{12}$ cm^{-3} (xxx). (iii) $n_P = 9.0 \times 10^{12}$ cm^{-3} (+++). (iv) $n_P = 2.26 \times 10^{13}$ cm^{-3} (∗∗∗). The lines are the calculated curves based on RMSA and using Z_P^* and κ^* from the PBC model. Taken from W. Hartl and H. Versmold, J. Chem. Phys. **88**, 7157 (1988).

(DFT) of Ramakrishnan and Yussoff[102] predicts that if S_{max} decreases, S_n must increase for freezing to occur. The experimentally observed values[132] of S_{max} and S_n are consistent with the predictions of the DFT. In addition to our measurements, we find that the calculated values of S_{max} (using RMSA) for the parameters corresponding to the experimentally determined freezing phase boundary in Fig. 7 are also less than 2. Thus, the important question of S_{max} at the colloidal liquid–crystal freezing transition thus remains unresolved.

12. POLYDISPERSITY EFFECTS

There is an important and unavoidable difference between pure atomic or molecular materials and colloidal systems. In the former, all particles are identical, whereas in the latter there is inevitably some distribution of particle size or surface charge. It is therefore natural to ask how polydispersity affects the character of the crystal–fluid phase transition.

The effect of polydispersity on the equation of state of an ordered electrostatically stabilised colloidal suspension has been examined based on the model of Evans and Napper.[134] These simple and rather qualitative calculations predict that the polydispersity becomes important when the Debye screening length κ^{-1} is much less than the mean particle radius $\bar{a}_P$ (i.e., $\kappa \bar{a}_P \gg 1$). This is because the repulsive potential is steep for $\kappa \bar{a}_P \gg 1$, and the suspension behaves like an assembly of close-packed hard spheres in which particles of radius larger than $\bar{a}_P$ cannot be accommodated without dislocating the ordered structure. On the other hand, for systems with $\kappa \bar{a}_P \ll 1$, the softness of the repulsion makes the system less sensitive to particle size. It has been suggested[134] that the stability of dilute ordered suspensions ($\phi \lesssim 0.01$)

---

[134] E. Dickinson, J. Chem. Soc. Faraday Trans. II **75**, 466 (1979).

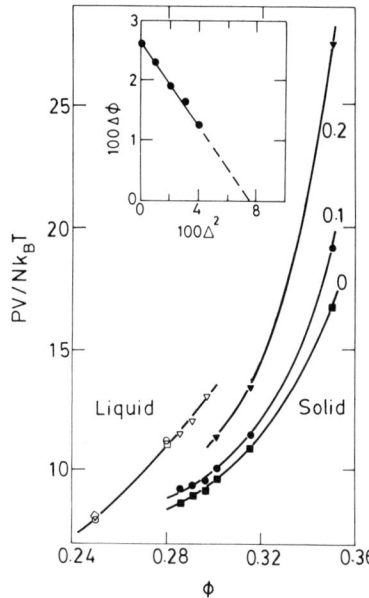

FIG. 24. Calculated equation of state for different degrees of polydispersity. The open points are for the liquid state, and the filled symbols refer to the ordered phase. The values of Δ are marked on the curves. The inset shows $\Delta\phi$ versus Δ^2, extracted from the simulation data. Taken from E. Parkinson, R. Parker, and M. Lal, Chem. Phys. Lett. **79**, 578 (1981); E. Dickinson and R. Parker, J. Phys. Lett. **46**, L229 (1985).

of low ionic strength ($\sim 10^{-6}$ mole/litre) is not expected to be affected by polydispersity, whereas the structures of moderately concentrated suspensions ($\phi \gtrsim 0.3$) at high ionic strengths are very sensitive to it.

Molecular dynamics simulations[36] on the charged colloidal suspensions with high ionic strength ($\kappa \bar{a}_P > 3$) have shown that polydispersity has a negligible effect on the osmotic pressure in the liquidlike state. On the other hand, for the crystalline state, the osmotic pressure increases significantly with polydispersity. The simulations were done on a rather small system ($N = 108$), taking the interaction potential to be a sum of $U_A(r)$ [Eq. (8.14)] and $U_R(r)$ [Eq. (8.11a)], suitably modified for the polydisperse system. The particle radii were distributed according to a symmetrical triangular distribution:

$$\mathscr{P}(\bar{a}_P) = \frac{1}{(\bar{a}_P \Delta)^2}(\bar{a}_P \Delta - |a_P - \bar{a}_P|), \quad \text{for } 1 - \Delta < \frac{a_P}{\bar{a}_P} < 1 + \Delta,$$

$$= 0, \quad \text{for } 1 - \Delta \geqslant \frac{a_P}{\bar{a}_P}, 1 + \Delta \leqslant \frac{a_P}{\bar{a}_P}. \quad (12.1)$$

The width of the distribution Δ is related to the root-mean-square standard deviation, $\delta = \Delta/\sqrt{6}$. Figure 24 shows the computed compressibility factor $PV/Nk_B T$ versus ϕ for $\Delta = 0$, 0.1, and 0.2. These data have been used to estimate the width of the coexisting regions, $\Delta\phi = \phi_{\text{crystal}} - \phi_{\text{liquid}}$, by adopting a Maxwell "equal area" construction.[37] The inset in Fig. 24 shows

$\Delta\phi$ as a function of Δ^2. By extrapolating, $\Delta\phi = 0$ gives the critical value of $\Delta_c = 0.27 \pm 0.01$; that is, $\delta_c = 0.11$ at which the freezing transition disappears.[37]

The mechanical and thermodynamic stability of hard spherical colloids has been investigated[38] by using the density functional theory of freezing.[102] It is found that the polydisperse colloidal crystals are not stable beyond a critical degree of polydispersity $\delta_c \sim 0.067$. The value of δ_c is the same for both the triangular and rectangular distributions of hard sphere particle radii. Pusey has proposed the following simple physical picture to estimate the δ_c for suppressing the crystallization of hard spherical colloids. Keeping the number density and the mean size constant, let the polydispersity be increased in a crystal having an average separation of the centres of nearest-neighbour particles equal to d. The crystalline state is expected to be stable when most of the particles have radii less than $d/2$; that is,

$$\bar{a}_P(1 + x\delta) < d/2, \tag{12.2}$$

where x is a number of order 1. The volume fraction ϕ is

$$\phi = 4\pi\overline{a_P^3}n_P/3. \tag{12.3}$$

For a narrow distribution, $\overline{a_P^3} \simeq \bar{a}_P^3$. For closed-packed structures,

$$0.74 = 4\pi(d/2)^3 n_P/3. \tag{12.4}$$

From Eqs. (12.2)–(12.4), the criterion for a stable crystalline phase is

$$x\delta < (0.74/\phi)^{1/3} - 1. \tag{12.5}$$

For $\phi = 0.545$, the volume fraction at which the monodisperse hard sphere crystal melts, $x\delta_c < 0.11$. Choosing $x \sim 1$ gives $\delta_c \sim 0.11$, fortuitously the same as the computer simulation results.[37] So far, there is no systematic experimental study of the effect of polydispersity on the crystalline order formation in the colloids.

Beyond the critical degree of polydispersity, $\delta > \delta_c$, it has been speculated[38] that the solid solution could form either a colloidal glass or show phase separation into crystals with different compositions. The latter possibility would require the diffusion of particles over distances comparable to the size of a crystallite and hence would be much slower than the crystal formation in an equivalent monodisperse suspension. The case of collodial mixtures of particles with two or more diameters is different from that of the polydisperse suspension with a monomodal particle size distribution. It is possible that the smaller particles can fit into the open spaces of the crystal consisting of the larger particles. Therefore colloidal suspensions with multimodal particle size distribution can form colloidal alloys[135,136] that can be substitutionally disordered or ordered (like AB, AB_2, AB_4, etc.), glasses[12,13] or show phase separation.[12]

13. COLLOIDAL ALLOYS IN BINARY MIXTURES

Order formation is not restricted to one-component monodisperse colloids but can also occur in binary mixtures of colloidal particles of two different diameters. This was first observed by Sanders and Murray[137] from their analysis of structures in naturally occurring opals consisting of two kinds of silica spheres. In charged polyball colloidal suspensions, several crystalline compound structures have been observed by optical microscopy in concentrated binary mixtures ($\phi \sim 0.3$) with particle diameters ranging from 0.2 to 0.8 μm.[135,136] The structures observed are of the type found in the intermetallic compounds AlB_2, $NaZn_{13}$, $CaCu_5$, and $MgCu_2$. Another structure that does not have its counterpart in the atomic systems is the AB_4 type with $P6_3$/mmc symmetry. It is instructive to mention briefly the main features of the experiments. To start with, deionised, monodisperse suspensions ($\phi \sim 0.1$) of two different diameters, each one showing crystalline order, were mixed. The particle number ratio n_A/n_B was mostly in the range 10 to 20. Just after mixing, the mixture did not show any iridescence and the particles did not move much, suggesting a glassy state for the mixture. In order to hasten the ordering process, the glassy mixtures were diluted so that the system phase separated into crystalline and liquidlike phases. It took about 10 to 50 h (and sometimes several weeks) for the ordered structures to form. The ordered phase consisting of up to two different structures coexisted with the disordered phase. The exact particle number ratios were not specified by Yoshimura and Hachisu, possibly because of sedimentation of larger particles. An attempt has been made[135,138] to analyze the observed structures on the basis of packing of effective hard spheres,[139] as was done for the silica spheres.[137] The main idea of the packing model is that the type of structure is determined by the state of maximum packing density for a given hard sphere diameter ratio σ_B/σ_A.

For a given total particle density n_P, three independent parameters are

[135]S. Yoshimura and S. Hachisu, *J. Phys. (Paris) Colloq.* **46**, C3-115 (1985).
[136]S. Yeshimura and S. Hachisu, *Nature* **283**, 188 (1980); S. Hachisu and Y. Yoshimura, in "Physics of Complex and Super-molecular Fluids" (S. A. Safran and N. A. Clark, eds.), p. 221. Wiley, New York, 1987.
[137]B. J. Sanders and M. J. Murray, *Nature* **275**, 201 (1978); *Phil Mag.* **42**, 721 (1980); J. V. Sanders, *Phil. Mag.* **42**, 705 (1980).
[138]Yoshimura and Hachisu have used the semiempirical formula for $\phi(r, n_A/n_B)$ from Ref. 139. We find that the curves in Figs. 8 and 12 of Yoshimura and Hachisu (Ref. 135) are not consistent with the equation given for $\phi(r, n_A/n_B)$.
[139]S. Yerazunis, S. W. Cornel, and B. Winter, *Nature* **207**, 835 (1965).

needed to characterize a binary mixture of hard spheres: (i) size ratio $\gamma = \sigma_B/\sigma_A$, (ii) the particle density ratio x_A/x_B, where $x_A = n_A/n_P$, and n_A is the number density of the Ath component, and the total particle density $n_P = n_A + n_B$, and (iii) the partial volume fraction $\phi_i = \pi\sigma_i^3 n_i/\sigma$ ($i = $ A or B), $\phi = \phi_A + \phi_B$. A lower limit to the packing fraction of the binary system for the formation of ordered structures is 0.74, when the two sets of particles separate into two monodisperse phases. There is no unique method to determine the structure that maximizes the packing fraction for specified γ and x_A/x_B. Sanders and Murray[137] have determined the possible structures for given γ ($\gamma < 1$; i.e., $\sigma_B < \sigma_A$) and x_A/x_B by testing which structure out of a number of given structures has the highest packing fraction. For example, cubic AB_{13} ($\phi = 0.760$), hexagonal AB_2 ($\phi = 0.782$), and cubic AB ($\phi = 0.793$) are calculated to be the structures of close-packed binary alloys for $\gamma = 0.566$, 0.527, and 0.414, respectively.

In order to use the hard sphere packing model to understand the structures in binary mixtures of charged polyball suspensions, one has to determine experimentally the effective hard sphere diameters of the particles. This is usually not easy. The procedure adopted by Yoshimura and Hachisu[135] is not fully satisfactory. There is no theory to analyze the structures in colloidal alloys of soft sphere colloidal particles such as polyballs. Therefore, instead of using an effective hard sphere model for the polyballs, it is better to study experimentally binary mixtures of suspensions of the hard sphere colloidal particles. Such studies have been reported recently,[61] which used sterically stabilized suspensions of PMMA particles with diameters $\sigma_A = 3350$ Å and $\sigma_B = 2030$ Å (i.e., size ratio $\gamma = 0.61$). Their results are quite interesting, and the four distinct solid phases observed are shown in Fig. 25 as a function of the partial volume fractions ϕ_A and ϕ_B. (For details, see van Megen et al.[61])

14. COLLOIDAL GLASSES

The glassy "phase" in colloids can be characterized as follows: (i) lack of positional long-range order and, hence, absence of iridescence of visible light seen from colloidal crystals, (ii) finite rigidity to low-frequency shear to distinguish the glassy state from the liquid phase, (iii) much slower particle diffusion than in the liquid state, and (iv) distortion or split in the second peak of the $S(Q)$ or $g(r)$, similar to what has been observed in metallic glasses. The first two are the essential defining features of glass.

a. *Glassy State in Binary Mixture*

In multicomponent systems ordering of the particles can be frustrated, leading to an amorphous state. This has been amply seen in conventional atomic systems where multicomponent solid solutions are good glass

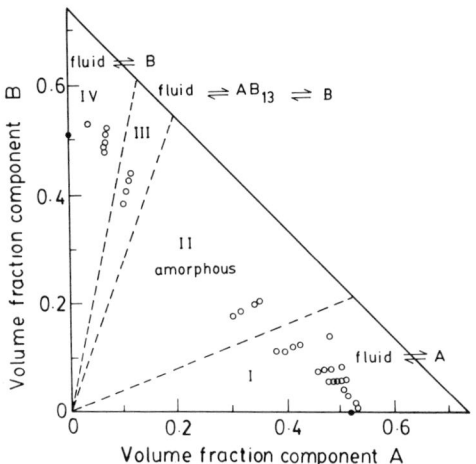

FIG. 25. Constant-volume phase diagram of the binary mixtures of two different particles of PMMA: $\sigma_A = 0.335\,\mu m$, $\sigma_B = 0.23\,\mu m$. The different phases are marked. Taken from W. van Megen, P. N. Pusey, and P. Bartlett, Phase Transition. **21**, 207 (1990).

formers.[140] We shall now discuss the experiments demonstrating glassy behaviour in binary colloids.

(1) *Mechanical Experiments.* The first evidence of glasslike behaviour in the polyball suspensions came from experiments on binary mixtures of 0.109- and 0.22-μm-diameter spheres.[12] It was observed that these dilute mixtures ($\phi < 0.08$) showed no Bragg scattering but had a nonzero shear modulus. The former observation rules out the occurrence of the crystalline phase, and the latter distinguishes the glass "phase" from the liquid state on an appropriate time scale. The question of time scale enters because even a liquid can support shear when $\omega\tau \gg 1$, where τ is the typical relaxation time and ω is the frequency at which shear modulus is being measured. The time scale of the shear–modulus measurements[12] was 0.1 to 1 s.

In the Lindsay and Chaikin experiments, the colloidal glass (also called Wigner glass) was formed by rapidly agitating mixtures of equal volumes of colloidal crystals of 0.109- and 0.220-μm particles. These samples had a shear modulus ~ 10–100 dynes/cm^2, similar to that of the colloidal crystals and the

---

[140] J. F. Sadoc and C. N. J. Wagner, in "Glassy Metals II" (M. Bech and H. J. Guntherödt, eds.), p. 51. Springer-Verlag, Berlin, 1983; M. Kimura and F. Yonezawa, in "Topological Disorder in Condensed Matter" (F. Yonezawa and T. Ninomiya, eds.), p. 80. Springer-Verlag, Berlin, 1983.

value decreased to zero by addition of HCl, thereby indicating the melting of the glass. When the colloidal glass samples were left undisturbed for several months, crystalline colloidal alloys were formed, suggesting that the glassy phase is metastable. Further, to see if the phase-separated state is the stable state of the mixture, colloidal crystals of 0.109-μm polyballs were added on the top of 0.22-μm polyball crystals. It was found that within a few days, the phase-separated crystalline phases merged to give the glassy state. These observations led to the conclusion that the ground state of the binary mixture is perhaps a crystalline compound. It takes very long ($\sim$ several months) for glass to reach the stable state. This is understandable because the diffusion coefficient D of the colloidal particles in the solid phase is very small ($\sim 10^{-12}$ cm^2/s)[99] and they have to move a distance $L \sim$ a few microns (time $\sim L^2/D$).

(2) *Scattering Experiments.* The light scattering experiments of Kesavamoorthy et al.[13] on a large number of dilute binary colloidal suspensions of various compositions, particle diameter ratios, and ionic strengths support the conclusion of a glassy state in the colloids. The glasslike "phase" was identified by its characteristic split in the second peak in $S(Q)$ and much slower diffusion of particles than that in the liquidlike ordered suspensions. The splitting in the second peak in $g(r)$ and, consequently, in $S(Q)$ is known to occur in conventional amorphous systems due to two inequivalent positions for the second-nearest neighbours.[140] In these studies, the polyball diameters range from 0.091 to 0.160 μm, and the diameter ratios lie between 0.92 and 0.69. The total volume fraction is less than 0.002. The ionic strength of the suspension is controlled by using a mixed bed of ion exchange resins. The two monodisperse suspensions, each one not deionised to the extent of showing crystalline order, were mixed in a sample cell that also contained at the bottom a mixed bed of ion exchange resins. Just after mixing ($t = 0$), the $S(Q)$ had smooth first and second peaks similar to those in monodisperse liquidlike suspensions. After some time a small distortion in the second peak appeared, which evolved continuously into a split second peak. Figure 26 shows careful measurements of the structure factors for binary mixtures of two monodisperse suspensions with particle diameters 0.091 and 0.109 μm in the concentration ratios 56:44 and 15:85. The time at which measurements were made after mixing are indicated (in hours) on the curves. The curve s in Fig. 26 refers to the measurement just after shear melting the sample corresponding to the curve c. It is clear that the distortion in the second peak in $S(Q)$, though small, is definitely present. The radial distribution function $g(r)$ obtained by Fourier-transforming these data also showed a split second peak.

In addition to this, Kesavamoorthy et al.[13] also monitored the fluctuations of near-forward scattered light by direct visual observations. Since the

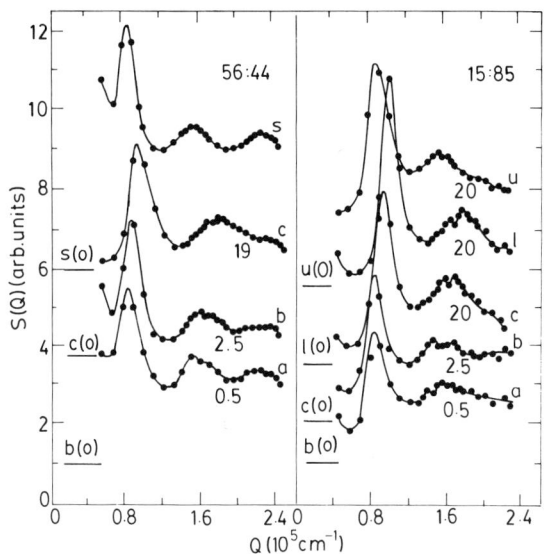

FIG. 26. Measured $S(Q)$'s for binary mixtures of polyballs with particle diameters 0.091 and 0.109 μm in ratios marked 56:44 and 15:85. The times at which measurements were made after mixing are indicated below the curves. The curves are displaced vertically for clarity. The solid lines drawn through the data points are only guides. Taken from R. Kesavamoorthy, A. K. Sood, B. V. R. Tata, and A. K. Arora, *J. Phys. C* **21**, 4737 (1988).

glasslike state and liquidlike suspensions coexisted in the sample, with the glasslike state lying near the bed of the ion exchange resin, the measurements of $S(Q)$ and τ_F were made as a function of height.[13] Figure 27 presents $S(Q)$ for the mixture of 0.11- and 0.12-μm polyballs in the ratio 42:58 with $n_P = 2.83 \times 10^{12}$ cm^{-3}, at two different heights: $h = 1.0$ mm and $h = 3.8$ mm in the sample cell ($h = 0$ refers to the ion exchange resin bed–colloid interface). The fluctuation times τ_F measured at various heights in the sample cell are plotted in the inset of Fig. 27. It is clear from Fig. 27 that τ_F is much larger at those heights in the sample cell where a split second peak is seen. These two observations together demonstrate the existence of the glassy behaviour in the dilute binary suspensions.

It is surprising that the diameter difference of $\sim 9\%$ is able to produce a glassy state. Since the average interparticle spacing in these dilute suspensions is ~ 10 average diameters, the physical diameter of the polyballs should not play a major role. It is the charge difference that is perhaps responsible for the observed glass "phase." Now comes an important question: May this glassy state be thought of as nearly the equilibrium state, or have we

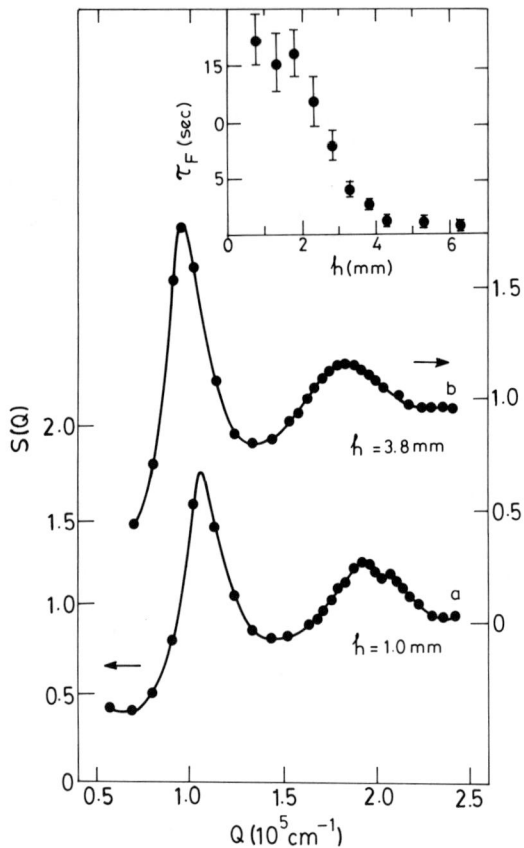

FIG. 27. Structure factor of a binary mixture of 0.11- and 0.12-μm particles in ratios 42:58 at different heights h above the ion exchange resins. The inset shows the intensity fluctuation time τ_F at different heights. The liquid–glass interface is at $h \sim 2.5$ mm. Taken from R. Kesavamoorthy, A. K. Sood, B. V. R. Tata, and A. K. Arora, *J. Phys. C* **21**, 4737 (1988).

effectively quenched the system to produce it? At first glance the deionizing times of many hours (~ 20 h) noted in Fig. 26 may seem to suggest that it is not a quench. However, one should compare the time scales in the atomic systems and the colloids. In the former, the time scale for crystallization is typically $\sim 5 \times 10^{-12}$ s, whereas in the latter it is the time taken by a polyball to diffuse some fraction of an interparticle distance a_s (cf. Section 7) $\sim a_s^2/D_0 \sim 5 \times 10^{-2}$ s. Therefore the ratio of the time scales is $\sim 10^{10}$, and, hence, a deionization time of 20 h is equivalent to a quench time of $\sim 10^{-5}$ s in conventional systems.[18]

(3) *Computer Simulations.* Molecular dynamics simulations have been carried out[141] for a 1:1 mixture of binary charged colloidal suspensions of polyballs with particle diameters $\sigma_A = 2220$ Å ($Z_A = 600$) and $\sigma_B = 1090$ Å ($Z_B = 300$). The interaction potential is taken to be of the DLVO form. In the simulations, it is assumed that no salt is present, and, hence, the screening is only due to the counterions. Different phases (liquid, colloidal alloy, and colloidal glass) are formed as the total volume fraction ϕ is increased. The different phases are characterized by $g(r)$ and the mean square displacement $\langle u^2 \rangle$. In the liquid phase, $\langle u^2 \rangle$ varies more or less linearly with time, whereas in a solid, be it crystal or glass, $\langle u^2 \rangle$ saturates asymptotically to a constant value because of the restricted motion of the particles. The crystal and glass are easily distinguished by their $g(r)$. The results are as follows:[141] (i) The mixture is a liquid when $\phi < 0.0075$. When ϕ is increased to 0.01, the system is found to freeze spontaneously into the AB type bcc structure. For $\phi > 0.01$, the mixture forms a glass. An interesting result is the inference of a preferred length scale ξ in glasses. This has been done by the analysis of the local potential energy profiles, calculated by using the Hessian (dynamical matrix). The value of ξ is estimated to range between $2n_P^{-1/3}$ and $3n_P^{-1/3}$ for $\phi = 0.1$, which contains approximately 100 particles. A precise physical meaning of this length is unclear.

b. *Glassy Phase in Monodisperse Suspensions*

The experimental phase diagram of monodisperse charged polyball suspensions determined by small-angle x-ray scattering (Fig. 7) shows clearly the occurrence of a glass "phase" at high volume fractions.[18] Monovoukas and Gast[47] have also remarked in passing, without structural identification, upon the formation of a glassy state in monodisperse polyball suspensions of $\phi > 0.14$. As in binary mixtures, the glass was characterized[18] by the similarity of $S(Q)$ and $g(r)$ to that of metallic glasses. Figure 28 shows the $g(r)$ obtained from the Fourier transform of measured $S(Q)$ [cf. Eq. (11.3)] for (a) the liquid and (b) the glass.[18] The arrows mark the positions of four nearest neighbours in the fcc structure; these are at r_m, $r_m\sqrt{2}$, $r_m\sqrt{3}$, and $2r_m$, where r_m is the position of the first maximum in $g(r)$. In the random close-packed (rcp) model of metallic glasses, the two parts of the split peak in $g(r)$ occur at $r_m\sqrt{3}$ and $2r_m$, suggesting that the rcp model of effective hard spheres may be able to explain the formation of colloidal glass in monodisperse concentrated suspensions of charged polystyrene spheres.

Another example of the glass "phase" in colloids is shown in Fig. 8 for monodisperse, nearly hard spherical colloidal particles of PMMA.[15,16,61]

---

[141]R. O. Rosenberg, D. Thirumalai, and R. D. Mountain, *J. Phys. Condens. Matter* **1**, 2109 (1989).

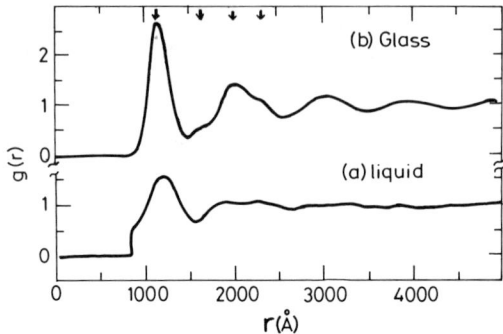

FIG. 28. $g(r)$ obtained from Fourier transform of measured $S(Q)$. Curve a: liquid; curve b: glass. The arrows mark the positions of four nearest neighbours in the fcc structure. Taken from E. B. Sirota, H. D. Ou-Yang, S. K. Sinha, P. M. Chaikin, J. D. Axe, and Y. Fuji, *Phys. Rev. Lett.* **62**, 1524 (1989).

The glass was identified by (i) the absence of iridescence[15] and (ii) a much slower decay of the dynamic structure factor $F(Q, \tau)$,[16,61] defined in Eq. (11.13). In all the samples studied, $F(Q, \tau)$ had two components: first a rapid initial decay and then a slower decay at longer times.[16,142] The fast initial decay arises from the local motion of the particles within a "cage" formed by near neighbours, whereas the slower-decaying component is associated with the hindered diffusion of the particle over long distances. The time constant of the slower decay changed from 0.1 s for a fluid phase (effective volume fraction $\phi_E = 0.48$) to a value of $> 10^3$ s for the glass phase ($\phi_E > 0.554$), thereby implying "structural arrest" in the latter. It was observed[61] that the nondecaying component (time constant $> 10^3$ s) occurs for all values of Q, both below and above the first peak position in $S(Q)$. This is an important observation, which proves that there is no preferred length scale in the density fluctuations that are partially frozen in at the liquid-to-glass transition. This result seems to be in contrast with the prediction of a preferred length scale in glasses by MD simulations of polyball binary mixtures.[141]

VII. Colloidal Crystals in Two Dimensions

The physics of low-dimensional systems is in principle very rich, but it is not always practical to realise a true two-dimensional crystal in conventional

---

[142]Theoretical analysis of dynamic light scattering by nonergodic media such as glasses has been recently reported in P. N. Pusey and W. van Megen, *Physica A* **157**, 705 (1989).

solids,[143] except for the adsorbed atomic monolayers on substrates such as graphite.[144] The macroscopic size of polyballs makes it relatively easy to confine them between two solid surfaces, thereby achieving a two-dimensional colloidal crystal.[145-147] Solid surfaces such as glass plates appear perfectly smooth on the colloidal length scale. Two-dimensional colloidal monolayers can also be formed at water–air interfaces[148-150] and water–solid interfaces.[151] Trapping of polyballs at the water–air interface occurs because of a finite contact angle between water and polystyrene.

15. Structure of 2D Crystals

Figure 29 shows the first layer at the interface of the colloidal crystal–glass surface in the sample cell.[81,152] The colloidal crystal with an fcc structure was formed in the monodisperse aqueous suspensions of polystyrene spheres of 0.53 μm diameter. The (111) plane of the fcc crystal was oriented parallel to the glass plate. It was inferred from the observations that a deep potential well exists near the glass plate (at a distance of $0.8 \times$ interparticle separation). The depth of the potential minimum decreases with distance from the glass plate. One can say that the first layer near the water–glass interface is quasi-two-dimensional.

A better example of the study of two-dimensional colloidal crystals and the transition from two to three dimensions is the work of Pansu et al.[145] Here the colloidal crystals are confined between two glass surfaces forming a wedge (wedge angle $\sim 10^{-2}$ rad). In this wedge geometry, a continuous transition from two to three dimensions is realised. Direct observations with an optical microscope show the following sequence of structures along the length of the wedge (i.e., increasing separation between the plates):[145]

$$1\Delta \to 2\square \to 2\Delta \to \cdots \to n\square \to n\Delta \to (n+1)\square \to \cdots, \qquad (15.1)$$

where n specifies the number of crystalline layers, and the symbol Δ or $\square$

---

[143]We mean only crystals formed by atoms or molecules. These do not include the possible formation of a two-dimensional ordered arrangement of electrons on liquid helium or in an inversion layer in semiconductor heterostructures.

[144]The adsorbed monolayers can interact with the substrate and, hence, may not be truly two-dimensional systems.

[145]B. Pansu, P. Pieranski, and L. Strzelecki, J. Phys. (Paris) **44**, 531 (1983).

[146]C. A. Murray and D. H. Van Winkle, Phys. Rev. Lett. **58**, 1200 (1987).

[147]Y. Tang, A. J. Armstrong, R. C. Mockler, and W. J. O'Sullivan, Phys. Rev. Lett. **62**, 2401 (1989).

[148]P. Pieranski, Phys. Rev. Lett. **45**, 569 (1980).

[149]G. Y. Onoda, Phys. Rev. Lett. **55**, 226 (1985).

[150]A. J. Armstrong, R. C. Mockler, and W. J. O'Sullivan, J. Phys. Condens. Matter **1**, 1707 (1989).

[151]S. Yoshimura, K. Takan, and S. Hachisu, in "Polymer Colloids II" (R. Fitch, ed.). Plenum, New York, 1980.

[152]R. Kesavamoorthy and B. V. R. Tata (unpublished), 1989.

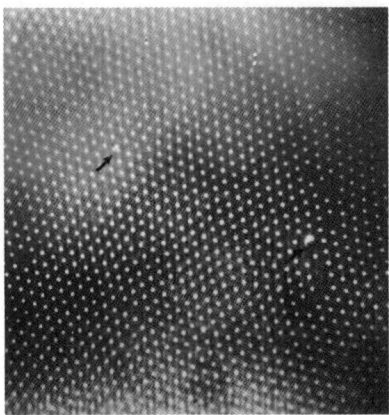

FIG. 29. Micrograph of the first layer [(111) plane] of the fcc colloidal crystal near the glass plate. The polyball diameter is 0.53 µm, and the interparticle separation is 3.2 µm. The arrows mark the aggregate and some inclusions. One can also note some low-angle grain boundaries. Courtesy: R. Kesavamoorthy, 1989.

denotes the coordination of particles in individual layers. For example, 2☐ indicates the stacking of two square layers. Stacking sequences such as ABC or ABA for 3Δ are not specified in the sequence given in Eq. (15.1). The structural transitions marked in Eq. (15.1) have been explained in terms of maximization of packing density in a system of hard spheres bounded by two plane surfaces.[153]

16. MELTING OF 2D COLLOIDAL CRYSTALS

Experiments using optical microscopy and digital imaging have been performed to study the melting transition in a 2D colloidal crystal of charged polystyrene particles.[146,147,150] These experiments address the question whether the melting transition in 2D is first order[154] or continuous, as proposed by Kosterlitz–Thouless–Halperin–Nelson–Young (KTHNY).[155] In the KTHNY theory, there is an intermediate hexatic (H) phase between the crystal (C) and liquid (L) phases. The hexatic phase is characterized by short-range translational order and quasi-long-range orientational order. The melting occurs in two stages: C–H transition characterized by divergence in translational order correlation length, and H–L transition characterized by divergence in orientational correlation length.

[153] B. Pansu, Pi. Pieranski, and Pa. Pieranski, *J. Phys. (Paris)* **45**, 331 (1984).
[154] T. V. Ramakrishnan, *Phys. Rev. Lett.* **48**, 541 (1982); H. Kleinert, *Phys. Lett. A* **95**, 381 (1983).
[155] J. M. Kosterlitz and D. J. Thouless, *J. Phys. C* **6**, 1181 (1973); B. I. Halperin and D. R. Nelson, *Phys. Rev. B* **19**, 2457 (1979); A. P. Young, *Phys. Rev. B* **19**, 1855 (1979).

The experiments of Murray and Van Winkle[146] were done on a colloidal monolayer in a small-wedge geometry, whereas melting in a freely expanding monolayer was studied by Tang et al.[147] The latter method has the advantage that the particle number density gradient present in the wedge method is eliminated. These experiments[146,147] give the positions of the centers of mass of the particles from the digitized and processed images. The translational correlation function $g(r)$, the bond-orientational correlation function $g_6(r)$, and the defect structure are computed from the positions of the particles. The $g_6(r)$ is defined as[150]

$$g_6(r) = \langle \psi_6(r)\psi_6^*(0)\rangle/g(r), \quad (16.1)$$

where

$$\psi_6(r) = \frac{1}{N}\sum_{i=1}^{N} \delta(r - r_i)\frac{1}{6}\sum_{j=1}^{6} \exp(i\delta\theta_{ij}), \quad (16.2)$$

and θ_{ij} is defined as the angle between the bond joining the ith and jth particles and some chosen axis. The correlation lengths ξ and ξ_6 are estimated by assuming exponential decay envelopes $\sim \exp(-r/\xi)$ and $\exp(-r/\xi_6)$ for $g(r)$ and $g_6(r)$, respectively. Similarly, the power-law exponents η and η_6 are determined: $g(r) \sim r^{-\eta}$ and $g_6(r) \sim r^{-\eta_6}$. These correlation lengths and power-law exponents are shown in Fig. 30a and b as a function of the particle number density.[147] Figure 30c shows the fraction of six coordinated particles. The number densities at the two transitions C–H and H–L are marked by n_a and n_b in Fig. 30. The results of Tang et al. shown in Fig. 30 are similar to those of Murray and Van Winkle[146] and are consistent with the results of isothermal–expansion melting experiments in a colloidal monolayer at the water–air interface.[150] The data support the two-stage melting theory of KTHNY. However, the direct observations of defect creation and evolution between n_a and n_b indicated the presence of grain boundaries possibly supporting the model of a first order melting transition by the spontaneous creation of low-angle grain boundaries.[147] Armstrong et al.[150] observed that the melting behaviour of the 2D colloidal monolayer depends on the particle size. The melting of the crystal consisting of 2.88-μm-diameter polyballs is consistent with the existence of a hexatic phase, whereas the 0.10-μm-diameter particle system appeared to melt via a first-order transition. This difference has been attributed to the fact that less energy is required to create bound dislocation pairs in smaller-diameter particle crystals, resulting in the presence of numerous multiparticle defects in the former. These multiparticle defects may lower the core creation energy of dislocations, thereby enhancing their production.[150]

To sum up, the picture of 2D melting in the colloidal monolayers is not fully understood. More work is required to gain an insight into the intricacies of melting and other phase transitions in 2D systems.

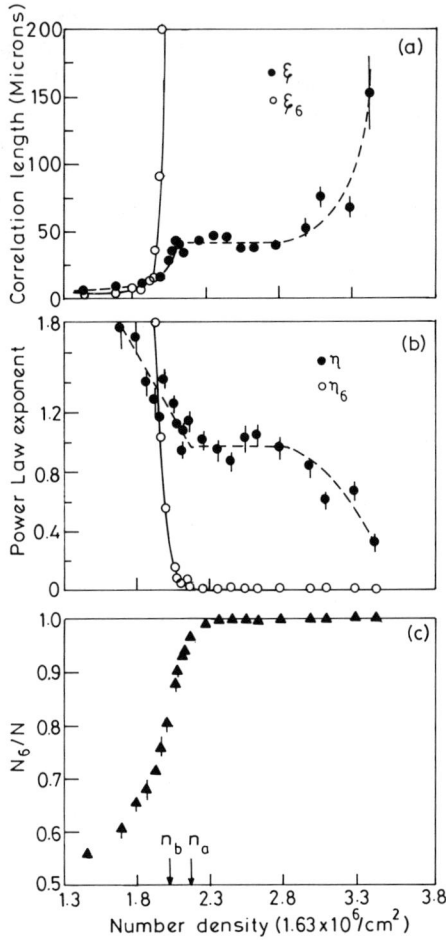

FIG. 30. (a) ξ and ξ_6. (b) η and η_6. (c) The fraction of particles with sixfold coordination. The estimated number densities at the two transitions are shown by n_a and n_b. The lines drawn are guides. Taken from Y. Tang, A. J. Armstrong, R. C. Mockler, and W. J. O'Sullivan, *Phys. Rev. Lett.* **62**, 2401 (1989).

VIII. Elastic Properties and Shear Flow

17. Elastic Properties

As mentioned in Section 3, the typical elastic constants (Young modulus Y, bulk modulus B, or shear modulus G) of the colloidal crystals lie in the range 1 to 1000 dynes/cm^2. The value obviously depends on the strength of the interparticle interaction, which can be controlled by the particle number density n_p, ionic strength n_α, effective surface charge Z_P^*, and the dielectric constant of the medium. It may be recalled that the large difference ($\sim 10^9$–10^{12}) between the values of the elastic constants of the colloids and conventional solids arises due to a similar order of magnitude difference in their particle densities.

Interestingly, the small value of Young's modulus results in the deformation of the colloidal crystal under its own weight. Since the density of polystyrene particles (1.05 g/cm^3) is different from that of water (1.0 g/cm^3), the particles exert a downward gravitational force proportional to their effective density ($=0.05$ g/cm^3). Though the gravitational force exerted is quite small, it produces significant elastic deformations due to the very small value of elastic moduli of the colloidal suspensions. This fact has been used to estimate Young's modulus for the polyball colloidal crystals[29] and the bulk modulus of the liquidlike suspensions.[156]

The viscoelastic behaviour at 40 kHz and steady flow properties of polyball suspensions were determined through the fluid–crystal phase transition.[157] The rigidity in the crystalline phase[158] was about 10^3 dynes/cm^2, and it reduced to almost zero in the disordered state. Low-frequency ($\omega < 20$ Hz) shear moduli of the polyball colloidal crystals have been measured as a function of ϕ and n_α by determining the frequency of standing shear modes.[12] To see the dependence of G on the parameters (such as κ) of the interaction potential, one should recall that G is proportional to the force constant, which depends on the second derivative of the potential with respect to the particle displacement. For the DLVO or Yukawa potential, it has been shown that[159]

$$G = [N_n n_P (\kappa d)^2 U_R(r = d)]/18, \qquad (17.1)$$

where N_n is the number of nearest neighbours (8 for bcc and 12 for fcc crystals), and d is the nearest-neighbour distance. Equation (17.1) has been used recently to estimate the effective charge Z_P^* of the polyball from the measured value of G.[18] The theory has been generalised to apply to a two-component colloidal glass.[160]

---

[156]R. Kesavamoorthy and A. K. Arora, *J. Phys. A* **18**, 3389 (1985).
[157]S. M. Mitaku, T. Ohtsuki, and K. Okano, *Jpn. J. Appl. Phys.* **19**, 439 (1980).
[158]Recall that the complex viscosity $\eta^* = \eta - iG/\omega$, where η and G are the viscosity and the rigidity, respectively. If the viscosity is shear viscosity, then G is called the shear module. Also η is related to the shear rate: η (chordal) $= \sigma/\dot{\gamma}$ or η (differential) $= (\partial \sigma/\partial \dot{\gamma})$, where σ is the shear stress.
[159]R. A. Johnson, *Phys. Rev. B* **6**, 2094 (1972); J. F. Joanny, *J. Colloid Interface Sci.* **71**, 622 (1979).
[160]R. Kesavamoorthy and A. K. Arora, *J. Phys. C* **19**, 2833 (1986).

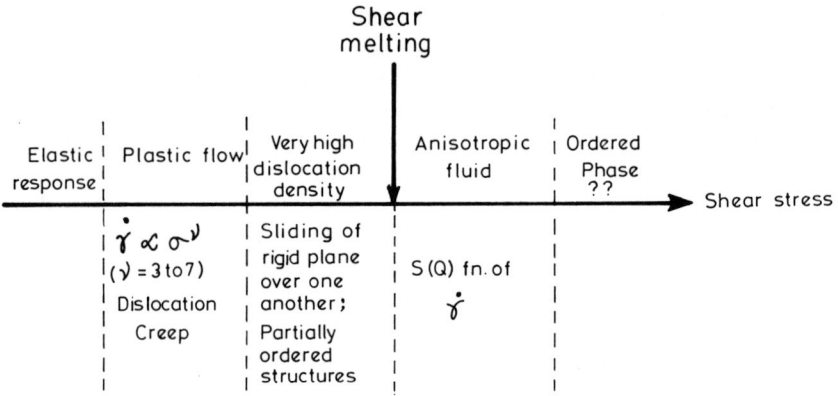

FIG. 31. A schematic illustration of various stages in the flow behaviour of the colloidal crystal as a function of applied shear stress. The transition between different regimes need not be sharp, and hence the dotted lines only help identify the different behaviours.

18. Shear Flows

In the laboratory it is very easy to apply a shear stress, $\sigma_{app} \sim 10^3$ dynes/cm². This applied stress is a very small fraction of the shear modulus G of the conventional solids ($G \sim 10^{12}$ dynes/cm²), $\sigma_{app}/G \sim 10^{-9}$. However, in colloidal crystals, $\sigma_{app}/G \sim 1$ to 10^2. In solids at low applied stress ($\sigma_{app} \lesssim 10^{-4} G$), the response is elastic. As stress increases to $10^{-3} G$ or so, plastic flow sets in. The possibility of achieving $\sigma_{app} \sim 10^2 G$ for the colloidal crystals opens up a new regime of highly nonlinear flow behaviour of condensed matter. In fact, a number of intriguing nonlinear flow behaviours have been observed for the colloidal crystals.[30-34] Figure 31 outlines one such scenario.[30] In the plastic flow regime, the shear rate $\dot{\gamma}$ varies as a power law with applied stress as

$$\dot{\gamma} \propto \sigma_{app}^{\nu}. \tag{18.1}$$

The exponent ν ranges from 3 to 7, depending on the particle density, charge, and ionic strength. As σ_{app} increases, the colloidal crystal becomes unstable and melts into an isotropic fluid. This is known as shear melting,[32,33] and has no known counterpart in conventional solids since the required stresses would be too high. Before the shear-melting transition, the planes in the crystal can slide over one another. A variety of partially ordered phases can occur. After the shear-melting transition, the fluid shows anisotropy in the structure factor. The viscosity, either chordal ($\sigma/\dot{\gamma}$) or differential ($\partial\sigma/\partial\dot{\gamma}$), of the shear-melted crystal is higher than that of the solid.[31] At very high shear

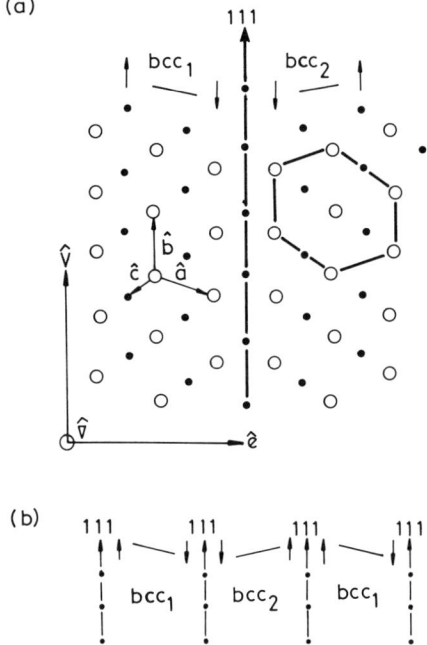

FIG. 32. (a) Twin structures bcc_1 and bcc_2. The open circles is the first layer, and the full dots is the second layer. $\hat{a}, \hat{b}$, and $\hat{c}$ are the basis vectors for the bcc lattice. The arrows show the direction of in-plane strain, which can change these layers into a distorted 2D hcp or distorted fcc structures. (b) Bcc twins stacked in the $\hat{e}$ direction. On application of finite stress, the (bcc) twin structures oscillate as indicated by the arrows and periodically convert into each other. Taken from B. J. Ackerson and N. A. Clark, *Phys. Rev. A* **30**, 906 (1984).

stress or shear rates, it is possible to see a reentrant ordered phase.[33] It is clear that the range of structures and behaviours seen in flowing colloidal crystals goes far beyond what we can hope to see in conventional crystals at ordinary stresses.

The partial ordering in sheared colloidal crystals is interesting and not fully understood.[34] It is summarized here. Before shear is applied, oriented bcc twin structures bcc_1 and bcc_2 coexist, as shown in Fig. 32. The $\langle 111 \rangle$ direction is parallel to that of the local solvent velocity $\hat{V}$. The (211) plane parallel to the $\hat{V}$-$\hat{V}$ plane is the common boundary between the twins, and the (110) planes are parallel to the $\hat{V}$-$\hat{e}$ plane ($\hat{\nabla}$ is the direction of the velocity gradient).[34] For a small nonzero shear rate, the structure changes from an oriented bcc twin to a distorted bcc twin because the structure can oscillate between the two configurations. The crystals orient with the closed packed direction $\langle 111 \rangle$ parallel to the velocity of the flow $\hat{V}$ and the closest-packed planes (110) perpendicular to the shear. As the shear rate is further increased, there is not enough time for the structure to relax from $bcc_1 \leftrightarrow bcc_2$ and the resultant structure becomes a distorted 2D hcp. At still higher shear rates, the (110) layers break up into strings of particles parallel to $\vec{V}$, which eventually break up at very high rates of shear, resulting in an amorphous or liquid structure. It is not clear how one should describe the shear-melted phase, which is far from equilibrium.

The shear experiments can be conveniently carried out in a Couette cell, wherein the colloidal suspension is contained in a narrow gap formed between two concentric cylinders. When the shear is produced by rotating the outer cylinder, only shear-melting instability occurs. But on the other hand, if the colloidal crystal is sheared by rotating the inner cylinder, the centrifugal force causes another instability, namely the Taylor instability (also observed in conventional fluids). Meglio et al.[31] have shown that the critical shear rate of melting is lowered when the Taylor instability occurs along with the shear melting. On the theoretical side, nonequilibrium generalizations of the density functional theory of freezing as well as some other dynamical approaches have been proposed to study shear melting.[161] A quantitative and clear understanding of the flow behaviour in colloidal crystals is far from being complete.

IX. Miscellaneous and Conclusions

19. SOME TOPICS NOT COVERED

We mention some other equally interesting topics that have not been covered in this review.

a. *Rod-Shaped Colloidal Particles*

So far we have considered colloidal particles of spherical shape for which orientational ordering is not relevant (except for the hexatic phase in two dimensions). The ordering in colloidal suspensions of rod-shaped particles gives rise to a whole class of anisotropic phases, such as nematic (N) and smectic (S) liquid crystals. Some examples of colloidal suspensions of rodlike particles are colloidal β-FeOOH,[162] γ-AlOOH,[163] and polytetrafluoroethylene.[164] All these particles are charged and therefore interact via repulsive interactions. So far most of the theoretical understanding is based

---

[161] S. Ramaswamy and S. R. Penn, *Phys. Rev. Lett.* **56**, 945 (1986); B. Bagchi and D. Thirumalai, *Phys. Rev. A* **37**, 2530 (1988); P. Harrowell and M. Fixman, *J. Chem. Phys.* **87**, 4154 (1987).

[162] Y. Maeda and S. Hachisu, *Colloids Surf.* **6**, 1 (1983); **7**, 357 (1983).

[163] J. Bugosh, *J. Phys. Chem.* **65**, 1791 (1961).

[164] T. Folda, H. Hoffmann, H. Chanzy, and P. Smith, *Nature (London)* **333**, 55 (1988).

on hard spherocylindrical particles. These particles are cylinders of length L, diameter D, and capped at each end with a hemisphere of the same diameter. The nature of the phase transition as a function of volume fraction depends on L/D. For small L/D (as in β-FeOOH particles), one can have isotropic (I) $\rightarrow$ S $\rightarrow$ crystal, whereas for large L/D (as in tobacco mosaic virus), the sequence of transition is I $\rightarrow$ N $\rightarrow$ S $\rightarrow$ crystal.[165] The spherical particle is a spherocylinder with $L/D = 0$, for which one observes an I $\rightarrow$ crystal transition.

b. *Colloidal Aggregates*

When the repulsive Coulomb barrier between the charged colloidal particles is greatly reduced either by increasing the ionic strength or by reducing the surface charge, the attractive London–Van der Waals forces come into play and the particles form aggregates. Understanding the stability of colloidal dispersions against aggregation has been one of the key issues in the field of colloidal science. Apart from its obvious commercial importance, it has been recently shown that aggregation is a prototype example of a complicated random process that can reveal such aspects as self-similarity, scaling, and universality.[166] The density–density correlation function $g(r)$ in colloidal aggregates is of power-law form $g(r) \sim r^{D_f - d_s}$ for values of r ranging between the particle size a_p and the average aggregate size R_g. Here D_f is the Hausdorff–Besicovitch fractal dimension relating the mass of the aggregate M to R_g: $M \sim R_g^{D_f}$. The value of D_f is less than or equal to the Euclidean dimension d_s and can be determined by scattering experiments.[166]

c. *Dynamics*

The interparticle interaction and hydrodynamic interactions with the surrounding fluid make the Brownian motion of the colloidal particles very intriguing. The colloidal suspension thus provides a "controllable" many-body system on which modern statistical mechanics and hydrodynamic theories can be tested. We refer to a review article by Pusey and Tough[124] and a recent paper[167] to give a feel for the issues involved in the dynamics of the colloidal particles.

20. A Few Unsolved Problems

The purpose of this review is definitely not to claim that ordering in colloids has been fully understood. In fact, there are a number of unsolved problems, some of which are here mentioned.

---

[165] H. N. W. Lekkerkerker, in "Phase Transitions in Soft Condensed Matter" (T. Riste and D. Sherrington, eds.). Plenum, New York, 1989.

[166] For reviews, see D. W. Schaefer, J. E. Martin, and K. D. Keefer, *J. Phys. (Paris) Colloq.* **46**, C3-127 (1985); R. Jullien, *Comments Cond. Mat. Phys.* **13**, 177 (1987).

[167] D. A. Weitz, D. J. Pine, P. N. Pusey, and R. J. A. Tough, *Phys. Rev. Lett.* **63**, 1747 (1989).

1. Now that all the normal phases (liquid, crystal, and glass) seen in conventional solids have been observed in model colloidal suspensions, one may ask whether the quasicrystalline phase with fivefold symmetry discovered recently in atomic systems[168] can be formed in colloids. The ordered $NaZn_{13}$-type structure[135,136] and AB_{13} phase[61] seen in the colloids may give some hope of observing a quasicrystalline phase in the colloidal suspensions under appropriate conditions.

2. The question of whether the electrostatic interaction between the polyballs is purely repulsive of the DLVO form [Eq. (8.12)] or has both repulsive and attractive parts [e.g., Eq. (9.2)] is still open. Some experimental observations do seem to indicate that the pure repulsive potential is not adequate.

3. Recent experimental results of Sirota et al.[18] on the phase diagram of the polyball suspension do not agree with the molecular dynamics simulations of the Yukawa system.[106] The discrepancy seems to be due to the omission of the geometrical factor in the Yukawa potential. This underlines the immediate need to carry out MD simulations of the complete phase diagram of the polyball suspensions interacting via the DLVO potential. Further there is no satisfactory density functional theory to explain the bcc–fcc and freezing transitions (cf. Fig. 17). It will also be interesting to see, both experimentally and theoretically, the bcc–fcc interfaces in colloidal crystals.

4. Though considerable attention has been paid in the literature to monodisperse suspensions, not much work has been done on binary mixtures. There is no satisfactory theory to explain the various ordered structures possible in the binary mixtures (e.g., as seen by Yoshimura and Hachisu[135]) of charged colloidal particles. The effect of size and charge polydispersity on the order formation in the dilute polyball suspensions is yet to be understood, both theoretically and experimentally.

5. The nonlinear flow behaviour in the colloidal crystals is far from being fully understood. The flow behaviour in the colloidal glasses is also worth investigating. The 2D colloidal monolayers offer a unique opportunity to probe the effect of shear on 2D melting.

21. SUMMARY

Even though the colloidal particles have spherical shape and the interaction is of a simple form, the collective behaviour of the particles is varied and conceptually rich. The mesoscopic structures observed in colloidal suspensions mimic the phases observed in atomic systems. Well-characterized colloids not only provide ideal model systems to study cooperative behaviour in condensed matter but also exhibit many features

---

[168]Paul J. Steinhardt and S. Ostlund, "The Physics of Quasicrystals." World Scientific, Singapore, 1987.

inaccessible in conventional solids. One such example is the occurrence of highly nonlinear exotic flow behaviours in the colloidal crystals. In conclusion, the field of colloids, though mature, is still full of excitement and promises to be so for many more years to come.

Acknowledgments

It is a pleasure to acknowledge an enjoyable collaboration with R. Kesavamoorthy, B. V. R. Tata, and A. K. Arora. I thank R. Kesavamoorthy for providing the micrograph in Fig. 29, and P. N. Pusey, N. Ise, J. Sogami, T. Okubo, and S. A. Asher for sending me their preprints. Thanks are due to Subrata Sanyal for his help in preparation of the manuscript. It is a pleasure to have had fruitful conversations with Sriram Ramaswamy on varied aspects of complex fluids. I also thank him for a critical reading of the manuscript. We thank the following publishers and the corresponding authors for granting permission to reproduce/adopt the figures: Seradyn Inc. (Fig. 3); Academic Press (Figs. 5, 6, and 18); the American Institute of Physics (Figs. 7, 9, 12–16, 23, 28, 30, and 32); Macmillan Magazines Ltd. (Fig. 8); Indian Academy of Sciences (Figs. 19 and 21); Elsevier Science Publishers (Figs. 22 and 24); Gordon and Breach (Fig. 25); and the Institute of Physics (Figs. 26 and 27).

Note Added in Proof

Recently, Sengupta and Sood (to be published) have carried out density-functional theory incorporating the two-body and three-body direct correlation functions of the liquid. The calculated phase diagram reproduces extremely well the features of the phase diagram determined by Monovoukas and Gast (Fig. 6b).[47]

Crystal Nucleation in Liquids and Glasses

K. F. KELTON

Department of Physics
Washington University
St. Louis, Missouri

		Page
I.	Introduction	75
	1. Motivation	75
	2. Background	78
II.	Classical Theory: Steady-State Nucleation	83
	3. Basic Theory	83
	4. Cluster Translation and Rotation	91
III.	Classical Theory: Time-Dependent Nucleation	93
	5. Phenomena	93
	6. Analytical and Closed-Form Solutions	96
	7. Numerical Solutions	99
	8. Preexisting Cluster Distributions	102
IV.	Experimental Studies of Nucleation in Undercooled Liquids	103
	9. Undercooling of Polymorphically Crystallizing Liquids	103
	10. Nucleation Rate Measurements	117
V.	Experimental Studies of Nucleation in Glasses	123
	11. Experimental Methods and Analysis	123
	12. Crystallization of Metallic Glasses	127
	13. Silicate Glasses	135
	14. A Test of the Kinetic Model of Classical Nucleation Theory	151
VI.	Computer Simulations of Steady-State Nucleation	155
	15. Monte Carlo Calculations	155
	16. Molecular Dynamics	157
VII.	Beyond Classical Nucleation Theory	159
	17. Analysis of the Classical Theory	159
	18. Curvature and Anisotropy Effects on σ	161
	19. Field-Theoretic Models of Steady-State Nucleation	162
	20. Further Treatments of Non–Steady-State Nucleation	169
VIII.	Concluding Remarks	172
	Appendix A: Approximations to ΔG	173
	Acknowledgments	177

I. Introduction

1. MOTIVATION

Many liquids and glasses can be maintained out of equilibrium without crystallizing. As illustrated in Fig. 1, showing dilatometric measurements of

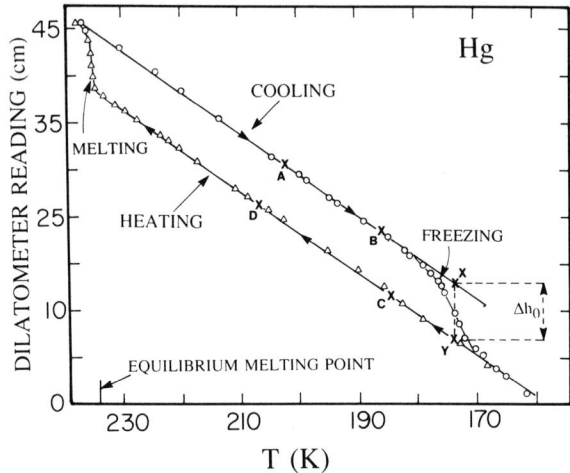

FIG. 1. Dilatometric measurements of an emulsion of Hg as a function of temperature, demonstrating the large undercooling possible in many liquids. Taken from D. Turnbull, *J. Chem. Phys.* **20**, 411 (1952).

liquid and solid mercury emulsions made by Turnbull,[1] the maximum undercooling attainable before solidification can be quite large. Here, the onset of freezing, marked by an abrupt change in slope corresponding to a densification upon crystallization, occurs at approximately 60 K below the equilibrium melting point. In other emulsions, Turnbull obtained undercoolings as large as 80 K, approximately one-third of the melting point.

Below the equilibrium melting temperature (the liquidus temperature for alloys), the Gibbs free energy of the liquid is higher than that of the crystal phases. Undercooled liquids are therefore thermodynamically less stable and should undergo a phase transformation. The large degree of observed undercooling without crystallization demonstrates the existence of a kinetic barrier to that transformation. This can be understood by assuming that the phase transformation is initiated by large-amplitude, localized fluctuations of some order parameter (here the density), leading to the appearance of small regions of the stable crystalline phase. If these regions are larger than some critical size, they will grow and eventually crystallize the liquid. This fluctuation process is called nucleation; the rate at which regions of the stable phase appear is the nucleation rate.

Homogeneous nucleation occurs randomly within the liquid. In the absence of spinodal transformation (an alternative route far from equilib-

---

[1] D. Turnbull, *J. Chem. Phys.* **20**, 411 (1952).

rium), it determines the ultimate limit on metastability. Homogeneous nucleation is material-dependent but does not depend on the specific sample (with the exception of quench rate effects leading to time-dependent nucleation in some glasses), the properties of the container, or the atmospheric conditions of the experiment. Frequently, phase transformations are initiated at specific sites. This heterogeneous nucleation generally dictates the practical limit on metastability. In some cases, it is possible to remove or render inactive the catalytic sites; the homogeneous rate can then be measured and compared with theoretical predictions. A similar study of heterogeneous nucleation is complicated by the difficulties of knowing and controlling the number of sites and the site activities.

While of extreme importance to physics, chemistry, and materials science, crystallization of a liquid is only one example of a nucleation-initiated first-order phase transformation. Other examples abound in diverse physical systems. These include the condensation of supersaturated water vapor (rain), phase separation in metallic alloys, polymers, liquids, and vapors, crystallization (devitrification) of glasses, orientational ordering in molecular crystals and nematic liquids, and domain formation in ferromagnetic systems. The subject of nucleation is therefore an extremely broad one that continues to receive a considerable amount of experimental and theoretical attention.

Experimental nucleation data are generally interpreted via a phenomenological, reaction rate theory, often referred to as the classical theory, first formulated by Volmer and Weber in 1926.[2] Since then, a sizable research effort has aimed at improving the theory, applying it to other first-order phase transformations, reevaluating fundamental assumptions, devising realistic computer simulations of phase transformations, and designing experiments to better test the theories. Despite this activity, the available theories are often in disagreement with each other and with the experimental data. This is due in part to the dependence of most theories of nucleation on unknown microscopic quantities, such as the interfacial energy and the atomic mobility at the interface.

In this review, experimental data and theoretical predictions for nucleation in condensed systems are examined. Since a plethora of articles discussing various aspects of nucleation already exists, and since there is insufficient space or time for an exhaustive treatment of nucleation, this discussion is limited primarily to homogeneous nucleation in undercooled liquids and glasses. Attention is focused on liquid metals and metallic and silicate glasses, since there the most complete kinetic and thermodynamic data exist. A collection of the measured maximum undercoolings and nucleation rates in those systems is presented and is used to examine the

[2]M. Volmer and A. Weber, *Z. Phys. Chem.* **119**, 227 (1926).

validity of the classical theory of nucleation. Previous reviews have generally restricted their discussions to steady-state, or time-independent, nucleation. Particularly for the devitrification of glassy systems, however, the magnitude of the nucleation rate depends on the annealing time. Time-dependent nucleation effects often enhance the phase metastability and may be important for glass formation. Consequently, time-dependent nucleation is reviewed extensively. Nucleation in silicate glasses is emphasized. A comparison between model predictions of the time-dependent nucleation rate and experimental data from those glasses has allowed recently a more critical examination of the kinetic model for nucleation in a condensed system than has been possible previously. Finally, in the last section the relevance of more recent theories for the description of nucleation in liquids and glasses is discussed.

There is no clear distinction between nucleation and growth, although no single theory that unifies both processes currently exists. Further, phase transformations can be initiated by the localized, large-amplitude fluctuations of nucleation or by the extended, small fluctuations of spinodal decomposition. A great body of literature deals with questions of spinodal transformations[3,4] and growth and coarsening under various conditions.[4-6] Those topics are not considered here. Nucleation from the vapor and nucleation in solids are only briefly mentioned; they are also covered in more detail elsewhere.[7]

2. BACKGROUND

Fahrenheit initiated the investigation of phase equilibrium and undercooling in 1714 when he conducted the first recorded systematic study of the crystallization of water.[8] He found that boiled water in a sealed, airtight container, could be kept overnight at the undercooled temperature of 15°F without crystallization. The introduction of small ice particles, however, initiated the crystallization, with the temperature of the ice–water mixture rising to 32°F, the melting temperature at atmospheric pressure. He noticed that a sudden jar to the container of undercooled water also initiated crystallization. As described by Dunning,[9] these observations were confirmed

[3] J. W. Cahn, *Trans. Metall. Soc. AIME* **242**, 166 (1968).
[4] J. D. Gunton, M. San Miguel, P. S. Sahni, in "Phase Transitions and Critical Phenomena" (C. Domb and J. L. Lebowitz, eds), Vol. 8, p. 267, Academic Press, New York, 1983.
[5] H. Biloni, in "Physical Metallurgy" (R. W. Cahn, and P. Haasen, eds.), third edition, p. 477. North-Holland, Amsterdam, 1983.
[6] R. D. Doherty in Ref. 5, p. 933.
[7] J. Feder, K. C. Russell, J. Lothe, and G. M. Pound, *Adv. Phys.* **15**, 111 (1966).
[8] D. B. Fahrenheit, *Phil. Trans. Roy. Soc.* **39**, 78 (1724).
[9] W. J. Dunning, in "Nucleation" (A. C. Zettlemoyer, ed.), p. 1. Marcel Dekker, New York, 1969, and references therein.

quickly and extended to other liquids.[10] Lowitz[11] first observed supersaturation in a salt solution and noted the analogy with undercooled water. Gay Lussac[12] showed that supersaturation is a general phenomenon and supported Fahrenheit's observation of the effects of motion by demonstrating that shaking, scratching, and rubbing could also induce crystallization.

Schröder, von Dusch, and Violette recognized that the observed variability of results was due largely to airborne particles and particles residing in the containers.[13,14] When these were partially eliminated, more consistent measurements were obtained.[15] A particular relationship with the crystallization product was necessary for a strong catalytic effect of the heterogeneity. Lowitz[11] found that seeding of supersaturated solutions or undercooled liquids with small crystallites of the stable phase led to rapid crystallization, while unrelated particles often had little effect. Ostwald[16] demonstrated that only very small seeds were required; 10^{-10} g was sufficient to crystallize sodium chlorate solutions.

The first evidence for a metastability limit was produced by de Boisbaudran,[17] who found that homogeneous nucleation occurred in highly supersaturated solutions, but did not occur in less supersaturated ones. De Coppet[18] measured the average time lag before crystallization in solutions of known supersaturation. Ostwald[16,19] defined two types of supersaturated solutions: (1) metastable, which in the absence of heterogeneous sites would remain unchanged indefinitely, and (2) labile, which will crystallize in a short time. Tammann[20] observed this boundary as a function of undercooling in piperine. In both regions, however, the transformation was initiated by a nucleation mechanism.

Gibbs gave the first detailed discussion of the criteria for phase equilibrium and proposed a sharp distinction between metastable and unstable regions.[21] According to Gibbs, phase transformations in the metastable region are initiated by nucleation; phase transformations in the unstable

---

[10]G. E. Fischer, *Geschichte der Phys.*, Goettingen **5**, 279 (1804); *Gehlers Phys. Wörterbuch* **1**, 678 (1789); via Ref. 9.

[11]J. T. Lowitz, *Crells Chemische Annalen* **1**, 3 (1795); via Ref. 9.

[12]J. L. Gay Lussac, *Ann de Chimie* **87**, 225 (1813); *Ann. Chim. Phys.* **11**, 296 (1819).

[13]H. Schröder and von Dusch, *Leibigs Annalen* **89**, 232 (1853); **109**, 35 (1859); via Ref. 9.

[14]C. Violette, *Compte. Rend.* **60**, 831 (1865).

[15]D. Gernez, *Compte. Rend.* **60**, 833; **61**, 71 (1865).

[16]W. Ostwald, *Z. Phys. Chem.* (Leipzig) **22**, 289 (1897).

[17]Lecoq de Boisbaudran, *Compte. Rend.* **63**, 95 (1866).

[18]L. C. de Coppet, *Ann. Chim. Phys.* **6**, 275 (1875); via Ref. 9.

[19]W. Ostwald, *Z. Phys. Chem.* **34**, 493 (1900).

[20]G. Tammann, *Z. Phys. Chem.* (Leipzig) **25**, 441 (1898); *Kristallisieren und Schmelzen* (Leipzig), 151 (1903); P. Othmer, *Z. Anorg. Allgem. Chem.* **91**, 200 (1915); via Ref. 9.

[21]J. W. Gibbs, *Trans. Connect. Acad.* **3**, 108 (1876); **3**, 343 (1878); "Scientific Papers," Vols. I, II. Longmans Green, London, 1906.

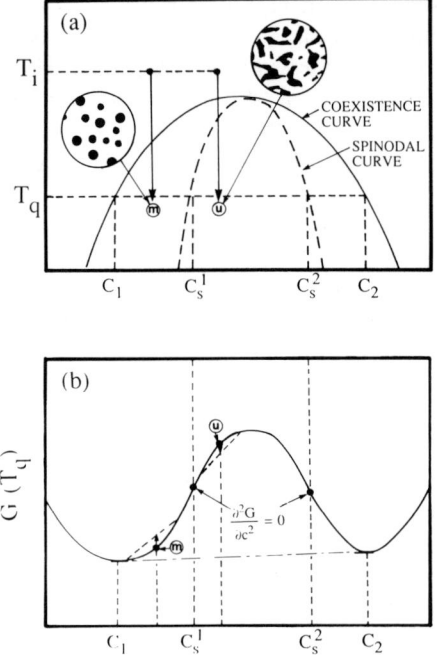

FIG. 2. (a) A schematic depiction of the morphology of a crystallizing liquid obtained by quenching into the metastable (m) or unstable (u) regions. The liquid–solid coexistence and the classical chemical spinodal curves are indicated. (b) A schematic diagram showing the relation of the boundaries of the spinodal and coexistence regions to the shape of the Gibbs free-energy curve.

region occur by a spinodal mechanism involving long-range fluctuations of infinitesimal amplitude. This is illustrated schematically in Fig. 2a for the case of phase separation of a binary alloy. The solid line is the coexistence curve as a function of alloy composition. Quench m is into the metastable region where transformation proceeds by nucleation and growth. The boundary of the spinodal region within which phase transformations proceed by the spinodal mechanism (quench u) is called the spinodal curve (here the chemical spinodal) and is noted in Fig. 2a. It is defined as the locus of points inside the coexistence curve for which the curvature of the free energy changes from convex to concave (Fig. 2b). From Fig. 2b, the energy increases with a spontaneous concentration fluctuation in the nucleation region, giving rise to an energetic barrier that stabilizes the system in the metastable state. No energetic barrier to phase separation exists in the spinodal region; there, the system is unstable to spontaneous fluctuations.

As illustrated schematically in Fig. 2, a nucleation-initiated phase transformation generally produces a droplet morphology; an interconnected structure results from a spinodal transformation. The phase morphology alone, however, is insufficient to identify unambiguously the transformation mechanism,[22-24] since an interconnected structure can also result from the superposition of many separately nucleated grains. The spinodal mechanism is best identified from small-angle x-ray scattering experiments.[25,26]

While remaining a useful concept, Gibbs' notion of a sharp dividing line between nucleation and spinodal transformation cannot be valid. The limit of metastability, defined as the point at which phase transformation is overwhelmingly probable, and determined from the undercooling limit for liquids or the supersaturation limit for vapors, is dependent on the time scale of the experiment. An operational, but not an intrinsic, limit on metastability is therefore determined.[27] This observation is supported by field-theoretic calculations demonstrating a gradual transition of the dynamical behavior of a quenched system as the concentration is varied in the vicinity of the classical spinodal curve.[4]

Nucleation and growth are thermally activated processes. Near equilibrium, the kinetic barrier is very large, and the probability of occurrence for a significant number of fluctuations leading to the stable phase is infinitesimal. At large departures from equilibrium, yet still within the metastability region, this barrier decreases to a few $k_B T$, defining a limit on metastability for which these fluctuations are present in appreciable numbers among the equilibrium fluctuations that describe the liquid state.

The significance of Gibbs' work to nucleation remained largely unrecognized until 1926, when Volmer and Weber[2] asserted that the metastability of a supersaturated phase was largely a question of kinetics. They constructed the first theory of nucleation by using Gibbs' formulation of the reversible work for the formation of a static cluster of the new phase, to calculate the equilibrium cluster density distribution. The nucleation rate was taken to be proportional to the rate of formation of critical-sized clusters in equilibrium. Szilard[28] and Farkas[29] formulated a detailed kinetic model for cluster

---

[22] J. E. Hilliard, in "Phase Transformations" (H. I. Aronson, ed.), American Society for Metals, Metals Park, OH, 1970.
[23] H. Herman and R. K. MacCrone, *J. Am. Ceram. Soc.* **55**, 50 (1971).
[24] C. M. F. Jantzen and H. Herman, in "Phase Diagrams: Materials Science and Technology" (A. A. Alper, ed.), Vol. 5, p. 127. Academic Press, New York, 1978.
[25] V. Gerold and G. Kostorz, *J. Appl. Crystallogr.* **11**, 376 (1978).
[26] K. Hono and Ken-Ichi Hirano, *Phase Transitions* **10**, 223 (1987).
[27] J. S. Langer, in "Lecture Notes in Physics, Systems Far from Equilibrium" (L. Garrido, ed.), Vol. 132, p. 12. Springer-Verlag, Berlin, 1980.
[28] L. Szilard, referred to in Farkas, Ref. 29.
[29] L. Farkas, *Z. Phys. Chem. A* **125**, 236 (1927).

evolution that became the basis for later treatments. Becker and Döring[30] argued that a steady-state distribution was more appropriate than the equilibrium distribution chosen by Volmer and Weber, and obtained an expression for the steady-state nucleation rate. Turnbull and Fisher[31] extended that formalism to the case of nucleation in a liquid. Collectively, these theories constitute what is commonly called the classical theory.

There are many proposed extensions of the classical theory. Lothe and Pound[32] argued that cluster translational and rotational degrees of freedom must be taken into account for the computation of the reversible work of cluster formation, particularly for vapor condensation. These factors change the predicted nucleation rate by approximately 17 orders of magnitude; no comparable discrepancy between predictions from the original formulation of the nucleation theory and experimental data appears. The proper inclusion of these terms continues to be debated, though they should be of negligible concern for liquid–solid nucleation. Cahn and Hilliard[33] used a density functional approach to consider the case of nucleation when the interface between the nucleating cluster and the original phase is diffuse instead of infinitely sharp, as had been assumed previously. Theirs was primarily a thermodynamic investigation; the reaction rate kinetic model was assumed. Langer and co-workers extended that analysis by taking account of fluctuations in cluster size and shape and formulating a hydrodynamic model for the kinetics near the critical point.[27,34–36] These more advanced theories have only recently been applied to nucleation in condensed systems and then only under very restrictive and idealized conditions.

Nucleation theories in liquids have been tested traditionally by comparing the predicted and measured values for the maximum undercooling and the magnitude and temperature dependence of the steady-state nucleation rate. The first systematic studies were performed by Turnbull,[37] who used a technique pioneered by Vonnegut[38] to increase considerably the degree of undercooling in liquid metals by dividing the liquid into a large ensemble of isolated drops within a suitable emulsion. With mercury, Turnbull obtained the first measurements of the magnitude of the nucleation rate as a function of temperature.[1] He found qualitative agreement with the temperature de-

---

[30] R. Becker and W. Döring, *Ann. Phys.* **24**, 719 (1935).
[31] D. Turnbull and J. C. Fisher, *J. Chem. Phys.* **17**, 71 (1949).
[32] J. Lothe and G. M. Pound, *J. Chem. Phys.* **36**, 2080 (1962).
[33] J. W. Cahn and J. E. Hilliard, *J. Chem. Phys.* **31**, 688 (1959).
[34] J. S. Langer, *Ann. Phys.* **41**, 108 (1967).
[35] J. S. Langer, *Ann. Phys.* **54**, 258 (1969).
[36] J. S. Langer and L. A. Turski, *Phys. Rev.* **A 8**, 3230 (1973); **22**, 2189 (1980).
[37] D. Turnbull, *J. Appl. Phys.* **21**, 1022 (1950).
[38] B. Vonnegut, *J. Colloid. Sci.* **3**, 563 (1948).

pendence predicted by the classical theory, but large discrepancies between the magnitudes. To date, quantitative measurements have been reported only for mercury and gallium.[39] A recent reanalysis of the Ga data, however, suggests that they do not constitute a true measure of the homogeneous nucleation rate.[40] Recently, there has been a developing interest in nucleation studies in silicate glasses, since both steady-state and time-dependent rates are measurable.[41-43] Measurements of time-dependent phenomena, in particular, offer new tests for nucleation theories.[44]

II. Classical Theory: Steady-State Nucleation

3. BASIC THEORY

a. *Homogeneous Nucleation*

Homogeneous nucleation in undercooled liquids and glasses is generally described by the classical theory, a phenomenological theory in which actual clusters of atoms or molecules in the configuration of the transformation product arise spontaneously. The simplicity and flexibility of the classical theory make it applicable to the study of a wide range of nucleation phenomena. In this section, the expression for the steady state, or time independent, nucleation rate is developed. Time-dependent solutions are discussed in Section III.

For simplicity, spherical clusters of crystal with the same composition as the liquid or glass, that form by fluctuations in the homogeneous initial phase are considered. Stress effects are ignored. Contributions to the free energy of the cluster due to stresses arising from volume changes on transformation are important for nucleation of one crystalline phase in a crystalline phase of different composition or structure.[45] Assuming, however, that the growth rate of a cluster scales with the fluidity (proportional to the inverse of the viscosity) of the surrounding phase, stresses at the interface of a small cluster growing in a liquid or glass should relax at approximately the same rate as

---

[39] Y. Miyazawa and G. M. Pound, *J. Cryst. Growth* **23**, 45 (1974).
[40] D. Turnbull, "Hume-Rothery Memorial Symposium" (E. W. Coolings, C. C. Koch, eds.), p. 3. TMS-AIME, New Orleans, LA, 1986.
[41] P. F. James, *J. Non-Cryst. Solids* **73**, 517 (1985).
[42] I. Gutzow, D. Kashchiev, and I. Avramov, *J. Non-Cryst. Solids* **73**, 477 (1985).
[43] P. F. James, in "Glasses and Glass-Ceramics" (M. H. Lewis, ed.), p. 59. Chapman and Hall, London, 1989.
[44] K. F. Kelton and A. L. Greer, *Phys. Rev. B* **38**, 10089 (1988).
[45] H. I. Aaronson and K. C. Russell, in "Solid–Solid Phase Transformations" (H. I. Aaronson, D. E. Laughlin, R. F. Sekerka, and C. M. Wayman, eds.), p. 371. TMS-AIME, Warrendale, PA, 1982; W. C. Johnson and J. K. Lee, p. 151.

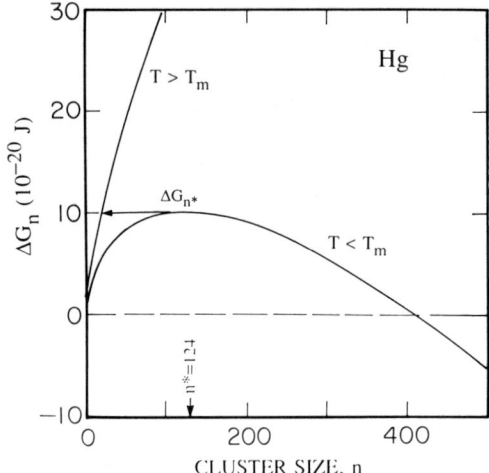

FIG. 3. The reversible work of formation of a cluster of n atoms, ΔG_n, computed using thermodynamic data for liquid mercury, 50 K above and 100 K below the equilibrium melting temperature, T_m. Below T_m a maximum is observed at a critical size n^*.

they develop. Finally, the volume changes associated with phase transitions in condensed systems are sufficiently small that either the Helmholtz or the Gibbs free energies can be used; the Gibbs free energy is used in this review.

Only the probability of obtaining a fluctuation of a given size in an equilibrium system can be computed within a purely thermodynamic approach to nucleation. From the theory of thermodynamic fluctuations,[46] the probability of obtaining a cluster of the new phase containing n atoms, P_n, depends on the minimum reversible work for cluster formation, ΔG_n:

$$P_n \propto \exp(-\Delta G_n/k_B T), \quad (3.1)$$

where k_B is Boltzmann's constant. This is the probability for sampling the microstates of the system that are available in equilibrium; it is valid when the homogeneous medium is equilibrated more rapidly than the formation time of the fluctuation. The equilibrium cluster size distribution per mole, N_n^e is readily obtained from Eq. (3.1):

$$N_n^e = N_A \exp[-\Delta G_n/k_B T], \quad (3.2)$$

where N_A is Avogadro's number of molecules.

[46]L. D. Landau and E. M. Lifshitz, "Statistical Physics," Chapter XII, Pergamon Press, Oxford, 1969.

Making the capillarity approximation where, following Gibbs, the interface between the cluster and the initial phase is sharp, ΔG_n can be expressed as the sum of volume and interfacial energy contributions:

$$\Delta G_n = n\,\Delta G' + (36\pi)^{1/3}\bar{v}^{2/3}n^{2/3}\sigma. \tag{3.3}$$

Here $\Delta G'$ is the Gibbs free energy per molecule of the new phase less that of the initial phase, $\bar{v}$ is the molecular volume, and σ is the interfacial energy per unit area. Within the classical theory, the macroscopic values for $\Delta G'$ and σ are used, even for small clusters. For undercooled liquids and glasses, $\Delta G'$ depends on the undercooling, $\Delta T = T_m - T$, where T_m is the equilibrium melting point and T is the temperature; $\Delta G'$ is positive above T_m and negative below. Figure 3 shows the dependence of ΔG_n on cluster size, assuming values for $\Delta G'$ and σ that are appropriate for mercury. Since σ is always positive, above T_m ΔG_n increases monotonically with increasing cluster size. Below T_m, ΔG_n also increases for small cluster sizes, due to the large surface to volume ratio, but decreases for large clusters where the volume free energy is dominant. A maximum in ΔG_n is then obtained,

$$\Delta G_{n^*} = \frac{16\pi}{3}\frac{\sigma^3}{\Delta G_v^2}, \tag{3.4}$$

corresponding to a critical cluster size, n^*, given by

$$n^* = \frac{32\pi}{3\bar{v}}\frac{\sigma^3}{|\Delta G_v|^3}. \tag{3.5}$$

Here ΔG_v is the Gibbs free energy difference per unit volume, $\Delta G'/\bar{v}$. At this maximum, a cluster of the final phase is in unstable equilibrium. Clusters smaller than the critical size, n^*, tend to shrink while clusters larger than n^* will on average grow. In the crudest sense then, nucleation is simply the production of postcritical clusters.

Volmer and Weber[2] first recognized the importance of including the kinetics of cluster growth and developed the first kinetic model for nucleation. The central ideas of that model have served as a basis for further development by Szilard,[28] Farkas,[29] Volmer,[47,48] Becker and Döring,[30] Zeldovich,[49] Frenkel,[50,51] Turnbull and Fisher,[31] and others.

---

[47] M. Volmer, *Z. Phys. Chem.* **25**, 555 (1929).
[48] M. Volmer, *Kinetik der Phasenbildung, Steinkopff, Dresden* 122 (1939).
[49] J. B. Zeldovich, *Acta Physiochim URSS* **18**, 1 (1943).
[50] J. Frenkel, *J. Chem. Phys.* **7**, 200 (1939).
[51] J. Frenkel, "Kinetic Theory of Liquids," Clarendon, Oxford, 1946.

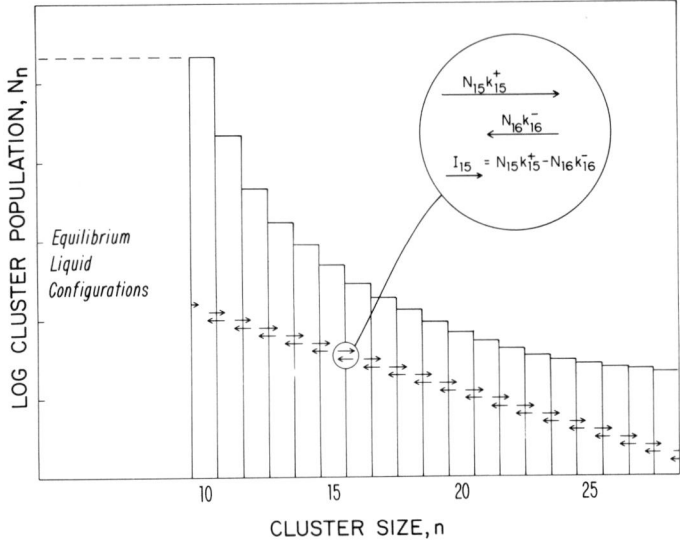

FIG. 4. A histogram of the population as a function of cluster size, n, showing the flux describing cluster growth. Below some lower limit (here $n = 10$), the clusters of the new phase are indistinguishable from equilibrium fluctuations in the liquid.

Clusters are assumed to evolve slowly in size by a series of bimolecular reactions:

$$E_{n-1} + E_1 \underset{k_n^-}{\overset{k_{n-1}^+}{\rightleftarrows}} E_n, \tag{3.6a}$$

$$E_n + E_1 \underset{k_{n+1}^-}{\overset{k_n^+}{\rightleftarrows}} E_{n+1}, \tag{3.6b}$$

where E_n represents a cluster of n molecules and E_1 a single molecule (or monomer). Here k_n^+ is the rate of monomer addition to a cluster of size n and k_n^- is the rate of loss. The time-dependent cluster density, $N_{n,t}$ is determined by solving a system of coupled differential equations of the form

$$\frac{dN_{n,t}}{dt} = N_{n-1,t} k_{n-1}^+ - [N_{n,t} k_n^- + N_{n,t} k_n^+] + N_{n+1,t} k_{n+1}^-, \tag{3.7}$$

which has the form of a master equation.

Figure 4 illustrates this model, showing the evolving cluster density and the reactions involved. It is difficult to decide which clusters have local configurations that are typical of the final phase and which arise naturally from equilibrium fluctuations within the liquid. This division is in fact artificial since, if the fluctuation approach is meaningful, all configurations must be microstates accessible from the metastable equilibrium state of the initial phase. For this figure, the lower limit is taken to be a cluster of 10 monomers. This is arguably valid, for example, in an fcc metal for which the smallest configuration clearly distinguishable from the range of configurations available in the liquid is that of two edge-sharing octahedra, consisting of 10 atoms.[52] From Fig. 4, the nucleation rate at a cluster size n, $I_{n,t}$, is the time-dependent flux of clusters past that size and is given by

$$I_{n,t} = N_{n,t}k_n^+ - N_{n+1,t}k_{n+1}^-. \tag{3.8}$$

Volmer and Weber[2] made the simplifying assumptions that the back flux is zero for cluster sizes larger than the critical size and the density $N_{n,t}$ is therefore set to zero for $n > n^*$. For $n \leqslant n^*$, $N_{n,t}$ is set to the equilibrium distribution, N_n^e [Eq. (3.2)]. The nucleation rate per mole, I, is then

$$I = N_{n^*}^e k_{n^*}^+ = k_{n^*}^+ N_A \exp\left(-\frac{\Delta G_{n^*}}{k_B T}\right). \tag{3.9}$$

Becker and Döring[30] argued that a steady-state distribution, of clusters, N_n^s, is more realistic than an equilibrium distribution, giving a constant nucleation rate, the steady-state nucleation rate, I^s,

$$I^s = N_n^s k_n^+ - N_{n+1}^s k_{n+1}^-. \tag{3.10}$$

For a true steady state, I^s is independent of the cluster size at which it is evaluated; this is not obtained experimentally, as will be discussed in greater detail in Sections III and V. In their treatment, Becker and Döring add $n^* + 1$ molecules to the initial phase for each nucleus removed to maintain a true steady-state condition. In practice, the number of molecules involved in forming the nuclei is sufficiently small that this generally has negligible effect,[52] at least in the initial stages of phase transformations.

If the initial phase were in equilibrium, the nucleation rate would be zero; Eq. (3.8) then gives

$$N_n^e k_n^+ = N_{n+1}^e k_{n+1}^-. \tag{3.11}$$

Solving Eq. (3.11) for the backward rate constant, k_{n+1}^-, we can write the steady-state nucleation rate [Eq. (3.10)] as

$$I^s = N_n^e k_n^+ \left[\frac{N_n^s}{N_n^e} - \frac{N_{n+1}^s}{N_{n+1}^e}\right]. \tag{3.12}$$

Cluster sizes u and v are chosen such that for $n \leqslant u$, $N_n^s = N_n^e$ and for $n \geqslant v$, $N_n^s = 0$. This assumption must be true as $u \to 0$ and $v \to \infty$. Fortunately the solution does not depend strongly on the values of u and v, provided that the energies corresponding to these cluster sizes are at least $k_B T$ lower than the

---

[52] K. F. Kelton, A. L. Greer, and C. V. Thompson, *J. Chem. Phys.* **79**, 6261 (1983).

energy at the critical size.[52] Summing Eq. (3.12) for all values of n between u and v and using the stated boundary conditions, we obtain

$$I^s \sum_u^v \frac{1}{N_n^e k_n^+} = \frac{N_u^s}{N_u^e} - \frac{N_{v+1}^s}{N_{v+1}^e} = 1. \tag{3.13}$$

I^s is obtained by evaluating the sum, making several assumptions: (1) the terms of the sum for values of n near n^* dominate since $1/N_n^e$ has a maximum at n^*; (2) k_n^+ is replaced by $k_{n^*}^+$; (3) N_A is taken to be a constant equal to the total number of molecules in the system (Avogadro's number per mole); (4) ΔG_n is replaced by the first two nonzero terms in a Taylor expansion about n^*; and (5) the sum is replaced by an integral from $n - n^* = -\infty$ to $n - n^* = \infty$, taking N_n^e to be a continuous function of n. With these approximations, the value for the steady-state nucleation rate is

$$I^s = N_{n^*}^e k_{n^*}^+ (|\Delta G'|/6\pi k_B T n^*)^{1/2} = N_{n^*}^e k_{n^*}^+ Z. \tag{3.14}$$

This differs from the Volmer and Weber expression (Eq. 3.9) by only a multiplicative factor, Z, called the Zeldovich factor,

$$Z = (|\Delta G'|/6\pi k_B T n^*)^{1/2}. \tag{3.15}$$

In most cases, $0.01 \leqslant Z \leqslant 0.1$. Knowing I^s, the steady-state distribution, we can obtain N_n^s from the relation

$$I^s \sum_n^v \frac{1}{N_n^e k_n^+} = \frac{N_n^s}{N_n^e}, \tag{3.16}$$

where $N_n^s = 0$ for $n \geqslant v$. With the same approximations as used to derive Eq. (3.14), $N_{n^*}^s = \frac{1}{2} N_{n^*}^e$.

As first shown by Zeldovich, the steady-state and time-dependent nucleation rates can be computed by constructing a Fokker–Planck equation that approximates the master equation.[49] Taking n to be continuous, we can combine Eqs. (3.7), (3.8), and (3.11) to give

$$\frac{\partial N_{n,t}}{\partial t} = \frac{\partial}{\partial n} \left\{ k_n^+ N_n^e \frac{\partial}{\partial n} \left(\frac{N_{n,t}}{N_n^e} \right) \right\}, \tag{3.17}$$

the Zeldovich–Frenkel equation.[49,51] This has the form of a diffusion equation where k_n^+ and $N_{n,t}$ are analogues to the diffusion coefficient and concentration, respectively.

For the case of nucleation from the vapor, k_n^+ and k_n^- are defined readily in terms of the molecular velocity in the vapor, the attachment probability to the developing cluster, and the evaporation rate. The appropriate mobilities and sticking probabilities for nucleation in a condensed phase are less obvious. Turnbull and Fisher[31] first applied the Becker–Döring formalism to nucleation in condensed systems. Assuming that with the addition or removal of a monomer to a cluster, the system passes through a configuration (the activated complex) that is higher in energy than either the initial or final state, the reaction rate theory[53] was used to calculate the forward and backward rate constants:

$$k_n^+ = 4n^{2/3} \gamma \exp[-\delta g_n/2k_B T], \tag{3.18a}$$

$$k_n^- = 4(n-1)^{2/3} \gamma \exp[+\delta g_{n-1}/2k_B T]. \tag{3.18b}$$

Here δg_n is the free energy of a cluster of $n+1$ molecules less that of a cluster containing n molecules, and γ is an unbiased molecular jump frequency at the cluster interface, taking the molecule from one local configuration to a new one. The factor of $4n^{2/3}$ is obtained from a consideration of the number of available attachment sites on the surface of a spherical cluster.[52] Expressed in terms of the attempt frequency, v,

$$\gamma = v \exp(-\delta g^*/k_B T), \tag{3.19}$$

where δg^* is the difference between the energy of the activated state and the average energies of the initial and final states. The jump frequency γ is often assumed to be the same as the jump frequency for bulk diffusion, D, giving

$$\gamma = 6D/\lambda^2, \tag{3.20}$$

where λ is the atomic jump distance.

With these approximations, the steady-state nucleation rate for a condensed system can be written as

$$I^s = \frac{24 D n^{*2/3} N_A}{\lambda^2} \left(\frac{|\Delta G'|}{6\pi k_B T n^*}\right)^{1/2} \exp\left(-\frac{\Delta G_{n^*}}{k_B T}\right). \tag{3.21}$$

The critical size n^* [Eq. (3.5)] and ΔG_{n^*} [Eq. (3.4)] decrease with increasing undercooling; Eq. (3.21) predicts a correspondingly sharp rise in the steady-state nucleation rate. I^s decreases with further undercooling since D drops sharply at lower temperatures, reflecting a lower atomic mobility. This predicted temperature dependence for I^s is observed experimentally (cf. Section V).

Equation (3.21) has the form

$$I^s = A^* \exp[-W^*/k_B T], \tag{3.22}$$

where W^* is the barrier to nucleation. The nucleation rate is then proportional to the thermodynamic probability of having a fluctuation leading to the formation of a critical cluster, and a dynamical factor describing the

---

[53] S. Glasstone, K. J. Laidler, and H. Eyring, "The Theory of Rate Processes," McGraw-Hill, New York, 1941.

rate at which that cluster grows, A^*. The value of A^* is central to the controversy over the validity of the classical theory of nucleation for describing nucleation in liquids and glasses. Experimentally determined values are typically many orders of magnitude larger than those calculated from Eq. (3.21) (cf. Sections IV and V).

b. *Homogeneous Nucleation in Multicomponent Systems*

Steady-state solutions for the homogeneous nucleation rate of a crystallite with a composition different from the initial phase have been obtained by a number of investigators, beginning with Reiss.[54] In that calculation, the composition of the parent phase was held fixed, but a range of cluster compositions was allowed. The cluster density was then specified by the number and type of atoms in each cluster. Using a bimolecular rate equation approach, they derived the steady-state nucleation rate,

$$I^s = A^*_{a,b} \exp\left[-\frac{\Delta G(n^*, n^*_a)}{k_B T}\right], \tag{3.23}$$

which has the same form as Eq. (3.22). The prefactor is a function of the rates for gaining or loosing a single a or b atom from a cluster during the reaction step, and the curvatures of the free-energy surface. $\Delta G(n^*, n^*_a)$ is the reversible work of formation of a critical cluster containing n^*_a a atoms and n^* total atoms. Tempkin and Shevelev[55] lifted the restriction of a constant matrix and introduced the role of diffusion to obtain a manifold of solutions that are considerably more complicated.

For a given liquid composition, there exists a solid with a composition that maximizes the free-energy change on transformation. Russell[56] restricted all clusters to this composition, but allowed a variation in the composition of the liquid. The cluster density, expressed in terms of the cluster size and the composition of the liquid shell immediately surrounding it, was described by a partial differential equation, similar to the Zeldovich-Frenkel equation (Eq. 3.17). The rate-limiting step is then the transfer of atoms from the parent phase to the liquid shell around the cluster.

Thompson and Spaepen simplified the problem considerably by assuming a fixed liquid composition and by restricting the nucleus composition to that which maximizes the free-energy change on transformation.[57] The problem is then reduced to cluster evolution along a single size axis. Ishihara *et al.* followed a similar route and assumed that the nucleus composition was given

---

[54]H. Reiss, *J. Chem. Phys.* **18**, 840 (1950).
[55]D. E. Tempkin and V. V. Shevelev, *J. Cryst. Growth* **52**, 104 (1981).
[56]K. C. Russell, *Acta Metall.* **16**, 761 (1968).
[57]C. V. Thompson and F. Spaepen, *Acta Metall.* **31**, 3021 (1983).

by that of the critical cluster, hence ignoring the saddle point of the two-dimensional free-energy surface.[58]

c. *Heterogeneous Nucleation*

Nucleation in undercooled liquids frequently occurs on the container surface, foreign particles, or other heterogeneities that catalyze the nucleation by reducing the cluster interfacial energy.[59] In solid–solid transformations, defects such as grain boundaries, edges or corners, and dislocations can serve as heterogeneous sites.[60,61] Within the classical theory, the thermodynamic barrier to nucleation, W^* is given by[62]

$$W^* = \frac{16\pi}{3} \frac{\sigma^3}{\Delta G_v^2} f(\theta), \qquad (3.24)$$

where

$$f(\theta) = \tfrac{1}{4}(2 + \cos \theta)(1 - \cos \theta)^2 \qquad (3.25)$$

and θ is the wetting angle.[5,6,62] For any wetting ($\theta \neq 180°$) the nucleation barrier is decreased, resulting in an increased nucleation rate. In practice, the quantity of heterogeneous sites are not known, and their catalytic effectiveness cannot be predicted,[63] making quantitative nucleation studies difficult to perform. Some existing investigations, however, appear to disagree with classical theory predictions.[64,65]

4. Cluster Translation and Rotation

The kinetic approach to nucleation presented in Section 3, has the advantage that it can describe nucleation for a wide range of conditions and it can predict the trajectory to steady state. Lothe and Pound,[32,66] and others,[67–69] however, have argued that the kinetic formulation ignores

---

[58] K. N. Ishihara, M. Maeda, and P. Shingu, *Acta Metall.* **33**, 2113 (1985).
[59] D. Turnbull, *J. Chem. Phys.* **18**, 198 (1950).
[60] J. W. Cahn, *Acta Metall.* **4**, 449 (1956); **5**, 169 (1957).
[61] R. B. Nicholson, in "Phase Transformations," p. 269. ASM, Metals Park, OH, 1970.
[62] J. W. Christian, "The Theory of Transformations in Metals and Alloys," second edition, Chapter 10. Pergamon Press, Oxford, 1975.
[63] W. A. Tiller, in "Physical Metallurgy" (R. W. Cahn, ed.), second edition, p. 403. North-Holland, Amsterdam, 1970.
[64] G. A. Chadwick, Proc. Int. Sympos. Chem. Met of Iron and Steel (1973), p. 207.
[65] B. Cantor and R. D. Doherty, *Acta Metall.* **27**, 33 (1979).
[66] J. Lothe and G. M. Pound, *J. Chem. Phys.* **48**, 1849 (1968).
[67] F. Kuhrt, *Z. Phys.* **131**, 185 (1952).
[68] K. Nishioka and G. M. Pound, in "Nucleation Phenomena, Advances in Colloid and Interface Science" (A. C. Zettlemoyer, ed.), Vol. 7, p. 205. Elsevier, New York, 1977.
[69] R. Kikuchi, in Ref. 68, p. 67.

important contributions to the free energy from cluster translations and rotations.

For vapor condensation, minimizing the Helmholtz free energy subject to the constraint of a fixed number of monomers, and ignoring pressure–volume work, the equilibrium concentration of nuclei is given by

$$N(n) = N_A \exp\left\{-\frac{1}{k_B T}\left[\Delta G_n - k_B T \ln\left(\frac{\lambda q_R}{N_A q_{rep}}\right)\right]\right\}, \quad (4.1)$$

where q_{rep} accounts for a decrease in entropy arising from the deactivation of six degrees of freedom when the cluster was reversibly removed from the bulk liquid. Here, λ is the translational partition function

$$\lambda = (2\pi n m k_B T)^{3/2}/h^3, \quad (4.2)$$

q_R is the rotational partition function

$$q_R = (2k_B T)^{3/2}(\pi I^3)^{1/2}/h^3, \quad (4.3)$$

I is the moment of inertia for the cluster, and $\hbar$ is Planck's constant h divided by 2π. Assuming reasonable values for vapor condensation, these additional terms contribute approximately $-45k_B T$ to the reversible work for critical cluster formation. The equilibrium cluster concentration and, therefore, the nucleation rate are increased by approximately 10^{17}. The requirement for an explicit inclusion of these terms in the free energy has been debated extensively.[68,70-73] Recent density functional calculations by Oxtoby and Evans[74] show reasonable agreement with classical theory predictions, suggesting that all cluster degrees of freedom are included properly in the classical theory formulation of the thermodynamics of nucleation. They also point out, however, that the calculational results are extremely sensitive to the range of the attractive potential used.

Supersaturation data for the vapor do not show such a large discrepancy with the classical theory. In any case, for nucleation in condensed phases, the external degrees of freedom for the cluster are primarily vibrational rather than translational and rotational; the additional partition functions, therefore, can contribute only slightly to the free energy. Cluster motion should then be of little consequence for nucleation in liquids and glasses and need not be considered. The reader interested in possible effects on vapor

---

[70] H. Reiss, in Ref. 68, p. 1.
[71] H. Reiss and J. L. Katz, J. Chem. Phys. **46**, 2496 (1967).
[72] H. Reiss, J. L. Katz, and E. R. Cohen, J. Chem. Phys. **48**, 5553 (1968).
[73] J. Lothe and G. M. Pound, J. Chem. Phys. **36**, 2080 (1962).
[74] D. W. Oxtoby and R. Evans, J. Chem. Phys. **89**, 7521 (1988).

condensation is referred to the vast literature on that subject; a good introduction can be found in the books by Abraham[75] and Zettlemoyer.[76]

III. Classical Theory: Time-Dependent Nucleation

5. PHENOMENA

Interpretations of nucleation experiments and formulations of most nucleation theories often assume a constant nucleation rate. While frequently true, this assumption is wrong in many cases. Time-dependent nucleation, also called transient nucleation, is important for many first-order phase transformations, including condensation from the vapor,[77-80] nucleation of a new crystalline phase within an existing crystalline phase,[81] liquid phase separation,[82] crystallization of undercooled liquids,[83] and devitrification (or crystallization) of a glass.[84,85] It may also be critical for glass formation in some systems.[86,87]

It is desirable to describe theoretically time-dependent nucleation under given experimental conditions. If properly understood, time-dependent nucleation studies can provide a probe of the microscopic models for nucleation. Recently, transient nucleation data were used to test the model for cluster evolution assumed by the classical theory.[44] In this section we present and evaluate the results of theoretical investigations of time-dependent nucleation within the classical theory.

Time-dependent nucleation is best introduced with a specific example. In silicate glasses, the number of small crystallites, which are related directly to the number of nuclei, appearing as a function of annealing time can be viewed by optical or electron microscopy. Assuming a steady-state nucleation rate I^s,

---

[75] F. F. Abraham, "Homogeneous Nucleation Theory," Academic Press, New York, 1974.
[76] "Nucleation and Nucleation Phenomena" (A. C. Zettlemoyer, ed.), Elsevier, New York, 1969, 1977.
[77] K. Oswatisch, Z. Angew Math. U. Mech. **22**, 1 (1942).
[78] A. Kantrowitz, J. Chem. Phys. **19**, 1097 (1951).
[79] W. J. Shugard and H. Reiss, J. Chem. Phys. **65**, 2827 (1976).
[80] F. J. Schelling and H. Reiss, J. Chem. Phys. **74**, 3527 (1981).
[81] D. Turnbull, Metals Tech., T.P. 2365 (1948) [Trans. AIME **175**, 774 (1948)].
[82] R. Pascova and I. Gutzow, Glastech. Ber. **56**, 324 (1983).
[83] I. Gutzow, Contemp. Phys. **21**, 121, 243 (1980).
[84] P. F. James, Phys. Chem. Glasses **15**, 95 (1974).
[85] A. M. Kalinina, V. N. Filipovich, and V. M. Fokin, J. Non-Cryst. Solids **38, 39**, 723 (1980).
[86] K. F. Kelton and A. L. Greer, J. Non.-Cryst. Solids **79**, 295 (1986).
[87] K. F. Kelton and A. L. Greer, in "Proc. 5th Int. Conf. on Rapidly Quenched Metals," (S. Steed and H. Warlimont, eds.), p. 223. North-Holland, Amsterdam, 1985.

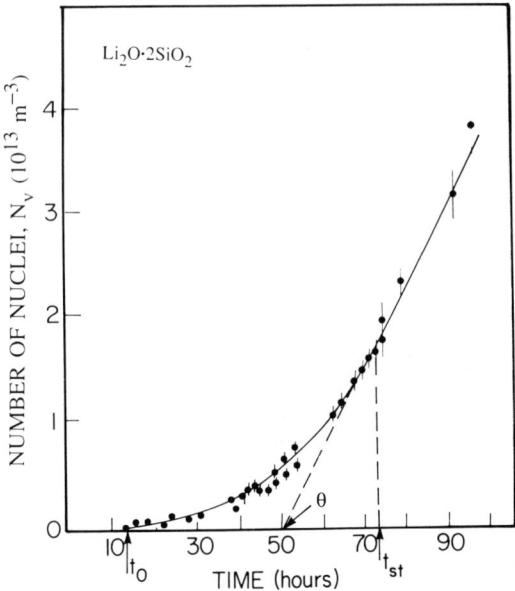

FIG. 5. The number of nuclei produced as a function of time at 703 K for lithium disilicate glass. The time for onset of nucleation, t_0, the time to steady state, t_{st}, and the effective time lag, θ, are indicated. Taken from V. M. Fokin, A. M. Kalinina, and V. N. Filipovich, *J. Cryst. Growth* **52**, 115 (1981).

the number of nuclei per unit volume, N_v would be

$$N_v = \int_0^t I(t)\,dt = I^s t. \tag{5.1}$$

A plot of N_v as a function of annealing time, t, should give a straight line with slope I^s Figure 5 shows the number of nuclei measured as a function of time for lithium disilicate glass; the nonlinearity of the curve demonstrates that the nucleation rate is not constant. From Eq. (5.1), $I(t)$, the time-dependent nucleation rate, is given by the local slope of N_v versus t. $I(t)$ is initially low and increases with time; it approaches I^s with long annealing times. For long annealing times, N_v can then be approximated by

$$N_v = I^s(t - \theta), \qquad t \gg \theta, \tag{5.2}$$

where θ is an effective time lag, also called an induction time. Experimentally, θ is obtained by extrapolating the number of nuclei produced as a function of time in the steady-state regime to the time axis, as illustrated in Fig. 5. The time before measurable nucleation is observed, t_0, and the time to reach

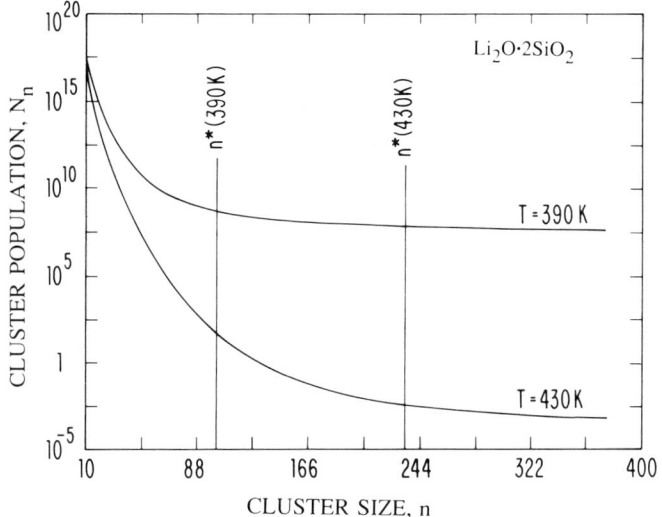

FIG. 6. The steady-state population of clusters calculated at two temperatures, using thermodynamic and kinetic parameters appropriate for lithium disilicate glass. The critical sizes at the two temperatures are indicated.

steady state, t_{st}, are also indicated. While all of these times are related to the relaxation rate to the steady state distribution, the time lag θ is most directly related to the transient time τ appearing in analytical expressions.

Transient nucleation is understood readily within the classical theory. For illustration, Fig. 6 shows the steady-state cluster distribution calculated from Eq. (3.16) for two different temperatures, using parameters that are appropriate for lithium disilicate glass. If the system were initially in steady state at T_1 and the temperature were instantly lowered to T_2, the distribution would evolve to the steady-state cluster distribution appropriate for the new temperature, following the bimolecular reactions described by Eq. (3.7). Accounting for the change in the critical size, the nucleation rate at the new temperature is proportional to the time-dependent critical cluster density at T_2 multiplied by the forward jump rate at n^*:

$$I_{n^*,t}(T_2) \propto N_{n^*,t}(T_2)k_{n^*}^+(T_2). \qquad (5.3)$$

For the case considered, the cluster density at the critical size for T_2 is initially lower than the appropriate steady-state density. A nucleation rate that is significantly lower than the steady-state prediction would thus be measured. As the cluster density evolves toward the steady-state distribution, the

[88]V. M. Fokin, A. M. Kalinina, and V. N. Filipovich, *J. Cryst. Growth* **52**, 115 (1981).

nucleation rate approaches the steady-state rate. Were the temperature change reversed after reaching steady state, the nucleation rate would be too large initially and would decrease with time. Analytical and numerical solutions for the time evolution of the nucleation rate are presented in the next two sections. For completeness, the time dependence of the mobility due to configurational relaxation of the glass should also be considered. This is discussed elsewhere.[89]

6. ANALYTICAL AND CLOSED-FORM SOLUTIONS

A viable theory of transient nucleation predicts the time-dependent nucleation rate as a function of time and temperature for different initial cluster distributions in terms of accessible materials parameters. Most treatments of transient nucleation have sought analytical solutions to the Zeldovich–Frenkel equation [Eq. (3.17)] by making approximations to the rate constants and free energy of cluster formation. The time-dependent nucleation rate at the critical size, $I_{n^*,t}$, is assumed to approximate the measured nucleation behavior. The transient behavior, however, depends strongly on the cluster size at which it is measured, and the time required for a cluster to grow from n^* to the measurement size must also be computed. This is discussed in Section V. The existing analytical solutions and the short time limits (when available) are collected in Table I.

The first treatment was due to Zeldovich.[49] He assumed that the clusters act as particles that are harmonically bound to $n = 0$; ΔG_n is then proportional to n^2. Taking $k_n^+ \approx k_{n^*}^+$, he found a time-dependent nucleation rate at the critical size,

$$I_{n^*,t} \approx I^s \exp(-n^{*2}/4k_{n^*}^+ t). \tag{6.1}$$

The most thorough analytical treatment is due to Kashchiev.[90] The free energy of cluster formation, ΔG_n, was approximated by the first two nonzero terms in a Taylor expansion about n^*, and k_n^+ was set to $k_{n^*}^+$ for all n. This gives

$$I_{n^*,t} = I^s \left[1 + 2 \sum_{m=1}^{\infty} (-1)^m \exp\left(-\frac{m^2 t}{\tau_K}\right) \right], \tag{6.2}$$

where I^s is the steady-state nucleation rate, and the transient time, τ, is given by

$$\tau_K = \frac{-24 k_B T n^*}{\pi^2 k_{n^*}^+ \Delta G'} = \frac{4}{\pi^3 k_{n^*}^+ Z^2}; \tag{6.3}$$

[89]K. F. Kelton, in "Materials Research Society Symposium Proceedings" (G. S. Cargill, F. Spaepen, and K. N. Tu, eds.), Vol. 57, p. 255. 1987.
[90]D. Kashchiev, Surf. Sci. **14**, 209 (1969).

TABLE I. ANALYTICAL EXPRESSIONS FOR THE TIME-DEPENDENT NUCLEATION RATE

AUTHOR	TIME-DEPENDENT NUCLEATION RATE	SHORT TIME APPROXIMATION	TRANSIENT TIME(n^*)
Zeldovich[49]	$I_{n^*,t} = I^s e^{-\tau/t}$	—	$\tau_Z = n^{*2}/4k_{n^*}^+$
Kashchiev[90]	$I_{n^*,t} = I^s \left[1 + 2\sum_{m=1}^{\infty}(-1)^m e^{-m^2 t/\tau}\right]$	$I_{n^*,t} = I^s(4\pi\tau/t)^{1/2} e^{-\pi^2\tau/4t}$	$\tau_K = 4/\pi^3 k_{n^*}^+ Z^2$
Collins[91]	$I_{n^*,t} = I^s \left[1 + 2\sum_{m=1}^{\infty}(-1)^m e^{-m^2 t/\tau}\right]$	$I_{n^*,t} = I^s(4\pi\tau/t)^{1/2} e^{-\pi^2\tau/4t}$	$\tau_C = 1/\pi^2 k_{n^*}^+ Z^2$
Kantrowitz[78]		$I_{2n^*,t} = 2N(k_{n^*}^+/\pi t)^{1/2} e^{-n^{*2}/k_{n^*}^+ t}$	
Wakeshima[92]	$I_{n^*,t} = I^s[1 - e^{-t/\tau}]$	—	$\tau_W = 1/8\pi k_{n^*}^+ Z^2$
Chakraverty[93]	$I_{n^*,t} = I^s[1 - e^{-t/\tau}]$	—	$\tau_{C'} = 1/2\pi k_{n^*}^+ Z^2$
Feder et al.[7]	$I_{n^*,t} = I^s[1 - e^{-t/\tau}]$	—	$\tau_F = 1/2k_{n^*}^+ Z^2$

Z is the Zeldovich factor [Eq. (3.15)]. In this treatment, the values of $N_{n,t}$ differ from the corresponding values of N_n^s only in the critical region where $\Delta G_{n^*} - \Delta G_n \leqslant k_B T$. The time dependence of small clusters is not modeled accurately, but since the nucleation rate evaluated at the critical size is dominated by the behavior of clusters in the critical region, $I_{n^*,t}$ should be described well. Collins[91] had derived Eq. (6.2) earlier, but he obtained a slightly different expression for the transient time (Table I).

Kashchiev related the experimentally obtained values for the time lag, θ, to τ, by integrating Eq. (6.2) to obtain the number of clusters passing the critical size, $\chi_{n^*,t}$, at long times:

$$\chi_{n^*,t} = I^s(t - \pi^2 \tau/6). \tag{6.4}$$

Assuming that all clusters larger than the critical size eventually appear as nuclei, N_v can be set to $\chi_{n^*,t}$. A comparison of Eq. (5.2) and Eq. (6.4) gives

$$\theta_{n^*} = \pi^2 \tau/6. \tag{6.5}$$

Wakeshima[92] solved Eq. (3.13) for the non-steady-state rate,

$$\int_1^0 d\left(\frac{N_{n,t}}{N_n^e}\right) = \int_u^v \frac{I_{n,t}}{N_n^e k_n^+} dn. \tag{6.6}$$

Making approximations that were similar to those made by Becker and Döring for the steady state case, he obtained

$$I_{n^*,t} = I^s[1 - \exp(-t/\tau)], \tag{6.7}$$

where τ is defined in Table I. This approximation is only valid when a significant population of clusters is established, and therefore is not valid for short annealing times.

Chakraverty assumed that the solution for $N_{n,t}$ was separable in n and t.[93] That expression for $N_{n,t}$ leads to a time-dependent nucleation rate that is identical to Eq. (6.7). A different transient time was obtained, however (Table I). Feder et al.[7] obtained Eq. (6.7) by calculating the time required for a cluster smaller than the critical size to disappear. The diffusion behavior for such small clusters, given by the Zeldovich–Frenkel equation, is dominated by the drift, driven by the potential gradient. By the principle of microscopic reversibility, this is also the time for a cluster to grow to the original size. The transient time is then given by that time plus the time required to diffuse through the critical region (Table I).

Several short-time solutions have also been presented. These are not discussed, but are listed in Table I.

[91] F. C. Collins, Z. Electrochem **59**, 404 (1955).
[92] H. Wakeshima, J. Chem. Phys. **22**, 1614 (1954).
[93] B. K. Chakraverty, Surf. Sci. **4**, 205 (1966).

The least restrictive treatments are due to Andres and Boudart,[94] Frisch and Carlier,[95] and Kelton, Greer, and Thompson.[52] Andres and Boudart simplified the problem considerably by solving for the temporal moments of $dN_{n,t}/dt$. Defining the effective time lag for cluster size n, θ_n, as

$$\theta_n = \int_0^\infty \left(1 - \frac{I_{n,t}}{I^s}\right) dt, \tag{6.8}$$

they showed that, with an initial distribution of clusters, $N_{n,0}$,

$$\theta_{v-1} = \sum_{n=u+1}^{v-1} \left[(N_n^s - N_{n,0}) \sum_{m=u}^{n-1} \frac{1}{k_m^+ N_m^e}\right], \tag{6.9}$$

with N_n^s calculated from Eq. (3.16).

Kelton *et al.* transformed the problem into frequency space to obtain

$$N_{n,t} = \sum_{m,j=u}^{v} N_{j,0} b_{nm} b_{mj}^{-1} \exp(\lambda_m t). \tag{6.10}$$

Here λ_m are the eigenvalues of the diagonalized matrix $\mathbf{A}$, defined as $\mathbf{A} = \mathbf{B}^{-1} \mathbf{KB}$, where $\mathbf{K}$ is a matrix of the coefficients in Eq (3.7), and b_{nm} are the elements of the matrix $\mathbf{B}$, whose columns are the normalized eigenvectors.

7. Numerical Solutions

Numerical solutions to the differential equations describing cluster evolution offer several advantages over analytical solutions. The numerical treatments simulate directly the discrete steps for cluster growth that are central to the classical theory. They do not make the continuum approximation and are free from the many, largely untestable, approximations that are required for the analytical solutions. The predictions of the numerical solutions can therefore be used to evaluate the expressions developed in the last section. Numerical treatments for time-dependent nucleation are also extended readily to include arbitrary cluster distributions,[52] nonisothermal annealing treatments and glass formation,[86,87] nucleation of a crystal of composition different from that of the liquid or glass,[96] and heterogeneous nucleation.[97]

Turnbull obtained the first numerical solution for time-dependent nucleation to explain recrystallization rates observed by others in cold-worked aluminum.[81] Considering two systems for which $n^* = 4$ and 25, the maximum

---

[94]R. P. Andres and M. Boudart, *J. Chem. Phys.* **42**, 2057 (1965).
[95]H. L. Frisch and C. C. Carlier, *J. Chem. Phys.* **54**, 4326 (1971).
[96]P. V. Evans, Ph.D. Thesis, University of Cambridge, Cambridge, UK, 1988 (unpublished).
[97]A. L. Greer, P. V. Evans, R. G. Hamerton, D. K. Shangguan, and K. F. Kelton, Proc. 9th Int. Cryst. Growth Conf., Sendai, Japan, 1989.

cluster size was taken to be n^*. Ignoring the back flux to $I_{n,t}$, Turnbull showed that for each cluster size, $N_{n,t}$ rose with a sigmoidal dependence on time from zero to N_n^s, and $dN_{n,t}/dt$ went through a maximum. The time to reach the maximum increased with n. $I_{n^*,t}$ also rose sigmoidally with time and approached the steady-state value. Christiansen[98] qualitatively confirmed these results by obtaining an exact solution for a system of four cluster sizes. In that analysis the backward fluxes were not set to zero, but k_n^+ and k_n^- were assumed to be independent of n with $k_n^+ \gg k_n^-$.

Courtney,[99] Abraham,[100] Kelton et al.,[52] and Volterra and Cooper[101] solved the coupled differential equations without making restrictive assumptions for the rate constants. Courtney, Kelton et al., and Volterra and Cooper used a finite difference method; Abraham used the Fowler–Warten algorithm,[102] which is well suited for systems of equations with a large spread in the time constants of the solution. Arbitrary upper and lower limits on the cluster size were required, yet the results were demonstrated to be rather insensitive to the exact values. The qualitative features of the earlier calculations were confirmed; $N_{n,t}$ rose sigmoidally with time, and $I_{n,t}$ showed a maximum for $n < n^*$ that was larger and occurred earlier for smaller n.

The finite difference calculation is most easily extended to situations other than isothermal annealing. The time is divided into a large number of small intervals, δt, and the number of clusters of size n at the end of the interval, $N_{n,t+\delta t}$, is calculated from

$$N_{n,t+\delta t} = N_{n,t} + \delta t \frac{dN_{n,t}}{dt}, \qquad (7.1)$$

where $dN_{n,t}/dt$ is given in Eq. (3.7). To compare the results of the numerical treatments with the predictions from the analytical expressions, it is necessary to assume a set of materials parameters. Kelton et al. chose parameters that closely approximated those for lithium disilicate glass, although a detailed fit to experimental data was not sought. Figures 7a and b compare the results from these numerical calculations for the time-dependent nucleation rate at the critical size $I_{n^*,t}$ with predictions from the analytical expressions listed in Table I. Although the Zeldovich expression [Eq. (6.1)] is probably the most widely used, these comparisons show that the shape of the curve and the predicted transient time are in poor agreement with the numerical solution. Predictions from the Kashchiev expressions [Eqs. (6.2) and (6.3)] for the time lag θ_{n^*} and $I_{n^*,t}$ agree best with the numerical calculation. All treatments

[98] J. A. Christiansen, Acta. Chem. Scand. **8**, 909 (1954).
[99] W. G. Courtney, J. Chem. Phys. **36**, 2009 (1962).
[100] F. F. Abraham, J. Chem. Phys. **51**, 1632 (1969).
[101] V. Volterra and A. R. Cooper, J. Non-Cryst. Solids **74**, 85 (1985).
[102] M. E. Fowler and R. M. Warten, IBM J. Res. Devel. **11**, 537 (1967).

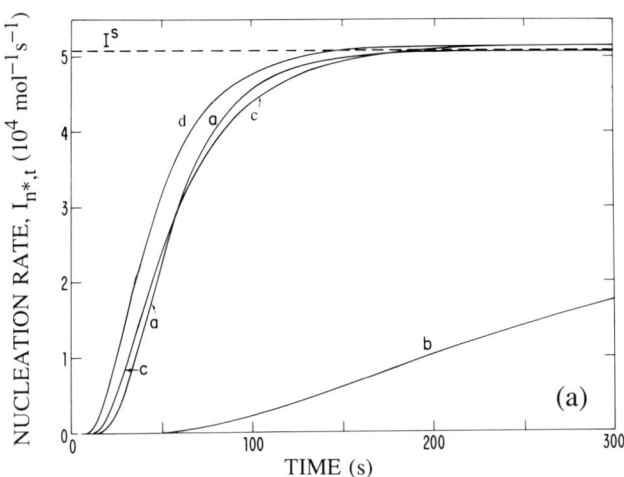

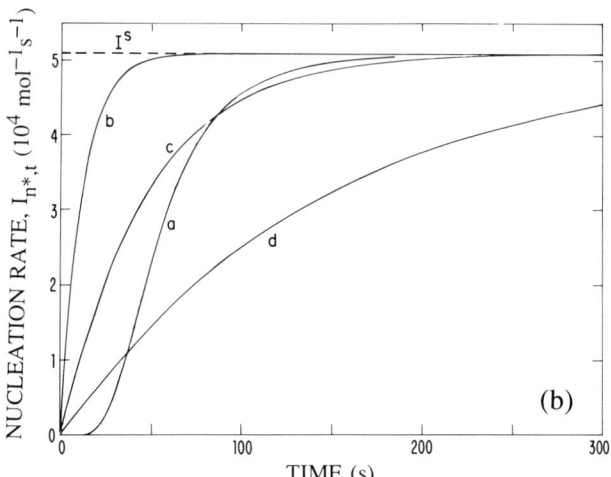

FIG. 7. (a) The time-dependent nucleation rate at the critical size, $I_{n^*,t}$, at 750 K using parameters that approximate those for lithium disilicate glass. The analytical expressions are listed in Table I: curve a—numerical calculation; curve b—Zeldovich; curve c—Kashchiev; curve d—Collins. (b) Curve a—numerical calculation; curve b—Wakeshima: curve c—Chakraverty; curve d—Feder et al. The effective time lags are indicated. Taken from K. F. Kelton, A. L. Greer, and C. V. Thompson, J. Chem. Phys. **79**, 6261 (1983).

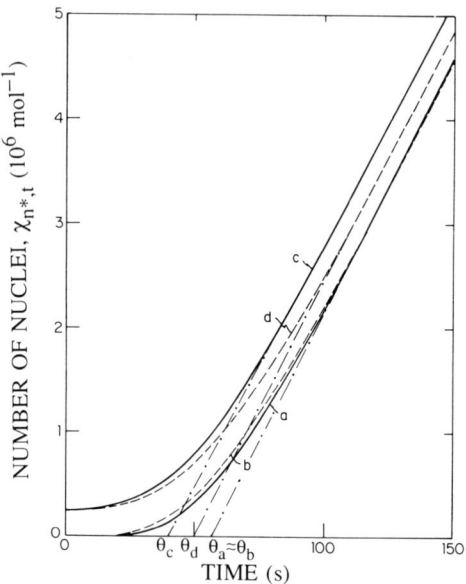

FIG. 8. The total number of clusters exceeding the critical size, $\chi_{n^*,t}$, as a function of time at 750 K using parameters that approximate those for lithium disilicate glass. With no initial distribution: (a) numerical calculation; (b) Kashchiev,[90] with an initial cluster distribution equal to the steady-state distribution at 800 K; (c) numerical calculation; (d) Kashchiev.[103] The effective time lags are indicated. Taken from K. F. Kelton, A. L. Greer, and C. V. Thompson, *J. Chem. Phys.* **79**, 6261 (1983).

predict an approximate scaling between the temperature dependencies of θ_{n^*} and the atomic mobility; the effective activation energies agree within $\approx 20\%$.

8. PREEXISTING CLUSTER DISTRIBUTIONS

An initial cluster distribution equal to zero for all cluster sizes greater than $n = 10$ was assumed in Section 7. In a rapidly quenched glass, the initial distribution is more likely to approximate the steady-state distribution at some higher temperature. The exact nature of that distribution can influence greatly the transient behavior.[88] Kashchiev developed an expression for the time-dependent nucleation rate for different initial cluster distributions.[103] Figure 8 shows the number of nuclei calculated as a function of time for lithium disilicate by Kelton *et al.*[52] and from the Kashchiev expression. Curve a is computed numerically, assuming no preexisting clusters. It compares well with curve b calculated from the Kashchiev expression for the same conditions. Curve c is computed numerically, assuming an initial cluster distribution that is the steady-state distribution at 800 K. The presence of preexisting clusters shortens the effective time lag but has no effect on the final

[103] D. Kashchiev, *Surf. Sci.* **18**, 389 (1969).

steady-state rate. Curve d is computed from the Kashchiev expression for the same distribution. The agreement is reasonably good, although the effect of a preexisting distribution appears to be underestimated by the Kashchiev treatment.

IV. Experimental Studies of Nucleation in Undercooled Liquids

9. Undercooling of Polymorphically Crystallizing Liquids

As discussed in the introduction, many liquids can be cooled far below their equilibrium melting temperature, T_m, without solidification. The amount of undercooling, ΔT, is defined as

$$\Delta T = T_m - T, \tag{9.1}$$

where T is the temperature. The classical theory of homogeneous nucleation predicts that the steady-state nucleation rate, I^s, rises sharply with undercooling; this is observed experimentally. In metals, for example, I^s increases by approximately a factor of 10 per degree of undercooling. Figure 9 shows experimental data for the nucleation rate as a function of undercooling for liquid mercury. The solid line is the predicted nucleation rate, calculated from Eq. (3.21).

For liquid metals, the growth velocities are sufficiently large in the nucleation regime that the time scale for crystallization is dominated by the time required to form nuclei. A maximum undercooling, ΔT_{max}, can therefore

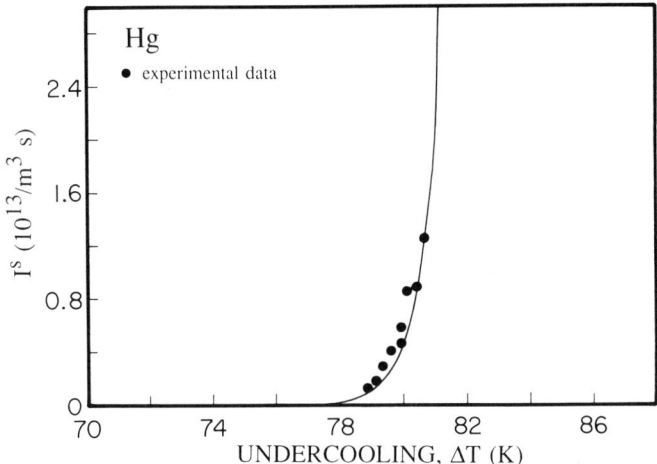

FIG. 9. The steady-state homogeneous nucleation rate as a function of undercooling of liquid mercury, demonstrating the existence of a sharp onset. The points are experimental data obtained from Ref. 1; the solid line is a fit to the classical theory of nucleation.

be defined and determined experimentally. The liquid is undercooled until the onset of crystallization is detected from some macroscopic property change, such as the volume (dilatometry),[1] heat evolution (calorimetry),[104,105] or electrical resistivity.[106] More microscopic probes, such as NMR, have also been used.[107].

Frequently, the observed ΔT_{max} is due to heterogeneous nucleation on container walls, on the surface of the liquid, or on structural impurities within the liquid. To obtain the undercooling limit imposed by homogeneous nucleation, these complications must be eliminated. In this section, techniques developed to remove or deactivate the heterogeneous sites are discussed. The maximum reported undercoolings for liquid metals are presented and discussed in light of classical theory predictions. Within the classical theory, estimates of the interfacial energy between the developing cluster and the initial phase can be computed from ΔT_{max}. These estimates are important for modeling the interfacial structure; they are computed for the liquid metals considered.

There have been several attempts to undercool bulk samples of liquid in clean crucibles, made from materials that only act weakly as catalytic sites for heterogeneous nucleation.[108,109] To eliminate surface oxide formation that can also lead to heterogeneous nucleation, the experiments are performed typically in vacuum or in a reducing atmosphere. With few exceptions, the undercoolings obtained are far less than those obtained by the techniques discussed here; they do not reflect the homogeneous limit and are not discussed further. The properties of the final product are strongly influenced by the microstructure, which is in turn determined to a large extent by the degree of undercooling prior to solidification.[110-112] Such studies are therefore of practical importance in designing microstructures, in grain refinement studies, and for the development of improved materials.

a. *Experimental Techniques*

Figure 10 illustrates schematically the most common methods used to eliminate heterogeneous sites.

[104] D. H. Rasmussen and C. R. Loper, Jr., *Acta Metall.* **24**, 117 (1976).
[105] G. A. Merry and H. Reiss, *Acta Metall.* **32**, 1447 (1984).
[106] I. S. Servi and D. Turnbull, *Acta Metall.* **14**, 161 (1966).
[107] T. Takahashi and W. A. Tiller, *Acta Metall.* **17**, 651 (1969).
[108] T. Z. Kattamis and M. C. Flemings, *Metall. Trans.* **1**, 1449 (1970).
[109] T. Z. Kattamis and M. C. Flemings, *Trans. Metall. Soc. AIME* **236**, 1523 (1966).
[110] J. H. Perepezko, D. H. Rasmussen, I. E. Anderson, and C. R. Loper, Jr., in "Sheffield International Conference on Solidification and Casting of Metals," p. 169. Metals Society, London, 1979.

Emulsion Technique In the droplet emulsion technique the liquid is dispersed into a large number of small droplets in an appropriate medium (Fig. 10a). If the dispersion is sufficiently fine (approximately 100 μm diameter), a significant number of droplets will contain no heterogeneous sites. Assuming that the droplet surface does not serve as a nucleating site, the homogeneously limited undercooling will be the maximum observed in the ensemble of droplets. To depress droplet coalescence in the emulsion and interactions with the container walls, the droplets are thinly coated. Organic fluids with organic or inorganic acid surfactants can be used for low-melting alloys ($T_m < 500°C$);[115] the recent use of molten salt mixtures has enabled higher-melting-point alloys to be studied.[116,117]

Vonnegut[38] pioneered the emulsion technique and measured the undercooling of liquid Sn. Turnbull[1,118–120] extended those studies to include Hg, Ga, Bi, and Pb. An almost universal reduced undercooling ($\Delta T_r = \Delta T_{max}/T_m$) of approximately 0.18 was reported. Perepezko and co-workers[121] and others[122] have subsequently applied the emulsion technique to study many pure metals and metallic alloys using differential scanning calorimetry (DSC) and differential thermal analysis (DTA). Calorimetry provides a relatively easy method for determining accurately the solidification temperature as a function of cooling rate, the volume fraction transformed, and the specific heat of the undercooled liquid.[123] In some cases, values for ΔT_r as large as 0.5 are reported (*cf.* Table III), suggesting that the values reported originally by Turnbull may have been heterogeneously limited. Care must be exercised when comparing these results, however, since calorimetric techniques are less sensitive to the initial stages of nucleation than are the dilatometric measurements. Further, the amount of undercooling critically depends on the

[111] B. A. Mueller, J. J. Richmond, and J. H. Perepezko, "Proc. 5th Int. Conf. on Rapidly Quenched Metals" (S. Steeb and H. Warlimont, eds.), p. 47. Elsevier, Amsterdam, 1985.
[112] G. Devaud and D. Turnbull, *Acta Metall.* **35**, 765 (1987).
[113] M. C. Flemings and Y. Shiohara, *Mat. Sci. Eng.* **65**, 157 (1984).
[114] J. H. Perepezko, *Mat. Sci. Eng.* **65**, 125 (1984).
[115] J. H. Perepezko and J. S. Smith, *J. Non-Cryst. Solids* **44**, 65 (1981).
[116] J. H. Perepezko, S. E. LeBeau, B. A. Mueller, and G. J. Hilderman, in "Rapidly Solidified Powder Aluminum Alloys" (M. E. Fine and E. A. Starke, eds.), p. 118. ASTM, Philadelphia, 1986.
[117] J. S. Paik and J. H. Perepezko, *J. Non-Cryst. Solids* **56**, 405 (1983).
[118] D. Turnbull, *J. Appl. Phys.* **20**, 817 (1949).
[119] D. Turnbull, *J. Chem. Phys.* **18**, 768 (1950).
[120] D. Turnbull, *Trans. AIME* **188**, 1144 (1950).
[121] J. H. Perepezko, B. A. Mueller, and K. Ohsaka, "Hume-Rothery Memorial Symposium" (E. W. Collings, C. C. Koch, eds.), p. 289. TMS-AIME, New Orleans, LA, 1986.
[122] D. H. Rasmussen and C. R. Loper, Jr., *Acta Metall.* **23**, 1215 (1975).
[123] J. H. Perepezko and J. S. Paik, *J. Non-Cryst. Solids* **61, 62**, 113 (1984).

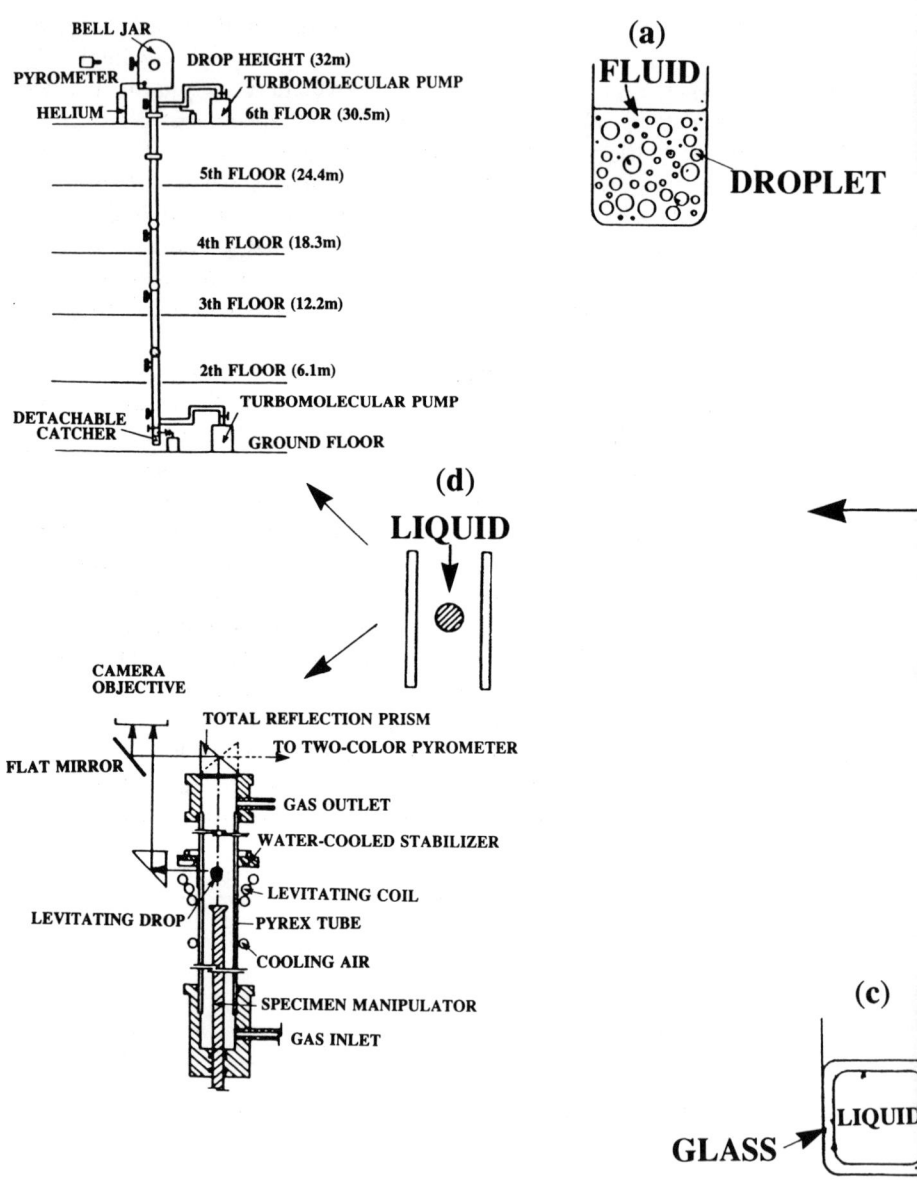

Fig. 10. A schematic illustration of the four most common methods for effectively removing heterogeneities from the melt to attain a homogeneously limited undercooling. A schematic diagram of the experimental apparatus used with each method is also provided. Adapted from M. C. Flemings and Y. Shiohara, *Mat. Sci. Eng.* **65**, 157 (1984); J. H. Perepezko, *Mat. Sci. Eng.* **65**, 125 (1984).

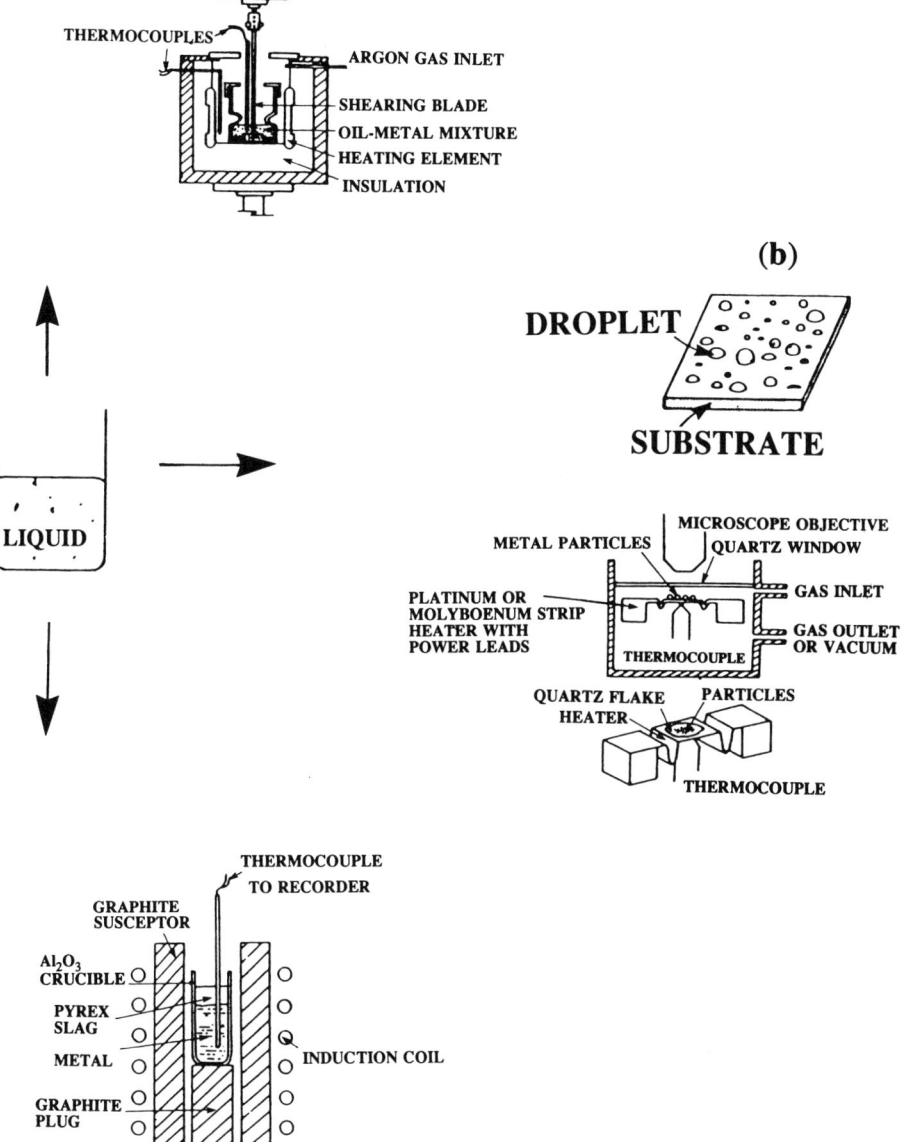

FIG. 10. (Continued)

TABLE II. UNDERCOOLING DATA FOR LIQUID METALS

METAL	EMULSION			SUBSTRATE			FLUXED				CONTAINERLESS		
	ΔT(K)	v (mol)	Q (K/s)	ΔT(K)	v (mol)	Q (K/s)	ΔT(K)	v (mol)	Q (K/s)	Flux	ΔT(K)	v (mol)	Q (K/s)
Ag	—	—	—	—	—	—	250[a,b]	4.63	—	Soda Lime Glass	—	—	—
Al	175[d]	~10^{-10}	500	48[c]	—	—	227[c](s)	~10^{-9}	~1	Pyrex	—	—	—
Au	—	—	—	230[c]	~10^{-9}	~1	130[c](s)	~10^{-8}	~1	NaOH	—	—	—
Bi	227[d]	~10^{-10}	0.3	90[c]	~10^{-10}	~1	221[c](s)	~10^{-9}	~1	Pyrex	—	—	—
Cd	110[d]	~10^{-10}	0.3	—	—	—	—	—	—	—	—	—	—
Co	—	—	—	330[c]	~10^{-9}	~1	310[e]	0.3	—	—	—	—	—
Cu	—	—	—	236[c]	~10^{-9}	~1	180[e]	0.3	—	Jena Glass	—	—	—
Fe	174[d]	~10^{-10}	0.3	295[c]	~10^{-8}	~1	250[f]	2.7	—	Jena Glass	—	—	—
Ga	—	—	—	—	—	—	—	—	—	Slag	420[g](l)	~0.03	—
Ge	88[d]	~10^{-10}	0.08	253[c]	~10^{-10}	~1	415[h]	4×10^{-5}	~4	B_2O_3	—	—	—
Hg	—	—	—	—	—	—	—	—	—	—	—	—	—
Hf	110[d]	~10^{-10}	0.3	—	—	—	—	—	—	—	450[i](d)	0.004	~140
In	—	—	—	—	—	1	—	—	—	—	—	—	—
Ir	—	—	—	—	—	—	—	—	—	—	340[i](d)	0.002	~175
Mn	—	—	—	—	—	—	308[g](s)	~10^{-9}	~1	Pyrex	—	—	—
Mo	—	—	—	—	—	—	—	—	—	—	520[i](d)	0.007	~200
Nb	—	—	—	—	—	—	—	—	—	—	525[j](d)	0.002	~300
Ni	—	—	—	319[k]	~10^{-8}	—	319[c](s)	~10^{-8}	~1	Pyrex	480[g](l)	~0.03	—
Pb	153[d]	~10^{-10}	0.3	240[l]	—	—	69[c](s)	~10^{-10}	~1	NaOH + KOH	—	—	—

Element													
Pd	—	—	—	—	—	—	—	—	—	—	—	—	—
Pt	—	—	—	332^c	—	—	310^e	—	—		380^i(d)	~10^{-3}	~125
Rh	—	—	—	400^m	~10^{-8}	~1	—	—	—	Jena Glass	450^i(d)	~10^{-3}	~140
Ru	—	—	—	426^m	~10^{-8}	—	—	—	—		330^i(d)	~10^{-3}	~80
Sb	210^d	0.3	—	135^c	~10^{-9}	~1	124^a	2.5	—		—	—	—
Se	—	—	—	—	—	—	—	—	—		—	—	—
Sn	191^d	0.3	~10^{-10}	25^c	~10^{-8}	~1	—	—	—	Soda Lime Glass	650^i(d)	~0.002	—
Ta	—	—	—	76.7^n	—	—	—	—	—		—	—	-500
Te	236^d	0.3	~10^{-10}	—	—	—	—	—	—		350^i(d)	~3×10^{-4}	~80
Ti	—	—	—	—	—	—	—	—	—		430^i(d)	~0.005	~125
Zr	—	—	—	—	—	—	—	—	—				

[a] Ref. 124
[b] Ref. 125
[c] Ref. 126
[d] Ref. 121
[e] Ref. 127
[f] Ref. 128
[g] Ref. 129
[h] Ref. 130
[i] Ref. 131
[j] Ref. 132
[k] Ref. 133
[l] Ref. 134
[m] Ref. 135
[n] Ref. 107

[124] G. L. F. Powell, *J. Aust. Inst. Metals* **10**, 223 (1965).
[125] G. L. F. Powell, *Trans. Metall. Soc. AIME* **239**, 1245 (1967).
[126] D. Turnbull and R. E. Cech, *J. Appl. Phys.* **21**, 804 (1950).
[127] J. Fehling and E. Scheil, *Z. Metallk.* **53**, 593 (1962).
[128] P. Bardenheuer and R. Bleckmann, *Stahl. U. Eisen* **58**, 49 (1941).
[129] D. W. Gomersall, S. Y. Shirashi, and R. G. Ward, *J. Aust. Inst. Metals* **10**, 220 (1965).
[130] G. Devaud and D. Turnbull, *Appl. Phys. Lett.* **46**, 844 (1985).
[131] W. H. Hofmeister, M. B. Robinson, and R. J. Bayuzick, *Appl. Phys. Lett.* **49**, 1342 (1986).
[132] Lacy, M. B. Robinson, T. J. Rathz, *J. Cryst. Growth* **51**, 47 (1981).
[133] R. E. Cech and D. Turnbull, *J. Metals* **3**, 242 (1951).
[134] M. J. Stowell, T. J. Law, and J. Smart, *Proc. Roy. Soc. A* **318**, 231 (1970).
[135] C. E. Mendenhall and L. R. Ingersoll, *Phil. Mag.* **15**, 205 (1908).

coating material and the film thickness.[136] It depends to a lesser degree on the droplet size, the initial temperature of the liquid, and the undercooling rate.[137,138]

Substrate Technique Mendenhall and Ingersoll[135] first noticed that liquid droplets of Ir, Pd, Pt, Rh, and Si could be undercooled several hundred degrees on an electrically heated substrate. For the substrate technique, small droplets of liquid, 10 to 100 μm in diameter, are placed on a heated, inert, glass substrate (Fig. 10b). If the droplets are sufficiently small, not contaminated during fabrication, and do not interact with the substrate, significant undercoolings can be obtained. Unlike the emulsion technique, high-melting-point alloys are easily studied. The nucleation temperature is obtained by cooling the substrate and using optical microscopy to observe the temperature at which the particle surface roughens, indicating solidification. For high-melting-point alloys the recalescence due to the evolution of the latent heat on solidification is measured. To prevent oxidation of the surface, experiments are often performed in vacuum or in a reducing atmosphere.

Turnbull and Cech[126] made the first quantitative measurements of undercooling for a wide range of elements using the substrate technique. An inspection of Table II shows that the undercoolings obtained with the substrate technique are generally not as low as those from the emulsion technique, presumably reflecting the catalytic activity of the substrate. The substrate method is, however, a superior method for electron microscopy investigations.[134] Drehman et al.[139] have used this technique to prepare bulk samples of amorphous $Pd_{40}Ni_{40}P_{20}$ by undercooling the liquid on a clean silica substrate in vacuum.

Fluxing Fluxing is a technique whereby the liquid is coated by or immersed in a material (typically amorphous) that isolates it from contact with the crucible walls and the atmosphere, dissolves impurities or changes their structure to render them less active, and does not itself provide heterogeneous sites (Fig. 10c). Typically an amorphous phase[140,141] or one with a complex crystal structure[114] is used for the flux. Fluxing is particularly

---

[136] J. H. Perepezko, in "Rapid Solidification Processing: Principles and Technologies II" (R. Mehrabian, B. H. Kear and M. Cohen, eds.), p. 56. Claitor's, Baton Rouge, LA, 1980.
[137] F. Spaepen and D. Turnbull, *Scripta. Metall.* **13**, 149 (1979).
[138] J. H. Perepezko and J. S. Paik, "Material Research Society Symposium Proceedings" (B. H. Kear, B. C. Giessen, and M. Cohen, eds.), Vol. 8, p. 49. 1982.
[139] A. J. Drehman, A. L. Greer, D. Turnbull, *Appl. Phys. Lett.* **41**, 716 (1982).
[140] T. Kattamis and M. C. Flemings, *Trans. AIME* **236**, 1523 (1966).
[141] J. L. Walker, in "Physical Chemistry of Process Metallurgy" (G. R. St. Pierre, ed.), p. 845. AIME, New York, 1961.

useful for cooling large samples; Bardenheuer and Bleckmann first used this method to undercool 150 g of Fe by 258°.[142] Turnbull and Cech[126] combined fluxing with the substrate technique. The maximum undercoolings obtained for pure metals with fluxing are listed in Table II. An s indicates that the results were obtained by using a combined flux and substrate method. Recently, Kui et al.[143] studied bulk glass formation in $Pd_{40}Ni_{40}P_{20}$, using an amorphous B_2O_3 flux. They demonstrated glass formation in a 4-g sample by cooling at only 1 K/s. In the flux, the glass could be cycled to within 50 K of the melting point without crystallization.

Containerless Solidification Since heterogeneous nucleation occurs frequently on container walls, there is a great deal of interest in containerless solidification. Two methods are illustrated in Fig. 10d. Metallic droplets can be both heated and levitated by an rf field.[144–146] Using this method, droplets of Fe and Ni have been undercooled consistently by ≈ 400 K below T_m. While rf levitation does provide a containerless environment, nucleation that is catalyzed by eddy "current-induced convective flow remains a problem. Also, rf levitation techniques can only be used to study metallic melts. Recently, acoustic levitation methods have been developed to study nonmetallic melts.[147]

The effect on nucleation of gravity-induced convection and segregation is also of interest.[148,149] Several microgravity studies have already been made on *Skylab* and on the space shuttle,[150–153] and experiments on the orbiting space station are planned. Similar studies can be conducted, for shorter times, in aircraft and rocket flights. Terrestrial experiments that simulate containerless solidification in a weightless environment can be performed less expensively in drop-tower experiments, where droplets solidify during free-

---

[142]P. Bardenheuer and R. Bleckmann, *Stahl u. Eisen.* **58**, 49 (1941).
[143]H. W. Kui, A. L. Greer, and D. Turnbull, *Appl. Phys. Lett.* **45**, 615 (1984).
[144]S. Y. Shirashi and R. G. Ward, *Can. Metal. Quart.* **3**, 117 (1964).
[145]D. W. Gomersall, S. Y. Shirashi, and R. G. Ward, *J. Aust. Inst. Metals* **10**, 220 (1965).
[146]G. J. Abbaschian and M. C. Flemings, *Met. Trans. A* **14**, 1147 (1983).
[147]K. Ohsaka and E. H. Trinh, *J. Cryst. Growth* **96**, 973 (1989).
[148]K. W. Benz and G. Nagel, in "Proc. 5th European Symp. on Materials Science under Microgravity," p. 157. ESA-SP 222, 1984.
[149]K. W. Benz, A. Danilewsky, B. Notheisen, and G. Nagel, in "Proc. 6th European Symp. on Materials Science under Microgravity Conditions," p. 345. Bordeaux, France, ESA SP-256, 1987.
[150]"Proc. 3rd Space Processing Symp.: Skylab Results," Vols. I, II, NASA Report M-74-5, 1974.
[151]"Materials Sciences in Space with Applications to Space Processing, Progress in Astronautics and Aeronautics" (L. Steg, ed.), p. 52. AIAA, New York, 1977.
[152]J. R. Carruthers, *J. Cryst. Growth* **42**, 379 (1977).
[153]R. J. Naumann, "Early Space Experiments with Materials Processing," NASA Report TM-78237, 1979.

fall in an evacuated or inert-gas-filled tube (Fig. 10d). Particularly long tubes (32 m or 105 m) exist at the NASA Marshall Space Flight Center, Huntsville, Alabama, USA,[131] while typical laboratory tubes are 3 to 6 m in length. Drop-tower experiments have been used to undercool Ti, Zr, Nb, Mo, Rh, Hf, Ta, and Pt by approximately 0.2 of T_m and Ir and Ru by 0.13 of T_m. Table II lists the maximum undercoolings obtained by levitation, l, and in drop-tower experiments, d.

Steinberg et al.[154] first observed metallic glass formation in a drop tube for 1.5-mm-diameter spheres of $Pd_{77.5}Cu_6Si_{16.5}$. Subsequently, glass formation has been reported by Drehman and Turnbull[155] ($Pd_{82}Si_{18}$), Drehman and Greer[156] ($Pd_{40}Ni_{40}P_{20}$), Kiminami and Sahm,[157] and Shong et al.[158] ($Ni_{58.5}Nb_{41.5}$).

Entrained Droplet The entrained droplet or "mush quenching" technique follows a different approach. Here, phase separation from the melt into a majority fraction of solid solution and a minority fraction ($<10\%$) of a second liquid phase occurs upon undercooling. This method is not shown in Fig. 10. Wang and Smith[159] used this technique to study nucleation of the entrapped liquid on further undercooling. Southin and Chadwick[160] investigated the undercooling in liquid droplets of simple eutectic and monotectic alloys. Studies in other system have been conducted by Boswell et al.[161,162] and Ramachandrarao et al.[163] The nucleation is probably heterogeneous; however, some solid solutions may act as only weak catalytic sites. The technique has the advantage that intimate contact between the liquid and the solid solution provides a good sink for heat evolution during crystallization. Difficulties arise, however from pressure effects, difficulties in estimating the true undercooling due to changes in the liquid composition, and the appropriate extrapolation of the liquidus temperature.

b. *Evaluation of Experimental Parameters*

Turnbull noted that the droplet coating was the dominant factor in

[154] J. Steinberg, A. E. Lord, L. L. Lacy, and J. Johnson, *Appl. Phys. Lett.* **38**, 135 (1981).
[155] A. J. Drehman and D. Turnbull, *Scripta Metall.* **15**, 543 (1981).
[156] A. J. Drehman and A. L. Greer, *Acta Metall.* **32**, 323 (1984).
[157] C. S. Kiminami and P. R. Sahm, *Acta Metall.* **34**, 2129 (1986).
[158] D. S. Shong, J. A. Graves, Y. Ujiie, and J. H. Perepezko, "Material Research Society Symposium Proceedings" (R. H. Doremus and P. C. Nordine, eds.), Vol. 87, p. 17. 1987.
[159] C. C. Wang and C. S. Smith, *Trans. AIME* **188**, 136 (1950).
[160] R. T. Southin and G. A. Chadwick, *Acta Metall.* **26**, 223 (1978).
[161] P. G. Boswell, G. A. Chadwick, R. Elliott, and F. R. Sale, "Solidification and Casting of Metals," p. 175. Metals Society, London, 1979.
[162] P. G. Boswell and G. A. Chadwick, *Acta Metall.* **28**, 209 (1980).
[163] P. Ramachandrarao, K. Lal, A. Singhdeo, and K. Chattopadhyay, *Mater. Sci. Eng.* **41**, 259 (1979).

determining the degree of undercooling.[1] Other, less important variables are the droplet size and the cooling rate. These conclusions are supported by extensive studies by Perepezko et al.[121] Using the factorial design method,[164] a statistical technique that allows an evaluation of the relative importance of a set of experimental variables, they demonstrated that the degree of undercooling can be optimized by using high-purity samples with a fine, narrowly distributed droplet size and surfactant treatments that promote the formation of thin noncrystalline coatings. These conclusions are also supported by the data reported in Table II. The undercoolings from the emulsion, fluxing and substrate techniques, where droplet contamination is probable, show a broad range of values. The droplets in drop-tower experiments are less susceptible to surface contamination and show the largest undercooling.

Assuming steady-state, homogeneous nucleation, the droplet size and cooling rate must be taken into account when comparing undercooling data. The maximum undercooling results from those droplets that are most resistant to nucleation, corresponding to the case where the probability of having no nuclei in a drop becomes small. For droplets of uniform size v (expressed in moles), the probability of having no nuclei in a droplet, given N nuclei per mole (or equivalently the volume fraction of droplets containing no nuclei) is

$$x = \exp(-vN). \tag{9.2}$$

If the cooling rate, Q, from the melting temperature, T_m, to the maximum undercooling temperature, T_u, is uniform,

$$x = \exp\left[-\frac{v}{Q}\int_{T_m}^{T_u} I^s(T)\,dT\right]. \tag{9.3}$$

If different techniques are equally sensitive to x and if heterogeneous nucleation is not a factor, the maximum observed undercooling temperature will then vary with the ratio of sample volume to quench rate, v/Q. Undercooling data must be corrected for different values of v/Q before they can be compared.

c. *Analysis of Undercooling Values*

The values for maximum undercooling, ΔT_{max}, the melting temperature, T_m, and when available, the values for v/Q are listed in Table III. The reduced undercooling values, $\Delta T_r = \Delta T_{max}/T_m$, are also listed. Arguing from existing data, Turnbull[37] first suggested a homogeneously limited value for ΔT_r

---

[164]G. E. P. Box, W. G. Hunter, and J. S. Hunter, "Statistics for Experimenters," Wiley, New York, 1978.

TABLE III. MAXIMUM UNDERCOOLING DATA FOR LIQUID METALS

Metal	ΔT_{max}	v/Q	T_m (K)	ΔT_r	σ (J/m²)	σ_g (kJ/mol)	ΔH_f (kJ/mol)[c]	n^*	r^*(Å)
Ag	227	10^{-9}	1234.0	0.18	0.128	5.11	11.4	481	12.5
Al	175	2×10^{-13}	933.3	0.19	0.108	4.25	10.5	340	11.0
Au	230	10^{-9}	1336.2	0.17	0.137	5.45	12.8	505	12.7
Bi	227	3×10^{-10}	544.2	0.42	0.088	5.74	10.9	68	8.3
Cd	110	3×10^{-10}	594.1	0.19	0.058	2.71	6.4	398	12.7
Co	330	10^{-9}	1765.2	0.19	0.238	7.14	15.5	501	11.0
Cu	236	10^{-9}	1356.2	0.17	0.178	5.58	13.0	500	11.2
Fe*	420	~0.003	1809.2	0.23	0.277	8.65	15.2	494	11.2
Ga	174	3×10^{-10}	302.9	0.57	0.077	3.35	5.59	38	5.6
Ge	415	10^{-5}	1210.2	0.34	0.300	14.4	32.2	74	7.3
Hg	88	10^{-9}	234.3	0.38	0.031	1.48	2.32	164	9.7
Hf	450	3×10^{-5}	2500.2	0.18	0.221	10.6	24.1	484	13.7
In	110	3×10^{-10}	429.6	0.26	0.036	1.86	3.27	370	13.2
Ir[a]	340	10^{-5}	2727.2	0.12	—	—	—	—	—
Mn	308	10^{-9}	1517.2	0.20	0.216	6.90	14.7	414	10.7
Mo[a]	520	4×10^{-5}	2888.2	0.18	—	—	—	—	—
Nb[a]	525	10^{-5}	2740.2	0.19	—	—	—	—	—
Ni[b]	480	~0.003	1728.2	0.28	0.3	10.40	17.7	317	9.4
Pb[b]	240	$~10^{-8}$	600.6	0.40	0.06	3.51	4.98	185	11.0
Pd	332	10^{-8}	1825.2	0.18	0.207	7.49	16.7	503	12.1
Pt	380	8×10^{-6}	2042.2	0.19	0.239	8.80	19.7	464	11.9
Rh	450	7×10^{-6}	2239.2	0.20	0.301	10.4	22.6	403	11.0
Ru	330	10^{-5}	2583.2	0.13	0.221	7.58	19.7	916	14.4
Sb	210	3×10^{-10}	903.7	0.23	0.130	7.58	19.9	148	10.2
Se	25	10^{-8}	490.2	0.05	0.021	1.13	6.28	1503	21.4
Sn	191	3×10^{-10}	505.1	0.38	0.075	4.10	7.08	120	9.2
Ta	650	4×10^{-6}	3253.2	0.20	0.301	12.5	24.7	540	13.2
Te	236	3×10^{-10}	723.2	0.33	0.125	7.89	17.6	87	8.9
Ti	350	4×10^{-6}	1940.2	0.18	0.202	8.23	18.8	480	12.6
Zr	430	4×10^{-5}	2125.2	0.20	0.193	9.49	20.1	426	13.3

[a]Heat of fusion not measured.
[b]v/Q estimated—data not available.
[c]Data taken from Smithells Metals Reference Book, sixth edition.

between 0.18 and 0.2 in metals. The more recent data in Table III demonstrate a broader range from 0.1 to 0.5, with an average value near Turnbull's original estimate. Assuming that the temperature of maximum undercooling is approximated by the temperature at the peak, recent calculations based on the classical theory predict $\Delta T_r \approx 0.56$–0.67,[165,166] far in excess of the measured values. The peak temperature, however, is generally a poor approximation to the maximum undercooling temperature. A better comparison might be made between values for W^*. Using values for the interfacial energy determined later in this section, estimating ΔG_v from Eq. (9.4), and correcting for droplet sizes and cooling rates, $W^* = (60 \pm 2)k_B T$, in the range expected from classical theory predictions for a homogeneous nucleation limited metastability limit.

The liquid–crystal interfacial energy σ can be estimated from undercooling data, assuming they are due to homogeneous nucleation. In the classical theory, Eqs. (3.21) and (3.22), the prefactor A^* contains the interfacial mobility, which is assumed to scale inversely with the viscosity, η. Measurements of η in the undercooled regime, however, are rare. Reasonable estimates for σ can still be obtained since (1) existing measurements of η show that it is rather independent of temperature in the range sampled by maximum undercooling measurements, and (2) the value of σ is relatively independent of the nature of the prefactor (and hence to the exact value for I_s and η) since it is raised to the third power in the exponential [Eqs. (3.21) and (3.4)]. When data are available, A^* may be calculated by extrapolating the measured viscosity at the melting point to the undercooled temperature to give $A^* \approx 10^{34 \pm 1}$/mol-s.

For metals, the growth velocity is generally large in the undercooled regime, and one nucleation event is sufficient to crystallize the droplets. Assuming this, the nucleation rate at the maximum undercooling is calculated from Eq. (9.3), taking $x \approx 0.05$. As demonstrated in Appendix A, for liquid metals the free-energy difference per unit volume, ΔG_v, can be approximated by the expression due to Turnbull:[1]

$$\Delta G_v = \Delta H_f \Delta T / T_m, \qquad (9.4)$$

where ΔH_f is the enthalpy of fusion per unit volume. Using this and calculating values for A^*, when data are available, values for the interfacial energy, σ, computed for the maximum undercooling limits, are presented in Table III.

The crystal–vacuum interfacial energy is related to the heat of sublimation in molecular crystals described by van der Waals interactions and to the

---

[165] H. B. Singh and A. Holz, *Solid State Commun.* **45**, 985 (1983).
[166] M. C. Weinberg, *J. Non-Cryst. Solids* **83**, 98 (1986).

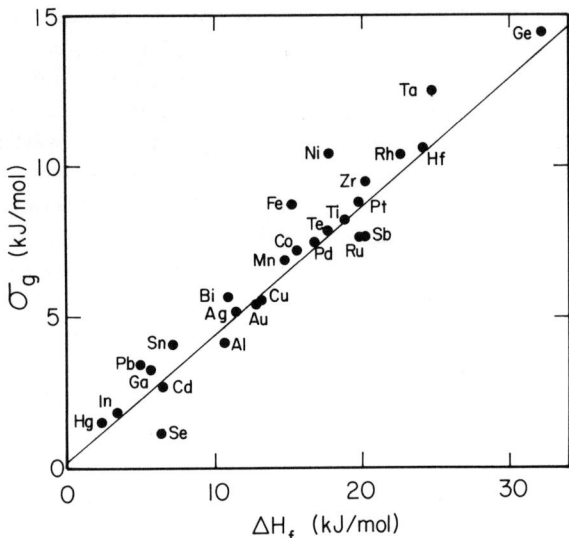

FIG. 11. The gram-atomic interfacial energy, derived from maximum undercooling experiments on elemental liquids, as a function of the gram-atomic heat of fusion. Due to uncertainties in v/Q, Fe, Ni, and Pb were excluded when determining the fit to the data.

crystal lattice energy in ionic crystals.[167] It is reasonable, therefore, to also expect a relation between the liquid–crystal interfacial energy and the heat of fusion, ΔH_f. Since ΔH_f is a molar quantity, Turnbull argued that the comparison should properly be made with the gram-atomic interfacial energy, defined by

$$\sigma_g = \sigma V^{2/3} N_A^{1/3}. \qquad (9.5)$$

The calculated σ_g's and the measured ΔH_f's are listed in Table III. Figure 11 shows a plot of σ_g versus ΔH_f. A linear dependence is obtained with a slope of 0.43. Turnbull found that while data for most metals fit on a line with slope 0.45, Ge, Sb, and Bi were best fit by a line with slope 0.32. Results from more recent measurements suggest that those anomalous data were incorrect.

The calculated number of atoms, n^*, and the radius, r^*, in the critical nucleus are also listed in Table III. A comparison with density functional calculations presented in Section 19 suggests that the interface is probably diffuse for all cases considered here. The computed values of σ are therefore questionable. Direct measurements of σ in Bi are approximately 15%

[167] R. A. Swalin, "Thermodynamics of Solids," second edition, pp. 230–237. Wiley, New York, 1972.

higher.[168] The significance of this for the observed correlation with ΔH_f is not clear.

Estimates of σ using different approximations for ΔG_v (Appendix A) give qualitatively similar results. The ratio, $\sigma_g/\Delta H_f$, varies between 0.4 and 0.48. It appears, therefore, that while the precise meaning of the value for σ is uncertain, it constitutes a parameter that can be determined for each element and profitably used to make predictions of the nucleation behavior.

For comparison, representative undercooling data on nonmetallic systems are listed in Tables IV and V. A range for ΔT_r similar to that for liquid metals is observed. Undercoolings obtained by the entrained droplet technique are listed in Table VI.

10. Nucleation Rate Measurements

A quantitative evaluation of nucleation theory is difficult to make from undercooling experiments. As indicated, the effects of heterogeneous sites are difficult to assess, an arbitrary value of the nucleation rate is assumed to analyze the undercooled limit, growth kinetics are neglected, and the statistics of nucleation in small volumes and cooling rates are often ignored. To more critically evaluate nucleation theories, it is necessary to obtain

TABLE IV. Maximum Undercoolings in Selected Alkali Halide Melts[a]

Salt	ΔT_{max}	$T_m(K)$	ΔT_r
CsBr	162	909.2	0.18
CsCl	152	919.2	0.17
CsF	153	955.2	0.16
CsI	206	894.2	0.23
KBr	162	1003.2	0.16
KCl	169	1049.2	0.16
KI	155	959.2	0.16
LiBr	94	820.2	0.11
LiCl	186	887.2	0.21
LiF	232	1115.2	0.21
NaBr	163	1028.2	0.16
NaCl	168	1074.2	0.16
NaF	281	1261.2	0.22
RbCl	163	988.2	0.16

[a] Data taken from Ref. 169.

[168] M. E. Glicksman and C. L. Vold, *Acta Metall.* **17**, 1 (1969).
[169] A. G. Walton, in "Nucleation" (A. C. Zettlemoyer, ed.), p. 225. Marcel Dekker, New York, 1969.

TABLE V. MAXIMUM UNDERCOOLINGS IN SELECTED ORGANIC LIQUIDS[a]

ORGANIC	ΔT_{max}	T_m (K)	ΔT_r
$Br \cdot [CH_2]_2 \cdot Br$	66.5	283.0	0.24
CBr_4	[a]82	362.8	0.23
	[b]82	362.8	0.23
CCl_4	[a]50.4	249.5	0.20
	[b]50.4	249.5	0.20
C_6H_6	70.2	278.6	0.25
$C_6H_5 \cdot CO_2H$	120	394.7	0.30
Diphenyl	86	344	0.25
H_2O	40.5	273.2	0.15
MeCl	55.6	174.0	0.32
Naphthalene	94.4	353.6	0.27
NH_3	40.3	195.6	0.21

a and b represent the two possible solid phases.
[c]Data taken from Ref. 169.

measurements of the magnitude and temperature dependence of the nucleation rate. These measurements are made typically on ensembles of droplets, in a manner similar to that for undercooling experiments. The experiments are difficult. It may not be possible to find an emulsion that effectively limits heterogeneous nucleation in the appropriate temperature range. The measurement of the change in fraction crystallized with time is both difficult and tedious. Quantitative measurements have therefore been made on only two metallic liquids and a similarly small number of organic liquids. These data present a disturbing discrepancy between the measured nucleation prefactor, A^*, and the value predicted from the classical theory.

Turnbull determined the nucleation rate in coated Hg droplets, using dilatometry to measure the volume fraction x. Those data, reduced to a master isotherm by scaling the time with temperature, are compared with predictions for different nucleation modes in Fig. 12 (see Refs. 1, 39). The sensitivity of nucleation to the surface coating is clear. Figure 12a shows the fraction solidified as a function of time for liquid droplets of Hg coated with mercury laurate. Figure 12b presents similar data for droplets coated with mercury acetate. The curves are calculations for a monodispersion of droplets and a polydispersion of droplets exhibiting volume nucleation or surface nucleation. A coating of mercury laurate is superior in suppressing the surface nucleation. A similar sensitivity to the surface coating was observed for undercooled Ga droplets.[39]

[170]M. J. Hunter and G. A. Chadwick, *J. Iron Steel Inst.* **210**, 707 (1972).
[171]M. J. Hunter, *J. Inst. Metals* **101**, 274 (1973).
[172]S. Chambey, Ph.D. Thesis, Department of Metallurgical Engineering, Institute of Technology, Banaras Hindu University, Varanasi, India, 1989 (unpublished).

TABLE VI. MAXIMUM UNDERCOOLINGS BY THE ENTRAINED DROPLET TECHNIQUE

Alloy	Primary Phase	ΔT_{max} (K)
Cu–wt.% Pb	Cu[a]	3
Cu–5 wt.% Bi	Cu[a]	22
Al–10 wt.% Sn	Al[a]	99
Ni–graphite	Ni[b]	775
Fe–graphite	Fe[b]	745
Ni–graphite*	Ni[b]	800
Fe–graphite*	Fe[b]	280
Fe–Fe$_3$C	Fe[b]	100
Mg–Mg$_2$Ni	Mg[c]	159
Ag–Cu	Ag[d]	126
	Cu[d]	235
Ag–Pb	Ag[d]	2
	Pb[d]	118
Al–Ge	Al[d]	291
	Ge[d]	88
Al–Si	Al[d]	176
	Si[d]	75
Bi–Cd	Bi[d]	91
	Cd[d]	82
Bi–Sn	Bi[d]	117
	Sn[d]	65
Cd–Zn	Cd[d]	32
	Zn[d]	2
Cd–Pb	Cd[d]	72
	Pb[d]	87
Pb–Sb	Pb[d]	55
	Sb[d]	41.5
Pb–Sn	Pb[d]	42
	Sn[d]	70.5
Sn–Zn	Sn[d]	62
	Zn[d]	116
Al–Cd	Al[d]	61
Al–In	Al[d]	9
Al–Pb	Al[d]	30
Al–Sn	Al[d]	91
Cu–Pb	Cu[d]	2
Zn–Bi	Zn[d]	146
Zn–Pb	Zn[d]	36
Zn–In	Zn[d]	39
Al–CuAl$_2$	Al[d]	57
Al–AgAl	Al[d]	29
Al–Cd	Cd[e]	60
Al–Bi	Bi[e]	157
Zn–Bi	Bi[e]	132
Al–In	In[e]	19

*Modified by addition of Mg.
[a]Ref. 159
[b]Ref. 170
[c]Ref. 171
[d]Ref. 160
[e]Ref. 172

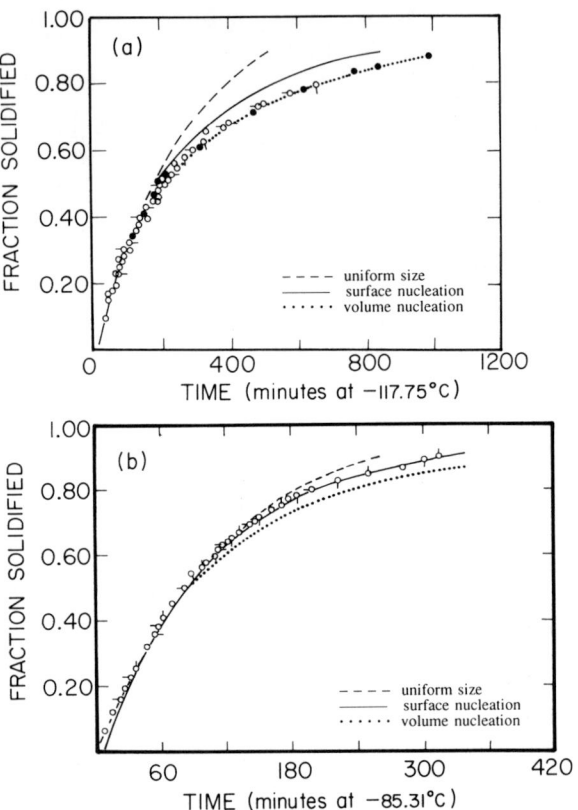

FIG. 12. (a) The fraction of mercury droplets coated with mercury laurate solidified as a function of time at $-117.75°C$, demonstrating volume nucleation. (b) The fraction of mercury droplets coated with mercury acetate solidified as a function of time at $-85.31°C$, demonstrating surface nucleation. The "uniform size" curves were calculated assuming a uniform distribution of droplet sizes; the other two curves assumed a distribution of droplet sizes. Adapted from D. Turnbull, *J. Chem. Phys.* **20**, 411 (1952).

The steady-state nucleation rate is obtained from Eq. (3.22), using ΔG_{n^*} from Eq. (3.4) and calculating ΔG_v from Eq. (9.4):

$$I^s = A^* \exp\left[-\frac{16\pi T_m^2}{3k_B \Delta H_f^2} \frac{\sigma^3}{T \Delta T^2}\right]. \tag{10.1}$$

A plot of $\log I^s$ versus $(T(\Delta T)^2)^{-1}$ should give a straight line with slope proportional to σ^3 and intercept equal to $\log A^*$. Plots of nucleation data for Ga and Hg show this behavior (Fig. 13). A large discrepancy, however, is observed between the data taken from two different measurements on Ga.

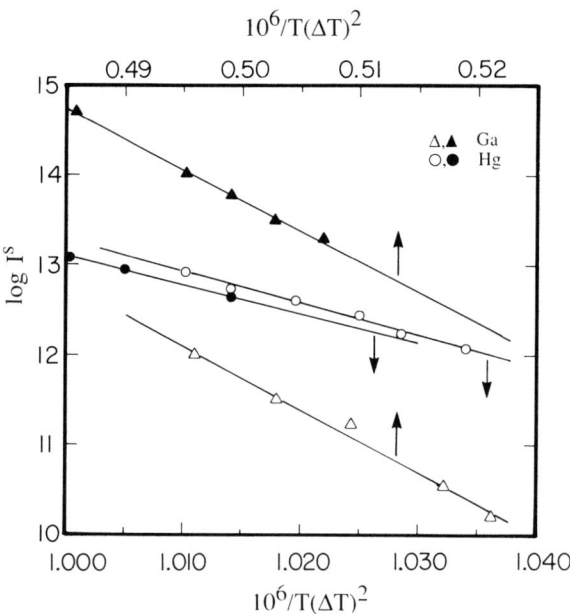

FIG. 13. The steady-state nucleation rate, I^s (m^{-3} s^{-1}), as a function of the undercooling in degrees Kelvin for liquid Hg (lower abscissa) and Ga (upper abscissa). Hg data obtained from D. Turnbull, *J. Chem. Phys.* **20**, 411 (1952); Ga data obtained from Y. Miyazawa and G. M. Pound, *J. Cryst. Growth* **23**, 45 (1974).

TABLE VII. PARAMETERS DERIVED FROM MEASUREMENTS OF $I^s(T)$ IN SELECTED SYSTEMS

SYSTEM	σ (J/m^2)	log A^* (m^{-3} s^{-1}) EXPERIMENTAL
Hg[a]	0.031	48.1
$C_{17}H_{36}$[b]	0.0072	36.5 ± 2
$C_{18}H_{38}$[c]	0.0096	37.3 ± 2
$C_{24}H_{50}$[c]	0.0082	30 ± 4
H_2O[c]	0.032	52.5

[a]Ref. 1
[b]Ref. 173
[c]Ref. 174

[173]D. Turnbull and R. L. Cormia, *J. Chem. Phys.* **34**, 820 (1961).
[174]G. R. Wood and A. G. Walton, *J. Appl. Phys.* **41**, 3027 (1970).

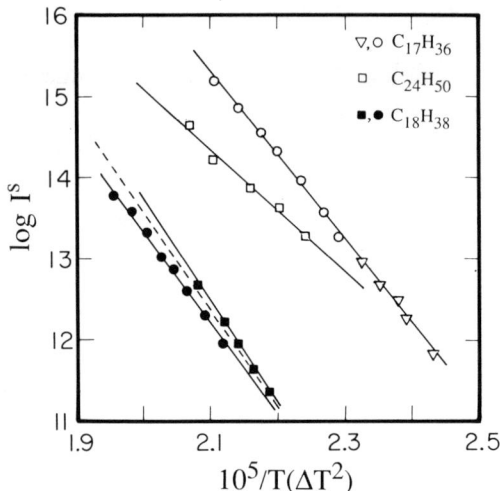

FIG. 14. The steady-state nucleation rate, I^s (m^{-3} s^{-1}), as a function of the undercooling in degrees Kelvin for several n-alkane liquids. The dotted line is the average of two sets of data for $C_{18}H_{38}$. Data obtained from D. Turnbull and R. L. Cormia, *J. Chem. Phys.* **34**, 820 (1961).

Also, Turnbull recently showed that the solidification rate reported for the Ga nucleation data scales with the average diameter of the droplet size ($\bar{d}$) rather than $\bar{d}^3$, as required for volume nucleation.[40] The relevance of the Ga data for homogeneous nucleation is therefore questionable, although the origin of the observed data scatter and the particle size dependence is unknown. The measured value of A^* for Hg, listed in Table VII, is approximately 10^7 larger than predicted from the classical theory ($A^* \approx 10^{41}$ m^{-3} s^{-1}). Turnbull first noted that the experimental values for A^* agree with theoretical predictions if σ increases linearly with temperature.[1] Turnbull[175] and Spaepen et al.[176,177] have argued that such a linear temperature dependence for metallic systems must be due to a sizable negative entropy (negentropy) near the interface arising from ordering in the liquid.

The measured nucleation rates for several n-alkane liquids[173] are plotted in Fig. 14. Straight lines are again obtained, however, the calculated values for A^* agree with the classical results (Table VII); no linear temperature dependence for σ is required. The calculated interfacial energies (Table VII)

[175] D. Turnbull, in "Physics of Non-Crystalline Solids" (J. A. Prins, ed.), p. 41. North-Holland, Amsterdam, 1964.
[176] F. Spaepen, *Acta Metall.* **23**, 729 (1975).
[177] F. Spaepen and R. B. Meyer, *Scripta Metall.* **10**, 257 (1976).

are also an order of magnitude smaller than for Hg. Turnbull and Spaepen[178] attribute these two effects to the nonrandom probability for occurrence of linear conformations in the melt. Based on the negentropic model, a small temperature dependence of the interfacial energy results since the localization of a small number of CH_2 groups forces the localization of a larger number of CH_2 groups in the same molecule without additional loss of entropy. As expected, the nucleation data for water also gives a prefactor that is approximately seven orders of magnitude higher than predicted.[174]

V. Experimental Studies of Nucleation in Glasses

Glasses are configurationally frozen liquids obtained by quenching the liquid on a time scale that is rapid compared with that required for crystallization. The cooling rates depend on the system. Silicate-based glasses can be produced by cooling at less than 1°C/s and occur naturally (obsidian, for example). Metallic liquids, however, must be cooled at rates from 1 to 10^{12}°C/s to produce a glass. No example of a naturally occurring metallic glass is known. Techniques for metallic glass formation were pioneered in the 1960s by P. Duwez and co-workers.[179] There are now several methods for producing metallic glasses, but most are made by quenching the liquid onto a rapidly rotating surface. A discussion of the quenching techniques is not provided; the reader is referred to other articles.[180]

Glasses transform on a longer time scale than undercooled liquids to a stable, generally crystalline phase. Direct measurements of nucleation and growth rates are therefore possible. These data allow more stringent tests of nucleation theories. For example, since the glass is more deeply undercooled than is possible with liquids, nucleation at large departures from equilibrium, where the classical theory might be expected to fail, can be studied. Time-dependent nucleation is also studied more easily in glasses than in liquids, because of the slower kinetics. Measurements of time-dependent nucleation as a function of temperature and sample thermal history can be tested against theoretical predictions and can provide a new probe of the cluster dynamics.

11. EXPERIMENTAL METHODS AND ANALYSIS

For metallic glasses, and for silicate glasses at temperatures well above the maximum in the nucleation rate, nuclei can be formed and developed by annealing at a single temperature. The crystallites are counted directly from

[178]D. Turnbull and F. Spaepen, *J. Polymer Sci.* **63**, 237 (1978).
[179]W. Klement, R. H. Willens, and P. Duwez, *Nature* **187**, 869 (1960); P. Duwez, R. H. Willens, and R. C. Crewdson, *J. Appl. Phys.* **36**, 2267 (1965); P. Duwez, *Trans. ASM* **60**, 605 (1967).
[180]Treatise on Materials Science and Technology, "Ultrarapid Quenching of Liquid Alloys" (Herbert Hermann, ed.), Vol. 20. Academic Press, New York, 1981.

prepared sections of the sample using optical microscopy,[84] scanning electron microscopy (SEM),[181] or transmission electron microscopy (TEM),[182,183] with standard stereological methods.[184,185] TEM hot-stage experiments can be used for in situ studies of the nucleation and growth. Those transformations, however, are frequently dominated by surface effects; more reliable data are obtained from sections of externally annealed samples.

Glasses are typically multicomponent systems. Crystallization of glasses, known as devitrification, occurs by one of three modes: (a) polymorphic crystallization, in which the glass transforms to a crystal of the same composition; (b) primary crystallization, for which there exists a compositional difference between the glass and the crystal; and (c) eutectic crystallization, which is a cooperative transformation to two intimately connected phases while maintaining the mean composition. If the growth characteristics are known for eutectically or polymorphically crystallizing glasses, it is possible to estimate the nucleation kinetics by measuring the crystallite size distribution.[186,187] Bulk nucleation can be either homogeneous or heterogeneous, and steady state or transient.[188-190] Figure 15 shows distributions calculated for several different mechanisms. These distributions are invalid if significant impingement of the crystallites has occurred and for systems that crystallize by a nonpolymorphic mechanism. Further, they assume that the growth velocity depends on the temperature only, and that the nuclei grow isotropically. The distribution shapes are characteristically different and can be used to identify readily the mode of devitrification of the glass.

For many silicate glasses, the peak in the steady-state nucleation rate occurs at a sufficiently low temperature that the growth of nuclei is slow. Nucleation rates can then be determined by a two-stage annealing treatment. The sample is first annealed at a temperature, T_N, where the nucleation rate is high and the growth velocity is small, to develop a population of nuclei.

---

[181] A. L. Greer, "Amorphous Metals and Semiconductors" (P. Haasen and R. I. Jaffe, eds.), p. 94. Pergamon, London, 1986.

[182] C. J. R. Gonzalez-Oliver and P. F. James, *J. Microscopy* **119**, 73 (1980).

[183] R. S. Tiwari, S. Ranganathan, and M. von Heimendahl, *Z. Metallk.* **72**, 563 (1981).

[184] E. E. Underwood, "Quantitative Stereology." Addison-Wesley, Reading, MA, 1970.

[185] R. T. Dehoff and F. N. Rhines, *Trans. Metall. Soc. AIME* **221**, 975 (1961).

[186] P. V. Evans, A. Garcia-Escorial, P. E. Donovan, and A. L. Greer, "Material Research Society Symposium Proceedings" (G. S. Gargill, F. Spaepen, and K. N. Tu, eds.), Vol. 57, p. 239. 1987.

[187] U. Köster, *Z. Metallk.* **75**, 691 (1984).

[188] U. Köster, and M. Blank-Bewersdorff, Mat. Res. Soc. Symp. Proc. (G. S. Cargill, F. Spaepen, K. N. Tu, eds.), Vol. 57, p. 115. 1987.

[189] H. Blanke and U. Köster, "Proc. 5th Int. Conf. on Rapidly Quenched Metals" (S. Steeb and H. Warlimont, eds.), p. 227. North-Holland, Amsterdam, 1985.

[190] U. Köster and H. Blanke, *Scripta Met.* **17**, 495 (1983).

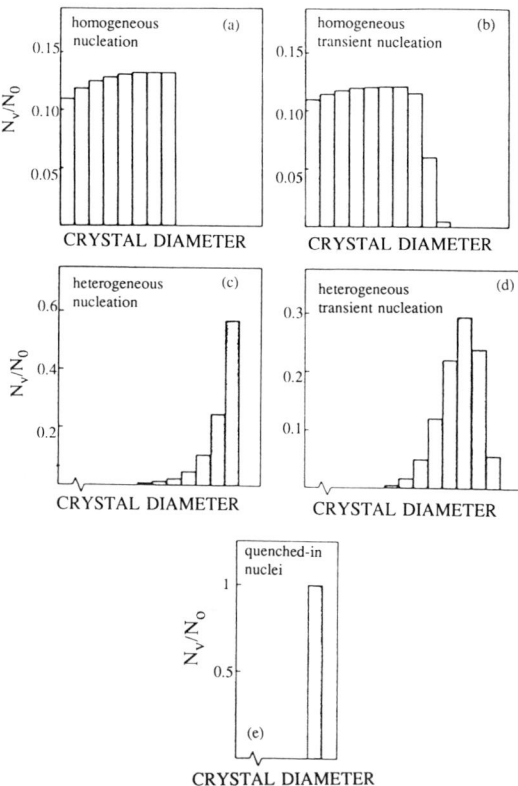

FIG. 15. Schematic histograms of the density versus crystal diameter for different nucleation mechanisms. Taken from U. Koster and M. Blank-Bewersdorff, *Mat. Res. Soc. Symp. Proc.* **57**, 115 (1987).

Those nuclei are grown subsequently at a higher temperature, T_G, where the growth velocity is large, but the nucleation rate is low enough to introduce no additional nuclei. Since the critical size increases with temperature, those nuclei between $n^*(T_N)$ and $n^*(T_G)$ will redissolve upon annealing at T_G, while those that are larger than $n^*(T_G)$ will grow to observable size. The number of nuclei between the critical sizes is small compared with those above $n^*(T_G)$ and does not significantly perturb the results. This has been verified experimentally[84,191,192] and theoretically.[87] In steady state, the nucleation rate is independent of the cluster size at which it is measured. The calculation of the nucleation rate at the critical size for the nucleating temperature should

[191] A. M. Kalinina, V. M. Fokin, and V. N. Filipovich, *Fiz. Khim. Stekla* **2**, 298 (1976).
[192] A. M. Kalinina, V. M. Fokin, and V. N. Filipovich, *Fiz. Khim. Stekla* **3**, 122 (1977).

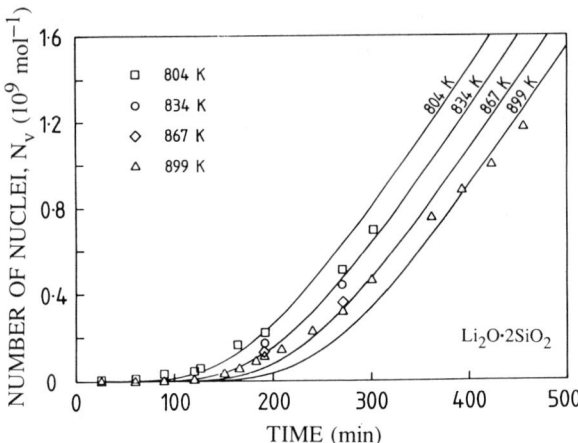

FIG. 16. The number of nuclei produced in lithium disilicate glass as a function of annealing time at 726 K, followed by a second anneal at four different growth temperatures. The experimental data are from Ref. 191; the lines are fits from a numerical calculation. Taken from A. L. Greer and K. F. Kelton, *J. Am. Cer. Soc.* (in press) by permission of the American Ceramic Society.

then give an accurate description of the measured rate. In practice, however, transient effects are often important, and the time-dependent nucleation rate must be calculated at the critical size for the growth temperature. As demonstrated in Fig. 16, if the nucleation temperature is held constant the time lag θ increases with increasing growth temperature. Since the critical size increases with increasing temperature, a longer time is required to diffuse from $n^*(T_N)$ to $n^*(T_G)$. The solid lines are predictions from a computation that is discussed in Section 14. The time-dependent nucleation rate is calculated typically at the critical size for nucleation. To compare with experimental data, the time required to grow from the critical size at the nucleating temperature to the critical size at the growth temperature must be computed. Using reaction rate theory, Kelton and Greer[86] found the average growth rate of a cluster,

$$\frac{dr}{dt} = \frac{16D}{\lambda^2}\left(\frac{3\bar{v}}{4\pi}\right)^{1/3} \sinh\left[\frac{\bar{v}}{2k_B T}\left(\Delta G_v - \frac{2\sigma}{r}\right)\right], \qquad (11.1)$$

where $\bar{v}$ is the molecular volume, ΔG_v is the free-energy decrease per unit volume on transformation, and σ is the interfacial energy per unit area. This expression becomes less accurate as the cluster size approaches the critical size. For cluster sizes $n \geqslant 1.5n^*$, predictions from this equation agreed with those calculated from a numerical solution of Eq. (3.7) to better than 3%.

Measurements of the devitrification kinetics are also used to estimate the nucleation and growth rates. The volume fraction transformed as a function of time, $x(t)$, with isothermal annealing is described theoretically by the Johnson–Mehl–Avrami (JMA) equation[194-196]

$$x = 1 - \exp(-(kt)^n). \quad (11.2)$$

Here k is an effective kinetic coefficient, and n describes the mode of transformation. For polymorphic crystallization on a fixed number of nuclei per unit volume N, n is equal to 3 and k is given by

$$k = (4\pi N/3)^{1/3} u, \quad (11.3)$$

where u is the growth velocity. For a constant volume nucleation rate I^s, $n = 4$ and

$$k = (\pi I^s u^3/3)^{1/4}. \quad (11.4)$$

Nonpolymorphic transformations generally give values for n between 1.5 and 2.5. The kinetics of crystallization are strongly dependent on the geometry of the sample and on the distribution of nuclei. A proper study therefore requires a knowledge of the microstructure and the transformation mechanism; otherwise, incorrect interpretations of the JMA analysis are likely.[197-199]

12. Crystallization of Metallic Glasses

a. *Transformation to Crystalline Phases*

The crystallization kinetics of metallic glasses have been studied extensively,[200,201] but few careful measurements of the nucleation rates exist. In addition to nuclei formed heterogeneously or homogeneously during annealing, nuclei are often formed during the quench. Devitrification occurs

---

[193] A. L. Greer and K. F. Kelton, *J. Am. Ceram. Soc.*, (in press).
[194] W. A. Johnson and R. F. Mehl, *Trans. Am. Inst. Min. Metall. Eng.* **135**, 416 (1939).
[195] M. Avrami, *J. Chem. Phys.* **7**, 1103 (1939).
[196] J. W. Christian, "The Theory of Transformations in Metals and Alloys," second edition, Chapter 12. Pergamon Press, Oxford, 1975.
[197] A. L. Greer, "Proc. 5th Int. Conf. on Rapidly Quenched Metals" (V. S. Steeb and H. Warlimont, eds.), p. 215. North-Holland, Amsterdam, 1985.
[198] K. F. Kelton and F. Spaepen, *Acta Metall.* **33**, 455 (1985); K. F. Kelton and J. C. Holzer, *Phys. Rev. B* **7**, 3940 (1988).
[199] M. C. Weinberg, *J. Non-Cryst. Solids* **72**, 301 (1985).
[200] U. Köster and U. Herold, in "Glassy Metals I, Topics in Applied Physics" (H. J. Güntherodt and H. Beck, eds.), Vol. 46, p. 225. Springer-Verlag, 1981.
[201] M. Scott, in "Amorphous Metallic Alloys" (F. E. Luborsky, ed.), p. 144. Butterworths, London, 1983.

frequently by growth on these quenched-in nuclei or by surface nucleation and growth.

Devitrification studies in $Fe_{40}Ni_{40}P_{14}B_6$[202,203] provide the best evidence for homogeneous nucleation in a metallic glass that transforms by a eutectic mechanism. Morris[202] used TEM to determine the number and size of crystallites that nucleated and grew during isothermal annealing. He showed that the crystallite size distribution increased with annealing time, as shown in Fig. 15a, with no evidence of site saturation, suggesting a homogeneous nucleation mechanism. No evidence for transient nucleation was reported. The nucleation rates were estimated using Eqs. (11.2) and (11.4). Tiwari[203] fit measured nucleation data for $Fe_{40}Ni_{40}P_{14}B_6$ to the classical theory to obtain an estimate for the interfacial energy between the crystal and the glass. Using a viscosity calculated from isothermal creep measurements[204] and a free energy calculated from Hoffman's expression [Eq. (A10)], he obtained $\sigma \approx 0.06$ J/m^2, which is lower than estimates for σ between crystalline Fe and Ni and their melts (Table III).

Assuming that the viscosity can be related to the diffusion coefficient via the Stokes–Einstein relation,[205]

$$D = k_B T/3\pi a \eta \tag{12.1}$$

where a is a length of order the atomic diameter, the classical expression for the steady-state nucleation rate (Eq. 3.22) can be written as

$$I^s = \frac{A}{\eta} \exp\left[-\frac{16\pi}{3k_B} \frac{\sigma^3}{T \Delta G_v^2}\right], \tag{12.2}$$

where $A = A^*\eta$. A plot of $I^s\eta$ versus $1/T \Delta G_v^2$ should give a straight line with slope proportional to σ^3 and intercept equal to the prefactor for the nucleation rate A. Figure 17 presents two different analyses of the combined data from Morris and Tiwari. The lower curve is constructed for a temperature-independent interfacial energy; the upper curve is constructed for σ increasing linearly with temperature. The viscosity data measured by Tiwari were used:

$$\eta = 3.3 \times 10^{-39} \exp(7.457 \times 10^4/T) \text{ Pa} \cdot \text{s}. \tag{12.3}$$

The Hoffman expression used by Tiwari is a poor description for ΔG_v in metallic glasses (Appendix A). The Singh and Holz expression [Eq. (A16)],

---

[202] D. G. Morris, *Acta Metall.* **29**, 1213 (1981).
[203] R. S. Tiwari, *J. Non-Cryst. Solids* **83**, 126 (1986).
[204] R. S. Tiwari, J. C. Claus, and M. V. Heimendahl, *Mater. Sci. Eng.* **55**, 1 (1982).
[205] A. L. Greer, C. J. Lin, and F. Spaepen, "Proc. 4th Int. Conf. on Rapidly Quenched Metals" (T. Masumoto and K. Suzuki, eds.), p. 567. Japan Institute of Metals, 1982.

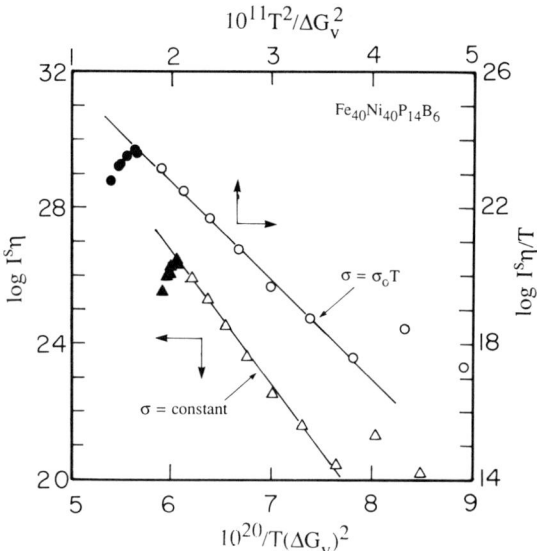

FIG. 17. The product of the steady-state nucleation rate, I^s (m^{-3} s^{-1}), and the viscosity, η (Pa·s), as a function of the volume free energy with undercooling, ΔG_v (J m^{-3}) for Fe$_{40}$Ni$_{40}$P$_{14}$B$_6$ metallic glass. The upper fit assumes a linear temperature dependence for the interfacial energy, σ, and refers to the upper abscissa and right ordinate. The lower curve takes σ to be constant and refers to the lower abscissa and left ordinate. Solid data R. S. Tiwari et al., Mater. Sci. Eng. 55, 1 (1982); open data are from D. G. Morris, Acta Metall. 29, 1213 (1981).

describes more accurately ΔG_v for undercooled melts and silicate glasses and is used here.

In contradiction with the predictions of classical theory, the curve is linear only over the central portion. This discrepancy was not rectified by using the viscosity data of Anderson and Lord.[206] Fitting the approximately linear portion of the curve gave $A \approx 10^{49.8}$, with $\sigma = 0.042$ J/m^2. Assuming $\sigma = \sigma_0 T$ gave $\sigma_0 = 8.2 \times 10^{-5}$ J/m^2·K and $A \approx 10^{30.6}$, in good agreement with the theoretical prediction for A from the classical theory. Given the uncertainty in the nucleation data and in the appropriate kinetic and thermodynamic parameters, however, this agreement is probably fortuitous.

Transient effects enhance the stability of some metallic glasses and may be important for glass formation in some systems.[86,89] Thompson et al.[207] studied the crystallization of (Al$_{100-y}$Cu$_y$)$_{77}$Si$_9$Ge$_{14}$, using DSC and TEM.

[206] P. M. Anderson and A. E. Lord, Jr., J. Non-Cryst. Solids 37, 219 (1980).
[207] C. V. Thompson, A. L. Greer, and F. Spaepen, Acta Metall. 31, 1883 (1983).

Nonisothermal DSC scans were analyzed using a Kissinger analysis[208,209] to obtain the growth velocity as a function of temperature. It was possible to fit the isothermal kinetic data to Eq. (11.2) only when non-steady-state nucleation was assumed.

Amorphous $Pd_{40}Ni_{40}P_{20}$ and $Pd_{77.5}Cu_6Si_{16}$ transform eutectically to give approximately spherical crystallites, indicating an isotropic growth velocity. These are among the best metallic glass formers, requiring a cooling rate from the melt of less than 100 K/s to prevent crystallization. Surface crystallization dominates the devitrification of small beads of $Pd_{40}Ni_{40}P_{20}$ that are quenched in a drop tower.[156] The bulk homogeneous nucleation rate was measured by a modification of the two-step annealing technique discussed in Section 11. Crystallites were nucleated in the glass by first annealing at 590 K. These nuclei were then grown to observable size by linearly increasing the temperature with time. The absence of bulk nuclei without the preannealing step suggests that nuclei are formed by homogeneous nucleation in that step; the independence of the nucleation behavior on the quenching conditions supports this conclusion. The number of nuclei increased linearly with time, indicating steady-state nucleation; no evidence for a transient was found. The maximum homogeneous nucleation rate was estimated as $10^6/m^3 \cdot s$, an extremely low rate for metallic alloys, explaining why this alloy so readily forms a glass. Kiminami and Sahm also measured a low nucleation rate of $10^7/m^3 \cdot s$ in $Pd_{77.5}Cu_6Si_{16}$ by a similar method.[157]

Devitrification often occurs via growth on nuclei that are formed during the quench (quenched-in nuclei). Greer found a high density of nuclei, $\approx 10^{18}/m^3$ in amorphous $Fe_{80}B_{20}$. Given steady-state nucleation, the number of nuclei, N_v, should decrease linearly with increasing quench rate $\dot{Q}$. Greer found that N_v was proportional to $\dot{Q}^{-2}$ to $\dot{Q}^{-4}$, suggesting important transient nucleation effects during the quench.[210]

Köster constructed statistical distributions of the number of nuclei as a function of size in $Fe_{65}Ni_{10}B_{25}$ after annealing at different temperatures (Fig. 18). The increasing number of nuclei of decreasing size and the site saturation suggest that devitrification is dominated by transient heterogeneous nucleation. By a similar analysis, $CoZr_2$ was shown to devitrify by transient homogeneous nucleation (Fig. 18).

Surface nucleation is important for devitrification of metallic glasses. It is generally favored over bulk nucleation because of a decreased surface energy penalty, higher diffusion coefficients, and a change in surface composition. The stresses resulting from the surface crystalline layer can effect the bulk properties and the stability.[211] Köster et al. found a crystallite size distribu-

[208] H. E. Kissinger, J. Res. Nat. Bur. Stand. 57, 217 (1956).
[209] D. W. Henderson, J. Non-Cryst. Solids 30, 301 (1979).
[210] A. L. Greer, Acta Metall. 30, 171 (1982).
[211] H. N. Ok and A. H. Morrish, Phys. Rev. B 23, 1835 (1981).

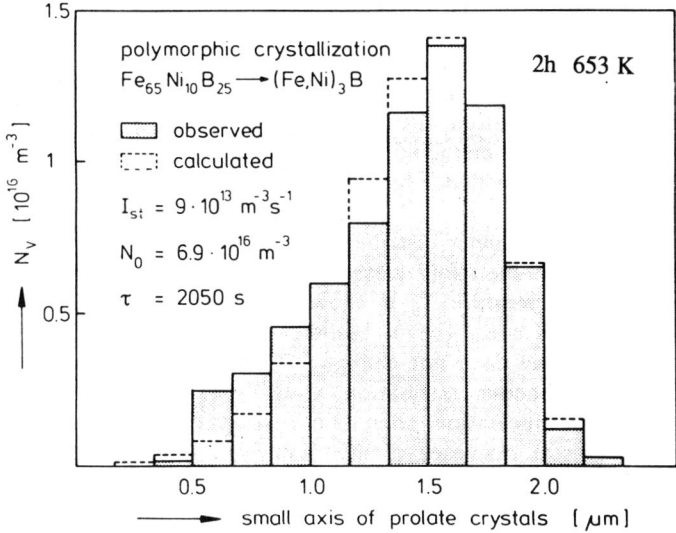

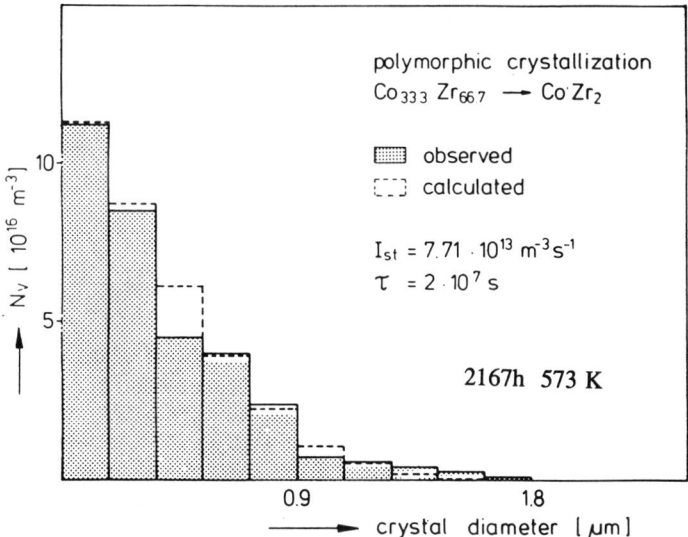

FIG. 18. (a) A comparison between the measured number of nuclei as a function of crystal size in polymorphically crystallizing $Fe_{65}Ni_{10}B_{25}$ glass and the distribution calculated assuming transient heterogeneous nucleation. (b) A comparison between the measured number of nuclei as a function of crystal size in polymorphically crystallizing $CoZr_2$ glass and the distribution calculated assuming transient homogeneous nucleation. Taken from U. Koster and M. Blank-Bewersdorff, *Mat. Res. Soc. Symp. Proc.* **57**, 115 (1987).

tion in a number of Ni–B and Co–B glasses that is very similar to that shown in Fig. 15d, suggesting heterogeneous transient nucleation on the surface.[212] Preferential oxidation at the surface, leading to a composition change, is frequently a cause of surface crystallization. This can be controlled partially with appropriate coatings and surface treatments. Chromium additions to $Fe_{79}B_{16}Si_5$ cause the formation of a protective layer of chromium oxide at the ribbon surface during annealing. Similarly a mild oxidation of $Pd_{40}Ni_{40}P_{20}$ inhibits surface crystallization because the formation of a thin layer of NiO prevents loss of phosphorus that can lead to crystallization.[213] The removal of an oxide layer resulting from the quench can frequently enhance surface crystallization. Ion beam cleaning followed by an immediate coating of the sample surface reduces the surface oxidation in some cases, but the results depend strongly on the coating material.[214]

b. *Transformation to Quasicrystals*

In 1984, Shechtman *et al.*[215] announced the discovery of a rapidly quenched alloy of aluminum and manganese that gave sharp diffraction patterns with an icosahedral symmetry, which is incompatible with translational periodicity. Similar phases have been discovered subsequently in other metallic alloys.[216]

The existence of icosahedral order in a condensed system was first proposed by Frank[217] to explain the large undercoolings observed in metallic melts. For a central potential (such as the Lennard-Jones potential), the icosahedron is the most stable configuration. According to Frank, metals will undercool because the replacement of that locally stable configuration with the crystalline configuration possessing a higher local energy involves a large energetic barrier. These qualitative arguments have been confirmed by computer calculations for particles interacting through a central potential that show a sharp increase in the number of icosahedra as the liquid is cooled.[218] Dense random packing models of metallic glasses contain a large

---

[212] B. Punge-Wittler and U. Köster, *Mater. Sci. Eng.* **97**, 343 (1988); U. Köster, *Mater. Sci. Eng.* **97**, 233 (1988).
[213] A. Garcia-Escorial and A. L. Greer, *J. Mat. Sci.* **22**, 4388 (1987).
[214] U. Herold, Ph.D. thesis, Dept. Mech. Eng., Ruhr-Universitat Bochum, 1982 (unpublished).
[215] D. Shechtman, I. Blech, D. Gratias, and J. W. Cahn, *Phys. Rev. Lett.* **53**, 1951 (1984).
[216] Z. Zhang, H. Q. Ye, and K. H. Kuo, *Philos. Mag. A* **52**, L49 (1985); K. F. Kelton, P. C. Gibbons, and P. N. Sabes, *Phys. Rev. B* **38**, 7810 (1988); M. D. Ball and D. J. Lloyd, *Scripta Metall.* **19**, 1065 (1985); P. Ramachandrarao and G. V. S. Sastry, *Pramana* **24**, L225 (1985); A. Tsai, A. Inoue, and T. Masumoto, *Jpn. J. Appl. Phys.* **26**, L1505 (1987).
[217] F. C. Frank, *Proc. Royal Soc. London A* **215**, 43 (1952).
[218] P. J. Steinhardt, D. R. Nelson, and M. Ronchetti, *Phys. Rev. Lett.* **47**, 1297 (1981); *Phys. Rev. B* **28**, 784 (1983); J. D. Honeycutt and H. C. Anderson, *J. Phys. Chem.* **91**, 4950 (1987); T. L. Beck and R. S. Berry, *J. Chem. Phys.* **88**, 3910 (1988).

number of polyhedra with five-sided faces.[219] Local icosahedral order is also found in many complex intermetallic crystalline structures; the Frank–Kasper phases[220] are best known.

The nucleation and growth of the icosahedral phase is poorly understood. Its formation is expected to be favored over competing crystalline phases, however, since less reconstruction of the polytetrahedral packing units presumably existing in the undercooled liquid is required. Several studies of the transformation of the icosahedral phase to more stable crystalline phases exist; these are reviewed elsewhere.[221] A qualitative analysis of the crystallization kinetics of icosahedral phase $Al_{86}Mn_{14}$ suggests a low interfacial energy between the icosahedral and crystalline phases, similar to that between a metallic crystal and its melt.[198] Only a few studies have been made of the amorphous to icosahedral phase transformation.

It has been suggested that some "amorphous" metals are actually composed of extremely fine grained quasicrystals. Bendersky and Ridder[222] argued that rapidly quenched droplets of $Al_{86}Mn_{14}$ solidify to a "microquasicrystalline" structure with grain sizes as small as 1 nm. Robertson et al.[223] explained the x-ray diffraction of a sputter-deposited amorphous $Al_{72}Mn_{22}Si_6$ film in terms of a 2.5-nm-grain-diameter microquasicrystalline structure. This was supported by DSC measurements of Chen and Spaepen,[224] demonstrating that sputter-deposited $Al_{82.6}Mn_{17.4}$ and $Al_{82-83}Fe_{17-18}$ films, which were apparently amorphous by diffraction, transformed by grain coarsening instead of by nucleation and growth. This implies a fine dispersion of icosahedral grains, which in turn suggests an extremely high nucleation rate from the liquid.

By contrast, Poon et al.[225] reported that rapidly quenched, amorphous $Pd_{58.8}U_{20.6}Si_{20.6}$ crystallized polymorphically to a single-phase icosahedral material. Drehman et al.[226] and Shen et al.[227] gave TEM and DSC evidence that the transformation proceeds by nucleation and growth. TEM studies of transformed samples showed a large density of small ($\approx 0.1\,\mu$) grains after

---

[219] G. S. Cargill III, in "Solid State Physics" (F. Seitz and D. Turnbull, eds.), Vol. 30, p. 227. Academic Press, New York, 1975; P. H. Gaskell, *J. Phys. C* **12**, 4337 (1979).
[220] D. P. Shoemaker and C. B. Shoemaker, in "Aperiodicity and Order" (M. V. Jaric, ed.), Vol. 1. Academic Press, Boston, 1987; A. K. Sinha, "Progress in Materials Science," Vol. 15, p. 81. Pergamon, New York, 1972.
[221] K. F. Kelton, *Phase Transitions*, **16/17**, 367 (1989).
[222] L. A. Bendersky and J. D. Ridder, *J. Mat. Res.* **1**, 405 (1986).
[223] J. L. Robertson, S. C. Moss, and K. G. Kreider, *Phys. Rev. Lett.* **60**, 2062 (1988).
[224] L. C. Chen and F. Spaepen, *Nature* **336**, 366 (1988).
[225] S. J. Poon, A. J. Drehman, and K. R. Lawless, *Phys. Rev. Lett.* **55**, 2324 (1985).
[226] A. J. Drehman, A. R. Pelton, and M. A. Noack, *J. Mat. Res.* **1**, 741 (1986).
[227] Y. Shen, S. J. Poon, and G. J. Shiflet, *Phys. B* **34**, 3516 (1986).

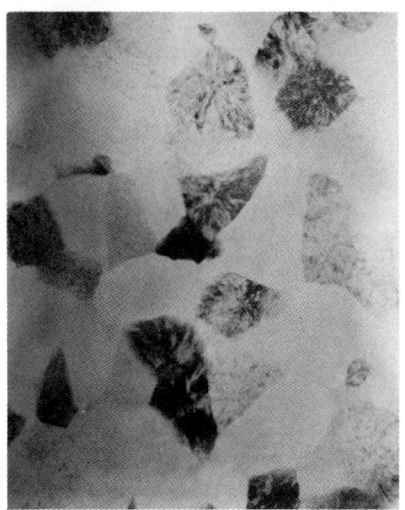 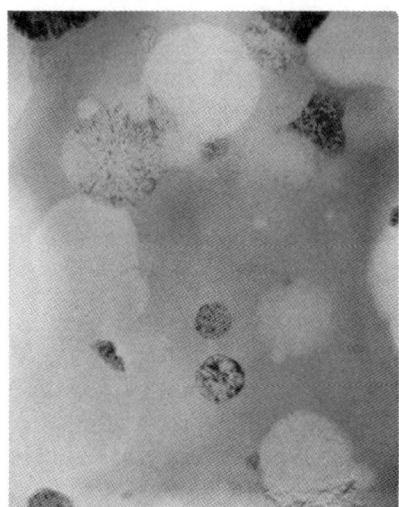

FIG. 19. A micrograph of rapidly quenched $Al_{75}Cu_{15}V_{10}$ showing the number of quasicrystal grains after partially transforming the glass (a) near the wheel side and (b) away from the wheel (less rapidly quenched), showing a nucleation and growth mechanism for quasicrystal formation from the glass.

devitrification. EELS measurements[228] confirmed that the icosahedral phase and the original glass have the same stoichiometry. Shen et al., however, reported an Avrami plot with $1.1 \leqslant n \leqslant 1.7$, which is inconsistent with bulk, polymorphic crystallization ($3 \leqslant n \leqslant 4$). On the basis of DSC measurements, Drehman et al. estimated a peak nucleation rate of $\approx 10^{18}/m^3 \cdot s$ between 730 and 760°C.

Tsai et al. reported DSC and TEM studies of the devitrification of amorphous $Al_{75}Cu_{15}V_{10}$ to a single-phase icosahedral material.[229] TEM studies showed clearly a homogeneous nucleation and growth process. A straightforward analysis of the crystallization kinetic data gave values for the Avrami exponent of $1.7 \leqslant n \leqslant 1.9$, similar to those found in $Pd_{58.8}U_{20.6}Si_{20.6}$; similar results were also reported for a rapidly quenched $Al_{55}Mn_{20}Si_{25}$ alloy.

Recent work by Holzer and Kelton[230] revealed a distribution of quenched in nuclei for rapidly quenched $Al_{75}Cu_{15}V_{10}$, showing fewer nuclei near the wheel side than away from the wheel (Fig. 19). Transient nucleation effects were also shown to be important. On the basis of TEM and DSC studies

---

[228] L. E. Levine, P. C. Gibbons, and K. F. Kelton, Phys. Rev. B **40**, 9338 (1989).
[229] A. P. Tsai, A. Inoue, Y. Bizen, and T. Masumoto, Acta Metall. **37**, 1443 (1989).
[230] J. C. Holzer and K. F. Kelton, Acta Metall., (in press).

of samples quenched at different rates, the nucleation and growth rates were determined to obtain an estimate of the interfacial energy between the icosahedral phase and the undercooled liquid or glass, $0.002 \leqslant \sigma \leqslant 0.015$ J/m^2. Such a low interfacial energy must result from a similar short-range order in the glass and the surface of the icosahedral phase. The "anomalous" Avrami coefficients observed previously are due to simultaneous nucleation and growth on a distribution of preexisting nuclei. They result from a misuse of the Avrami expression, which assumes that the transformation behavior is statistically the same for all regions of the samples, and is thus inapplicable here.

13. SILICATE GLASSES

Unlike metallic systems, silica-based melts are difficult to crystallize upon undercooling. Glass formation is then easy at relatively low, readily accessible, cooling rates. There is a considerable technological interest in the devitrification of silicate glasses, due in part to the importance of obtaining ceramics with a highly refined microstructure.[231] These glasses are also ideal for quantitative nucleation studies.

Silicate glasses are often less susceptible to heterogeneities and surface crystallization. They are prone to phase separation, however.[232-234] Many glasses show an extreme resistance to crystallization, and therefore only a limited number of glasses exhibit homogeneous nucleation. To crystallize other silicate glasses, it is necessary to add heterogeneous impurities as nucleating agents. In this section, homogeneous nucleation studies in several silicate based glasses are discussed. Heterogeneous nucleation and the role of nucleating agents are not discussed; references can be found elsewhere.[235-237]

a. *Steady-State Results*

The best studied glass is lithium disilicate ($Li_2O \cdot 2SiO_2$) since crystallization occurs by bulk, homogeneous nucleation, and measured data for ΔG_v and η exist as a function of temperature. Although $Li_2O \cdot 2SiO_2$ melts incongruently, the undercooled liquid or glass crystallizes polymorphi-

[231] P. W. McMillan, "Glass Ceramics," second edition. Academic Press, London, 1979.
[232] P. F. James, in "Advances in Ceramics" (J. H. Simmons, D. R. Uhlmann, and G. H. Beall, eds.), Vol. 4, p. 1. American Ceramic Society, Columbus, OH, 1982.
[233] A. F. Craievich, E. D. Zanotto, and P. F. James, *Bull. Mineral.* **106**, 169 (1983).
[234] P. F. James, in "Glasses and Glass-Ceramics" (M. H. Lewis, ed.), p. 59. Chapman and Hall, London, 1989.
[235] I. Gutzow, *Contemp. Phys.* **21**, 121, 243 (1980).
[236] G. H. Beall and D. A. Duke, in "Glass: Science and Technology" (D. R. Uhlmann and N. J. Kreidl, eds.), Vol. 1, p. 403. Academic Press, New York, 1983.
[237] A. I. Berezhnoi, "Glass Ceramics and Photositalls." Plenum Press, New York, 1970.

TABLE VIII. KINETIC AND THERMODYNAMIC PARAMETERS FOR SELECTED SILICATE GLASSES

Glass	V_m (10^{-6} m^3/mol)	Fulcher–Vogel Viscosity	ΔH_f (kJ/mol)	T_m (K)	T_g (K)	Remarks
BaO·2SiO$_2$	73.34[a]	$\log_{10}\eta = 1.83 + 1701.9/(T-795.6)$[b]	37.5[e]	1693[d]	961[d]	Congruent Melting[d]
Na$_2$O·2CaO·3SiO$_2$	126.6[c]	$\log_{10}\eta = -3.86 + 4893/(T-547)$[e]	91.3[f]	1562[e]	852[e]	Congruent Melting[e]
Li$_2$O·2SiO$_2$	61.2[g]	$\log_{10}\eta = -1.44 + 3370/(T-460)$[h] $\log_{10}\eta = 1.81 + 1346.6/(T-594.8)$[b]	57.4[g]	1306[i]	683[j]	Incongruent Melting within 1°C of the liquidus
2Na$_2$O·CaO·3SiO$_2$	129.7[k]	$\eta(T_{max}) \cong 10^{10}$ Pa·s[d]			743[d]	Incongruent Melting at 1422 K Liquidus 1473 K[l]
Na$_2$O·SiO$_2$	46.2[m]	$\eta(T_{max}) \cong 10^{10}$ Pa·s[d]	51.8[n] 26.1[q]	1362[i]	683[o]	Congruent Melting[d]
3BaO·5SiO$_2$	—	—	—	—	—	Incongruent Melting at 1696 K[p] Liquidus 1716 K
CaO·SiO$_2$	—	—	56[q]	1817 K[r]	1030[q]	Congruent Melting[q]
CaAl$_2$·Si$_2$O$_8$	100.8[s]	—	167[s]	1823[s]	—	—

[a]Ref. 242
[b]Ref. 243
[c]Ref. 244
[d]Ref. 41
[e]Ref. 245
[f]Ref. 246
[g]Ref. 238
[h]Ref. 247
[i]Ref. 248
[j]Ref. 249
[k]Ref. 250
[l]Ref. 251
[m]Ref. 252
[n]Ref. 253
[o]Ref. 254
[p]Ref. 255
[q]Ref. 256
[r]Ref. 257
[s]Ref. 258

cally;[238-240] the composition is completely liquid 1°C above the incongruent melting point. Since the congruent melting temperature must lie between these points, it can be approximated by the liquidus temperature. The composition of the glass is critical. If it is made lithia-rich (≥ 35.5 mole% Li_2O) some lithium metasilicate crystals appear;[84,241] if it is made lithia poor (≤ 32.0 mole% Li_2O), there is a metastable miscibility gap in the undercooled liquid leading to liquid–liquid phase separation. For glasses at exactly the $Li_2O \cdot 2SiO_2$ composition, however, crystallization occurs directly to lithium disilicate, without phase separation, and with no hint of other crystalline phases. Relevant thermodynamic and kinetic data are summarized in Table VIII.

Measurements of the temperature-dependent steady-state nucleation rates in $Li_2O \cdot 2SiO_2$ are presented with data from other silicate glasses in Fig. 20. The nucleation data for these glasses agree with the temperature dependence predicted by the classical theory. For $Li_2O \cdot 2SiO_2$, the range of observable nucleation lies between 698 and 803 K with the peak at 727 K. The peak nucleation rate varies, depending on the source of the data, from $\approx 7.8 \times 10^8$ to $4.25 \times 10^9 \ m^{-3} \ s^{-1}$. This scatter in nucleation rate may arise from sample purity variations. Certain metallic impurities are known to catalyze nu-

---

[238] K. Matusita and M. Tashiro, *J. Non-Crys. Solids* **11**, 471 (1973).
[239] E. G. Rowlands and P. F. James, *Phys. Chem. Glasses* **20**, 1, 9 (1974).
[240] M. C. Weinberg, G. F. Neilson, and D. R. Uhlmann, *J. Non-Cryst. Solids* **68**, 115 (1984).
[241] M. F. Barker, T. Wang, and P. F. James, *Phys. Chem. Glasses* **29**, 240 (1988).
[242] G. Oehlschlegel, *Glastech. Ber.* **44**, 194 (1971).
[243] E. D. Zanotto and P. F. James, *J. Non-Cryst. Solids* **74**, 373 (1985).
[244] E. G. Rowlands and P. F. James, in "The Structure of Non-Crystalline Solids" (P. H. Gaskell, ed), p. 215. Taylor and Francis, London, 1977.
[245] C. J. R. Gonzalez-Oliver and P. F. James, *J. Non-Cryst. Solids* **38, 39**, 699 (1980).
[246] C. Kröger and G. Kreitlow, *Glastech, Ber.* **29**, 393 (1956).
[247] K. Matusita and M. Tashiro, *Jpn. J. Ceram. Assoc.* **81**, 500 (1973).
[248] F. C. Kracek, *J. Phys. Chem.* **34**, 1583; 2641 (1930).
[249] M. Ito, T. Sakaino, and T. Moriya, *Bull. Tokyo Inst. Technol.* **88**, 127 (1968).
[250] R. W. G. Wyckoff and G. W. Morey, *Am. J. Sci.* **12**, 419 (1926).
[251] G. K. Moir and F. P. Glasser, *Phys. Chem. Glasses* **15**, 6 (1974).
[252] W. S. Donald and D. W. J. Cruickshank, *Acta Crystallogr.* **22**, 37 (1967).
[253] "JANAF Thermochemical Tables," second edition. U.S. Dept. of Commerce, National Bureau of Standards, Washington, D.C., 1971.
[254] J. E. Shelby, *J. Am. Ceram. Soc.* **66**, 754 (1983).
[255] R. S. Roth and E. M. Levin, *J. Res. Nat. Bur. Stand.* **62**, 193 (1959).
[256] E. D. Zanotto, *J. Non-Cryst. Solids* **89**, 361 (1987).
[257] B. Phillips and A. Muan, *J. Am. Ceram. Soc.* **42**, 414 (1959).
[258] D. Cranmer, R. Salomaa, H. Yinnon, and D. R. Uhlmann, *J. Non-Cryst. Solids* **45**, 127 (1981).

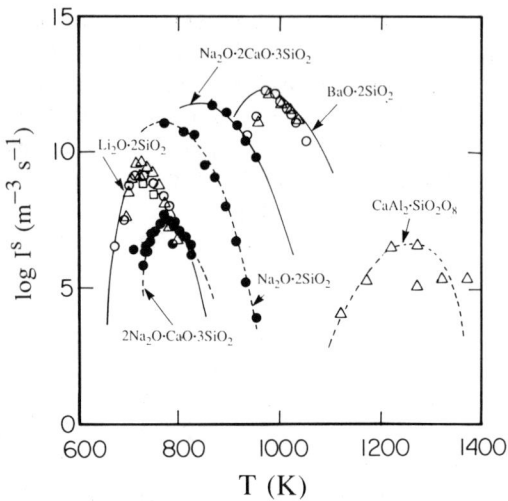

FIG. 20. The measured steady-state nucleation rate as a function of temperature for homogeneously nucleating silicate glasses: $Li_2O \cdot 2SiO_2$—○ (Ref. 259), △—(Ref. 84), □—(Ref. 247). $Na_2O \cdot 2CaO \cdot 3SiO_2$—● (Ref. 245). $2Na_2O \cdot CaO \cdot 3SiO_2$—● (Ref. 260). $Na_2O \cdot 2SiO_2$—● (Ref. 261). $BaO \cdot 2SiO_2$—○ (Refs. 262, 263), △—(Ref. 243). $CaAl_2 \cdot SiO_2O_8$—△ (Ref. 264). The solid lines are fits to classical nucleation, for those glasses where sufficient kinetic and thermodynamic data were available; the dashed lines are guides for the eye.

cleation in silicate glasses.[182,231,235,265-268] James et al.,[269] however, found that most impurities had no significant effect on the nucleation behavior of $Li_2O \cdot 2SiO_2$, but the frequency was enhanced markedly by the presence of water. This probably results from a change in atomic mobility since a similar decrease in the viscosity was reported.

Curve a in Fig. 21 is a plot for $Li_2O \cdot 2SiO_2$, using the viscosity measurements of Matusita and Tashiro[247] (Table VIII) and measured data

[259] V. N. Filipovich and A. M. Kalinina, "Inorganic Materials" (transl. of Izv. Akad. Nauk. SSSR Neorg. Mater.), Vol. 7, p. 1645. 1971.
[260] A. M. Kalinina, V. N. Filipovich, and V. M. Fokin, *J. Non-Cryst. Solids* **38, 39**, 723 (1980).
[261] L. C. Klein, C. A. Handwerker, and D. R. Uhlmann, *J. Cryst. Growth.* **42**, 47 (1977).
[262] E. G. Rowlands, Ph.D. Thesis, University of Sheffield, UK, 1976 (unpublished); via Ref. 40.
[263] P. F. James and E. G. Rowlands, in "Phase Transformations," Vol. 2, Section III, p. 27. Inst. of Metallurgists, Northway House, London, 1979.
[264] A. Hishinuma and D. R. Uhlmann, *J. Non-Cryst. Solids* **95, 96**, 449 (1987).
[265] R. D. Maurer, *J. Chem. Phys.* **31**, 444 (1959).
[266] S. D. Stookey, 5th Int. Glass Congress, *Glastech. Ber.* **32K**, paper VII (1959).
[267] C. S. Ray, W. Huang, and D. E. Day, *J. Am. Ceram. Soc.* **70**, 599 (1987).
[268] I. Gutzow and S. Toschev, in "Advances in Nucleation and Crystallization in Glasses" (L. L. Hench and S. W. Freiman, eds.), p. 10. American Ceramic Society, Columbus, OH, 1971.
[269] P. F. James, B. Scott, and P. Armstrong, *Phys. Chem. Glasses* **19**, 24 (1978).

TABLE IX. VALUES OF τ AND A FOR SILICATE GLASSES THAT CRYSTALLIZE BY HOMOGENEOUS NUCLEATION

	THEORETICAL	CALCULATED					
				$\sigma = \sigma_0 + \sigma_1 T$			
GLASS	A (Pa/m^3)	A (Pa/m^3)	σ(J/m^2)	σ_0(J/m^2 K)	σ_1 (J/m^2 K)	$\sigma(T_{max})$ This Work	(J/m^2 K) James[41]
Li$_2$O·2SiO$_2$	10$^{33.0}$	10$^{53.2}$a	0.139	0.138	2.1×10^{-5}	0.153	0.143
		10$^{60.1}$b	0.147	0.125	3.70×10^{-5}	0.152	0.147
Na$_2$O·2CaO·3SiO$_2$	10$^{32.6}$	10$^{60.8}$	0.131	0.103	3.14×10^{-5}	0.130	0.108
BaO·2SiO$_2$	10$^{32.9}$	10$^{55.3}$	0.100	0.077	2.77×10^{-5}	0.104	0.101

aUsing viscosity data from Ref. 247.
bUsing viscosity data from Refs. 243 and 270.

[270] C. J. R. Gonzalez-Oliver, P. S. Johnson, and P. F. James, *J. Mat. Sci.* **14**, 1159 (1979).

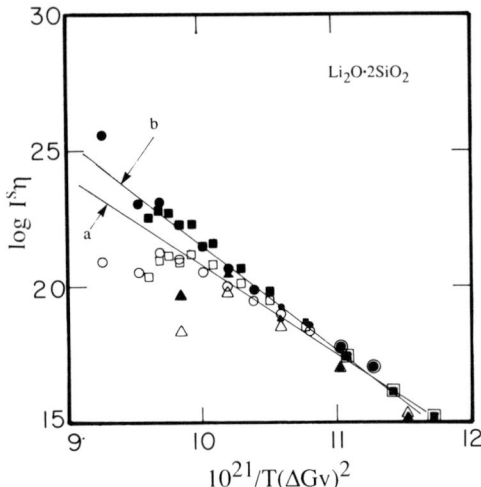

FIG. 21. The product of the steady-state nucleation rate, I^s (m^{-3} s^{-1}), and the viscosity, η (Pa·s), as a function of the volume free energy with undercooling, ΔG_v (J m^{-3}), for Li$_2$O · 2SiO$_2$ glass. Curve a assumes the viscosity data of Ref. 247; curve b assumes viscosity data of Ref. 243, 270. Nucleation data taken from Refs. 84, 247, 259.

for ΔG_v. The data deviate significantly from linearity at low temperatures. The calculated prefactor, A, is also about 20 orders of magnitude larger than the theoretical value (Table IX). These features were first noticed by Rowlands and James[239] and verified by others.[243,271] It is possible that the measured rates at low temperatures are not steady-state values, since there the transient times are large; however, the significant disagreement between the expected and calculated prefactor fit from higher-temperature data remains a problem.

Recently, the viscosity for Li$_2$O · 2SiO$_2$ was remeasured (Gonzalez-Oliver et al.[270] and Zanotto and James[243]) and fit to the Fulcher–Vogel expression (Table VIII). Zanotto and James also remeasured the nucleation rates and transient times and found good agreement with previous data. A plot constructed from available nucleation rates, using this new viscosity data is presented in Fig. 21 (curve b). Improved linearity is obtained, but the calculated values for σ and A are similar; A is over 25 orders of magnitude larger than the theoretical value. Also the use of the Matusita and Tashiro viscosity data gives the best agreement with the measured time lags (cf. Fig. 27), suggesting that the deviation at low temperatures is due to other factors.

Available data on other silicate systems are less abundant. Only two

[271] G. F. Neilson and M. C. Weinberg, J. Non-Cryst. Solids 34, 137 (1979).

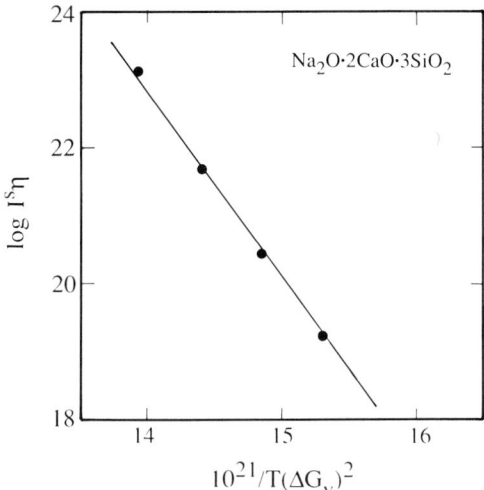

FIG. 22. The product of the steady-state nucleation rate, I^s (m^{-3} s^{-1}), and the viscosity, η (Pa·s), as a function of the volume free energy with undercooling, ΔG_v (J m^{-3}), for BaO·2SiO$_2$ glass. Nucleation data taken from Refs. 263 and 243.

glasses Na$_2$O·2CaO·3SiO$_2$ and BaO·2SiO$_2$, melt congruently, nucleate by a homogeneous, polymorphic mechanism, and have known thermodynamic and kinetic properties (Table VIII). The nucleation rates are plotted against temperature in Fig. 20. The range of observable nucleation for Na$_2$O·2CaO·3SiO$_2$ lies between 823 and 973 K with a peak rate of 5.6×10^{11} m^{-3} s^{-1} at 868 K. As for Li$_2$O·2SiO$_2$, transient nucleation was found at the lower temperatures. Barium disilicate (BaO·2SiO$_2$) shows observable nucleation between 935 and 1053 K with a slightly larger peak of 1.87×10^{12} m^{-3} s^{-1} at 973 K. Figures 22 and 23 present plots of log $I^s\eta$ versus $1/T \Delta G_v^2$ for BaO·2SiO$_2$ and Na$_2$O·2CaO·3SiO$_2$, respectively. As expected, linear behavior is observed. The assumption of a temperature independent σ again gives values for A that are much larger than the theoretical predictions (Table IX).

Nucleation data from several other glasses that devitrify polymorphically by volume nucleation are also plotted in Fig. 20. Unfortunately, there is insufficient thermodynamic data, or there are other problems that make quantitative comparisons with nucleation theory difficult. Taking the Singh and Holz approximation for the free energy [Eq. (A.16)], and using the thermodynamic data in Table VIII to analyze the nucleation data gives the expected linear behavior with A 20–30 orders of magnitude larger than the theoretical predictions. In many cases, these data are suspect. The data for

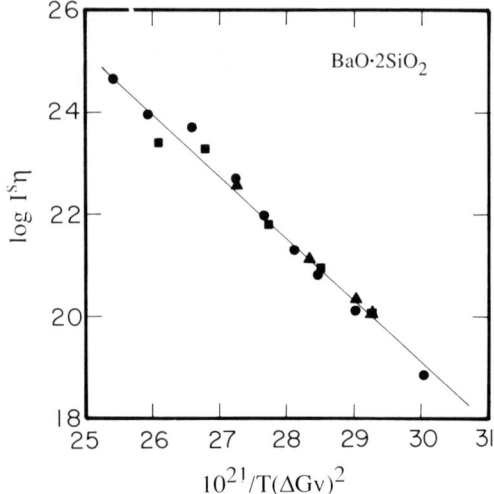

FIG. 23. The product of the steady-state nucleation rate, I^s (m^{-3} s^{-1}), and the viscosity, η (Pa·s), as a function of the volume free energy with undercooling, ΔG_v (J m^{-3}), for $Na_2O \cdot 2CaO \cdot 3SiO_2$ glass. Nucleation data taken from Ref. 245.

$Na_2O \cdot 2SiO_2$ were obtained indirectly from the rate of devitrification using the JMA equation and known crystal growth rates; the values of A were dependent on the volume fraction measured. Matusita and Tashiro[238] and Scott and Pask[272] observed only surface nucleation in $Na_2O \cdot 2SiO_2$. The bulk nucleation rate measured directly in $CaAl_2 \cdot Si_2O_8$ is approximately two orders of magnitude lower than that estimated from the devitrification measurements;[258,264] independent work found only surface nucleation.[41]

As for undercooled liquid metals, a consistent trend emerges: the data agree with the functional form of the classical nucleation theory, but the predicted prefactors are several orders of magnitude too large. This is apparently not due to heterogeneous nucleation.[239,240,269] Several explanations have been offered for these discrepancies: (1) a temperature-dependent interfacial energy;[239] (2) the nucleation of a metastable phase instead of the stoichiometric equilibrium compound;[264] or (3) a failure of the classical theory.

Following Turnbull's work in undercooled Hg[1], James et al. found improved agreement by assuming a linear temperature dependence for the interfacial energy, $\sigma(T) = \sigma_0 + \sigma_1 T$, to fit the $Li_2O \cdot 2SiO_2$ data.[239] A measure of the relative entropic contribution to the interfacial energy is given by the

[272]W. D. Scott and J. A. Pask, *J. Am. Ceram. Soc.* **44**, 181 (1961).

ratio σ_1/σ;[175-177] σ_0/σ gives the fraction of enthalpic contribution. The theoretical value for A was assumed, and $\sigma(T)$ was calculated by matching $I(T)$; σ increased from approximately 0.144 to 0.158 J m^{-2} over the nucleation range. A recent analysis by Greer and Kelton[193] of data obtained by others[191,192,259,273-275] gave $\sigma(T)$, in good agreement with James's values over a similar temperature range. The average values obtained by fitting these nucleation data and the values for σ at the maximum nucleation temperature are listed in Table IX. The values for the interfacial energy at the temperature of the maximum nucleation rate $\sigma(T_{max})$ agree qualitatively with those obtained by James. The discrepancies are due presumably to fits to different data sets and different assumptions for ΔG_v.

Hishinuma and Uhlmann[264] argued that the curvature in the nucleation plots and the large prefactors A for $Li_2O \cdot 2SiO_2$ may be due to the nucleation of new metastable intermediate phases in lieu of the stable phase. The problem may even be worse for other glasses, since there the free energy is unknown and must be approximated. The dependence of the analysis on the correct value for the free energy is illustrated in Fig. 24, where the nucleation data for lithium disilicate are analyzed using the viscosity data of Gonzalez-Oliver et al.,[270] assuming a temperature-independent σ and taking different approximations to ΔG_v. The sensitivity of the computed values for A to ΔG_v is staggering! The nucleation prefactor can vary from $A = 10^{54}$, using the Turnbull approximation [Eq. (A4)], to $A = 10^{106}$, assuming the Hoffman expression [Eq. (A10)]; the interfacial energy varies by approximately 5%. The use of the Singh and Holz expression [Eq. (A16)] gives the best agreement with the analysis using measured data for ΔG_v; this is expected from the fit to measured values for ΔG_v (Fig. A2). These results demonstrate that interpretations of nucleation data without available measurements for the free energy may be grossly in error. While this might explain the anomalously large values for A, curvature is not introduced; the use of the wrong free energy tends to shift the data while maintaining the linear dependence. Thus the low-temperature deviations in $Li_2O \cdot 2SiO_2$ are not explained.

Hishinuma and Uhlmann also argued that, although a temperature-dependent interfacial energy improves the values for A, the critical radius at the large undercoolings where the nucleation rates are measured is smaller than the unit cell of the growing crystal. They suggested that this signals a breakdown in the classical theory. Recent attempts to fit multiple-step annealing experiments also suggest a possible failure of the classical theory at

[273] V. M. Fokin, V. N. Filipovich, and A. M. Kalinina, *Fiz. Khim. Stekla* **2**, 129 (1976).
[274] V. N. Filipovich and A. M. Kalinina, "Inorganic Materials" (transl. of Izv. Akad. Nauk. SSSR Neorg. Mater.), Vol. 4, p. 1335. 1968.
[275] V. N. Filipovich and A. M. Kalinina, *ibid.*, **6**, 303 (1970).

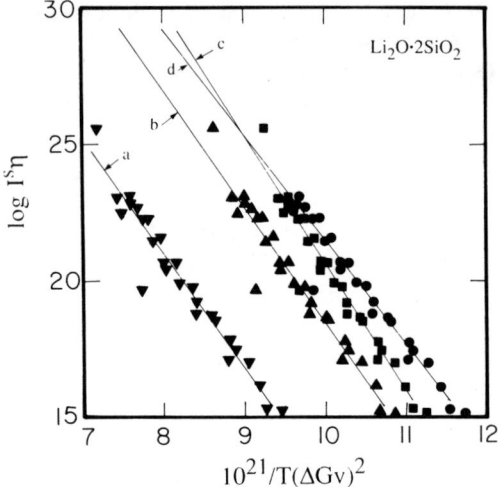

FIG. 24. The product of the steady-state nucleation rate, I^s (m^{-3} s^{-1}), and the viscosity, η (Pa·s), as a function of the volume free energy with undercooling, ΔG_v (J m^{-3}), for Li$_2$O·2SiO$_2$ glass, demonstrating the importance of using the correct form for the free energy to properly analyze nucleation data. Free-energy expressions: (a) Eqs. (A4); (b) Eqs. (A5) using ΔC_p at T_m; (c) measured value for ΔG_v; (d) Eq. (A16). The curve calculated using the Hoffmann expression would not fit on this scale.

large undercoolings (cf. Section 14). As will be discussed in Section 19, Harrowell and Oxtoby[277] showed that the form of the free energy with cluster size calculated from a density functional approach is similar to that predicted by the capillarity approximation, but the critical size is larger, possibly rectifying the problem of unphysically small critical sizes. The density functional calculations also predict the presence of an ordered layer near the interface giving rise to the negentropy required for a positive temperature dependence of the interfacial energy, in agreement with the assumptions of Turnbull[1] and Rowlands and James,[239] and with model calculations.[175-177]

James[41] recently compiled the nucleation data for all glasses that show volume nucleation and compared the temperatures for melting, T_m, glass transition, T_g, maximum nucleation rate, T_{max}, and a just-detectable nucleation rate of 1 cm^{-3} s^{-1}, T_d. He noticed that all glasses have T_{max}/T_m in the range from 0.54 to 0.59, in agreement with classical theory predictions,[165,166] and a relatively low reduced glass transition temperature T_g/T_m. Zanotto and Weinberg pointed out that volume nucleation occurs in those glasses with

[276] E. D. Zanotto and A. Galhardi, *J. Non-Cryst. Solids* **104**, 73 (1988).

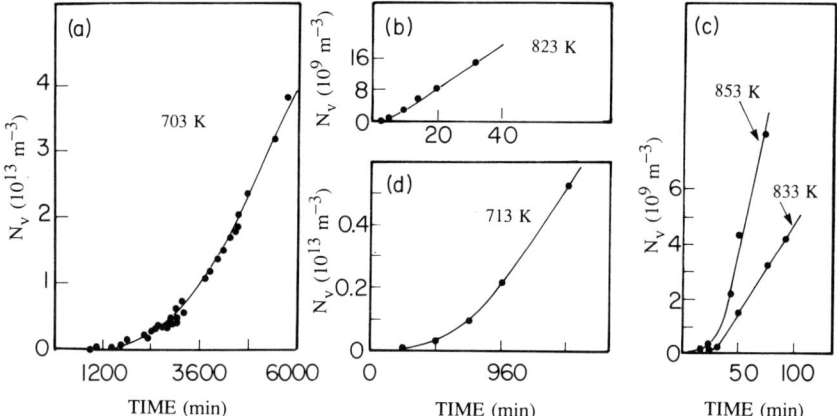

FIG. 25. The number of nuclei measured as a function of time for several silicate glasses: (a) $Li_2O \cdot 2SiO_2$; (b) $Na_2O \cdot BaO \cdot SiO_2$; (c) $Na_2O \cdot CaO \cdot SiO_2$; (d) $Li_2O \cdot 2SiO_2$, demonstrating the universality of time-dependent effects. Taken from I. Gutzow, D. Kashchiev, and I. Avramov, *J. Non-Cryst. Solids* **73**, 477 (1985).

$T_{max} \geqslant T_g$.[278] For glasses with $T_{max} \geqslant T_g$, transient times are long (due to the low atomic mobility), and the steady-state nucleation rate is suppressed. These glasses will either devitrify by another mechanism or will be resistant to crystallization.

b. *Transient Nucleation*

Since glasses are formed by cooling the melt on a time scale that is rapid compared with the time required for atomic rearrangements, the final cluster population depends on the details of the quench. Subsequent annealing treatments therefore will reveal a transient nucleation behavior as the distribution evolves to the steady-state distribution at the annealing temperature. Transient nucleation has been observed in a wide range of glass-forming systems. Gutzow et al.[279] presented the first evidence for transient nucleation in glass-forming melts, in $NaPO_3$. It was observed subsequently in metallic glasses,[207,280] glass–ceramic materials,[281] enamel,[282] and silicate glasses.[41,43] Transient nucleation has been studied most extensively in the silicate glasses. Figure 25 presents representative data for several glasses; in

[277] P. Harrowell and D. W. Oxtoby, *J. Chem. Phys.* **80**, 1639 (1984).
[278] E. D. Zanotto and M. C. Weinberg, *Phys. Chem. Glasses* **30**, 186 (1989).
[279] I. Gutzow, E. Popov, S. Toschev, and M. Marinov, in "Rost Kristallov," Proc. VII Int. Congress on Crystallography, Vol. 8, Part 2, p. 95. Moscow, July 1966 (Izv. Nauka, Moscow, 1968).
[280] U. Köster, *Phys. Stat. Sol.* **48**, 313 (1978).
[281] I. Gutzow, E. Zlateva, S. Alyakov, and T. Kovascheva, *J. Mat. Sci.* **12**, 1190 (1977).
[282] I. Penkov and I. Gutzow, *J. Mat. Sci.* **19**, 233 (1984).

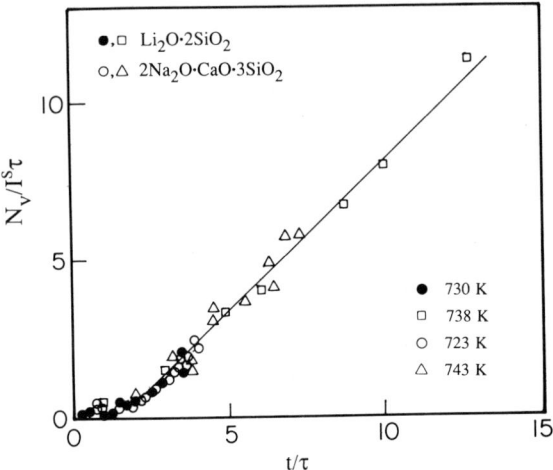

FIG. 26. The number of nuclei as a function of the time scaled by the product of the steady-state nucleation rate and the transient time, for $Li_2O \cdot 2SiO_2$ and $2Na_2O \cdot CaO \cdot 3SiO_2$ glass, demonstrating the validity of the Kashchiev approximation for the transient nucleation behavior. Adapted from V. M. Fokin, A. M. Kalinina, and V. N. Filipovich, *J. Cryst. Growth* **52**, 115 (1981).

all cases, the rate of production of nuclei is low initially and rises to a constant, steady-state rate with time.

The number of volume nuclei, N_v, produced as a function of time is best described by the Kashchiev expression [Eq. (6.2)]. Scaling the number of nuclei by the steady-state nucleation rate and the transient time at the annealing temperature, should produce a master curve.

$$\frac{N_v(t,\tau)}{I^s\tau} = \left[\frac{t}{\tau} - 2\sum_{m=1}^{\infty}\frac{(-1)^m}{m^2}\exp\left(\frac{-m^2 t}{\tau}\right) - \frac{\pi^2}{6}\right]. \quad (13.1)$$

Assuming this expression, nucleation data from $Li_2O \cdot 2SiO_2$ and $2NaO \cdot CaO \cdot 3SiO_2$ are plotted in Fig. 26. As predicted, the data are fit by a master curve that is independent of the glass composition and the annealing conditions. Since Eq. (13.1) was derived assuming the classical theory of nucleation, these results support that model.

Measured values for θ (defined in Fig. 5) in $Li_2O \cdot 2SiO_2$ and $2Na_2O \cdot CaO \cdot 3SiO_2$ are presented in Fig. 27a; the temperature dependence is clearly Arrhenius. James noted that in $Li_2O \cdot 2SiO_2$ the activation energy for θ is close to that for viscous flow in the same temperature range.[84] Kalinina *et al.* reported similar behavior for $2Na_2O \cdot CaO \cdot 3SiO_2$.[85] Those observations are taken to support the use of the bulk mobility to describe the interfacial dynamics. Values for the prefactor θ_0 and the activation energy ΔH_τ derived from the fits in Fig. 27a are listed in Table X with available activation energies

TABLE X. Transient Time for Three Silicate Glasses (Comparison with Viscosity and Crystal Growth)

$$\theta = \theta_0 e^{\Delta H_\mathrm{t}/k_\mathrm{B}T}$$

Glass	t_0 (s)	ΔH_t (kJ/mol)	Bulk Viscosity ΔH_η (kJ/mol)	Crystal Growth ΔH_u (kJ/mol)
$Li_2O \cdot 2SiO_2$	$10^{-28.1}$	447.0	585.8[a] 846.4[c]	282[b]
$2Na_2O \cdot CaO \cdot 3SiO_2$	$10^{-38.2}$	597.6	824.2[d]	426.8[d]
$Na_2O \cdot 2CaO \cdot 3SiO_2$	$10^{-10.6}$	229.8	600[e]	971[f]

[a]Ref. 88.
[b]Ref. 241.
[c]Estimated from Fulcher–Vogel expression (Table VIII).
[d]Ref. 85.
[e]Ref. 245.
[f]Estimates from two growth velocity measurements reported in Ref. 276.

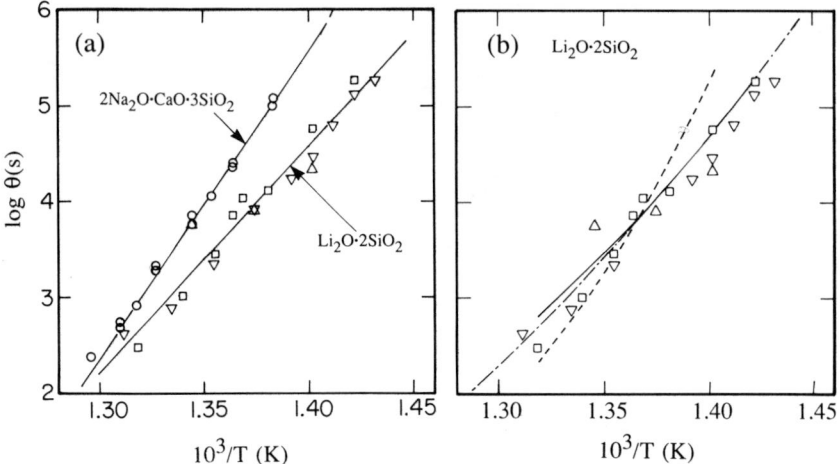

FIG. 27. (a) The induction time, θ, measured as a function of temperature for $Li_2O \cdot 2SiO_2$ glass (data from Refs. 84, 88, 241 and $2Na_2O \cdot CaO \cdot 3SiO_2$ glass (data from Ref. 85). The straight lines are Arrhenius fits to the data. (b) The results of a numerical calculation of θ for $Li_2O \cdot 2SiO_2$, using viscosity data from Ref. 247 (solid line) and Refs. 243, 270 (dashed line). The dot–dash line is a calculation from the Kashchiev expression, using the viscosity data from Ref. 247 and Eq. (11.1) to compute the time to grow from n^* at the nucleating temperature to n^* at the growth temperature.

for viscous flow and crystal growth velocities. The agreement between the activation energies for the transient time, the viscosity, and the growth velocity depends strongly on the chosen temperature range.

The time lag at the critical size θ_{n^*} is related to the transient time τ by Eq. (6.5). Measured values for θ, however, are often several times larger than τ. The measured θ is the sum of two terms: the true transient time, given by τ, and the time required for nuclei to grow from n^* at the nucleation temperature, T_N, to n^* at the growth temperature, T_G. Kalinina et al.[191] proposed that the growth time t_G is given approximately by

$$t_G = [r^*(T_G) - r^*(T_N)]/u \qquad (13.2)$$

where u is an effective growth rate for nuclei and $r^*(T)$ is the radius of the nucleus at temperature T.

Following Hillig,[283] the transient time is often taken to scale inversely with the diffusion coefficient at the crystal–liquid interface:

$$\tau(T) = \pi d^2/D, \qquad (13.3)$$

[283] W. B. Hillig, "Symposium on Nucleation and Crystallization in Glasses and Melts," p. 77. American Ceramic Society, Columbus, OH, 1962.

where d is the molecular diameter. A cluster is assumed to evolve to the critical size by accreting all monomers that diffuse to it by random walk.

Figure 27b compares calculated values for θ in $Li_2O \cdot 2SiO_2$ with measured data. The solid and dashed curves are computed by solving numerically the master equation [Eq. (3.7)] using the viscosity data of Matusita and Tashiro[247] and Gonzalez-Oliver et al.,[270] respectively. A critical size at the growth temperature, 899 K, of $n^*(T_G) = 65$ was assumed. The data are better fit by the Matusita and Tashiro viscosity values. Using that viscosity data, the Kashchiev prediction also fit the data well (dash–dot curve); the Hillig prediction was indistinguishable from the Kashchiev fit. That agreement reflects the long time required to grow from $n^*(T_N)$ to $n^*(T_G)$; the values for τ from the two expressions are quite different.

As discussed in Section 13a, volume nucleation in glasses can be suppressed by transient effects, leading to an enhanced stability of the glass. Assuming the Kashchiev expression and using experimental measurements for the viscosity and enthalpy of fusion for silicate glasses, Zanotto and Weinberg[278] calculated the transient times for several silicate glasses in the temperature range where the nucleation rates should be highest (Fig. 28). Of the glasses shown, only $BaO \cdot 2SiO_2$, $Li_2O \cdot 2SiO_2$, and $Na_2O \cdot 2CaO \cdot 3SiO_2$ are known to nucleate homogeneously. All others nucleate by another mechanism or are impossible to nucleate without using nucleating agents. These glasses also have the shortest transient times, supporting their hypothesis. Recently measured transient times were used to estimate better the interfacial mobility in $Li_2O \cdot 2SiO_2$; an analysis of the steady-state nucleation rates, however, still shows the anomalous behavior at low temperatures.[284]

c. *Compositional Dependence*

Most liquids and glasses crystallize to a phase with a different composition. As discussed briefly in Section 3b, the nucleation behavior can be modeled by assuming a compositional dependence for ΔG_v, σ, and η, although the actual dependence is generally not known. Silicate glasses are better suited for compositional studies than metallic systems since pseudobinary systems are readily constructed. Despite this and the common occurrence of nonpolymorphic devitrification, there are few quantitative experimental studies of compositional effects on the nucleation rate.

Burnett and Douglas examined Na_2O—BaO—SiO_2.[285] They found that volume nucleation rates peaked at a composition near the stoichiometric composition for $BaO \cdot SiO_2$. The onset temperature for nucleation decreased

[284]M. C. Weinberg and E. D. Zanotto, *J. Non-Cryst. Solids* **108**, 99 (1989).
[285]D. G. Burnett and R. W. Douglas, *Phys. Chem. Glasses* **12**, 117 (1971).

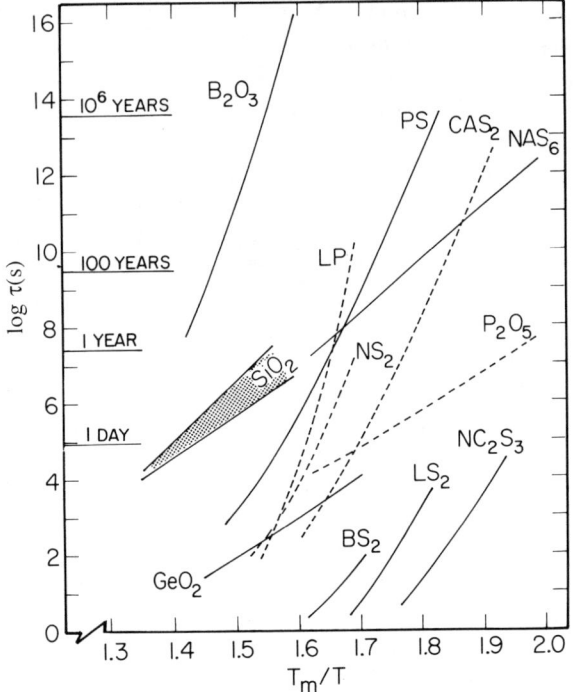

FIG. 28. The computed transient times as a function of reduced temperature for several silicate glasses, demonstrating that many glasses do not crystallize by volume nucleation and growth due to very long transient periods. LP($Li_2O \cdot P_2O_5$); PS(PbO·SiO_2); CAS_2(CaO·$Al_2O_3 \cdot 2SiO_2$); NAS_6($Na_2O \cdot Al_2O_3 \cdot 6SiO_2$); NS_2($Na_2O \cdot 2SiO_2$); NC_2S_3($Na_2O \cdot 2CaO \cdot 3SiO_2$); LS_2($Li_2O \cdot 2SiO_2$); BS_2(BaO·$2SiO_2$). Taken from E. D. Zanotto and M. C. Weinberg, *Phys. Chem. Glasses* **30**, 186 (1989).

as the composition moved away from BaO·SiO_2. This decrease followed a decrease in the liquidus temperature corresponding to a decrease in the free energy. Perepezko noted similar behavior in metallic alloys.[136]

Similar trends were observed in the pseudobinary systems Li_2O—BaO—SiO_2[286] and $Li_2O \cdot 2SiO_2$—BaO·$2SiO_2$.[232] An analysis of the nucleation data using available thermodynamic data suggested that the compositional dependence of ΔG_v was responsible primarily for the shift in nucleation rates. Strnad and Douglas[287] observed that for Na_2O—CaO—SiO_2, compositional regions near $Na_2O \cdot 2CaO \cdot 3SiO_2$ and

---

[286] P. F. James and E. G. Rowlands, in "Phase Transformations," Vol. 2, Section III, p. 27. The Institution of Metallurgists, Northway House, London, 1979.
[287] Z. Strnad and R. W. Douglas, *Phys. Chem. Glasses* **14**, 33 (1973).

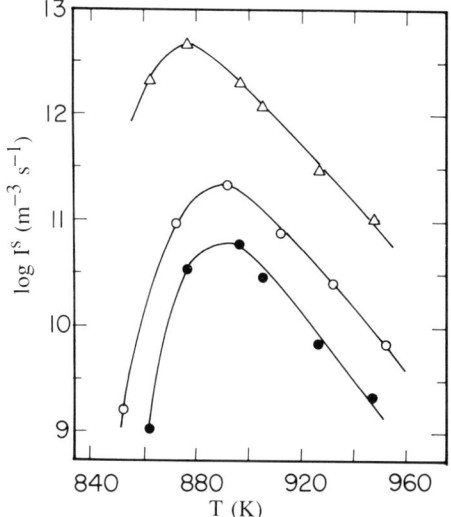

FIG. 29. The steady-state nucleation rate as a function of temperature for 16.7Na$_2$O—33.3CaO—50SiO$_2$ mol% (O), 17.7Na$_2$O—32.9CaO—49.4SiO$_2$ mol% (△), and 15.7Na$_2$O—33.7CaO—50.6SiO$_2$ mol% (●) glasses, demonstrating a strong dependence on concentration. Taken from P. F. James, in "Glasses and Glass-Ceramics" (M. H. Lewis, ed.), p. 59. Chapman and Hall, London, 1989.

2Na$_2$O · CaO · 3SiO$_2$ showed volume nucleation; outside that region, surface nucleation was dominant.

Gonzalez-Oliver and James[245] found that the nucleation rates in Na$_2$O · CaO · 3SiO$_2$ depend sensitively on the SiO$_2$ concentration (Fig. 29). The glasses with more or less SiO$_2$ than Na$_2$O · 2CaO · 3SiO$_2$ showed higher or lower nucleation rates, respectively, than the base glass. Evidence that ΔG_v and η were compositionally dependent made the interpretation of the nucleation data difficult.

These initial data present no unexpected trends. However, to probe properly the composition dependence of the nucleation rate, systematic measurements of ΔG_v, η (or D), I^s, and θ must be made as a function of composition. The paucity of available data makes it difficult to draw sweeping conclusions at this time.

14. A Test of the Kinetic Model of Classical Nucleation Theory

Although measurements of the nucleation rates in some liquids and glasses agree with the predictions of the classical theory, there are hints that aspects of that theory may be flawed. An analysis of a broader range of materials is hampered by many problems, including the lack of data for $\Delta G_v(T)$ and $\eta(T)$,

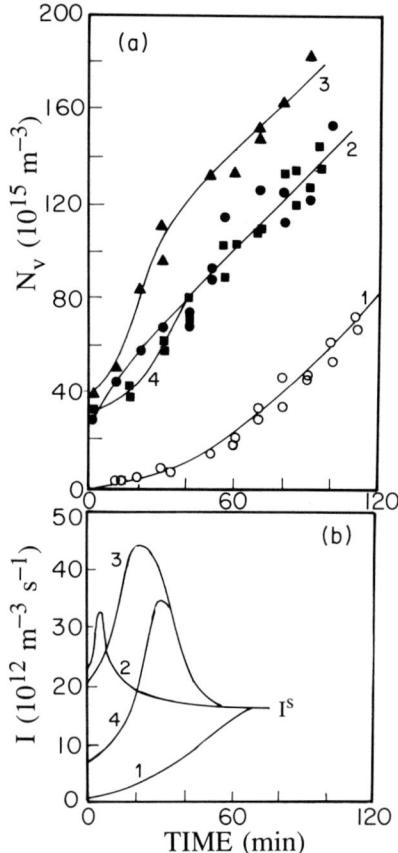

FIG. 30. (a) The number of crystal nuclei as a function of annealing time at 753 K for $2Na_2O \cdot CaO \cdot 3SiO_2$ glass (1) with no preannealing treatment, (2) after a 4-h anneal at 743 K, (3) after a 3-h anneal at 733 K, (4) after a 65-h anneal at 723 K. (b) The nucleation rate derived from the local slope of the data in (a). Taken from A. M. Kalinina, V. N. Filipovich, and V. M. Fokin, *J. Non-Cryst. Solids* **38, 39**, 723 (1980).

significant heterogeneous nucleation, prohibitively long transient times, and phase separation. Even when the nucleation is known to be homogeneous and data are available for $I(T,t)$, $\Delta G_v(T)$, and $\eta(T)$, the significance of agreement with nucleation theories is unclear. The interfacial energy σ is often known only from fitted nucleation data and, therefore, cannot be compared with values obtained by other techniques. Further, it is unclear whether the bulk atomic mobility derived from measurements of the shear viscosity appropriately describes atomic motion in the interface. Most theoretical attention has focused on improving the thermodynamic aspects of the nucleation theories. The kinetic model, central to the classical theory and many of its variations, however, has received less attention.

A proper schedule of annealing treatments for $Li_2O \cdot 2SiO_2$ glass causes the nucleation rate to go through maximum with time (Fig. 30). The intensity

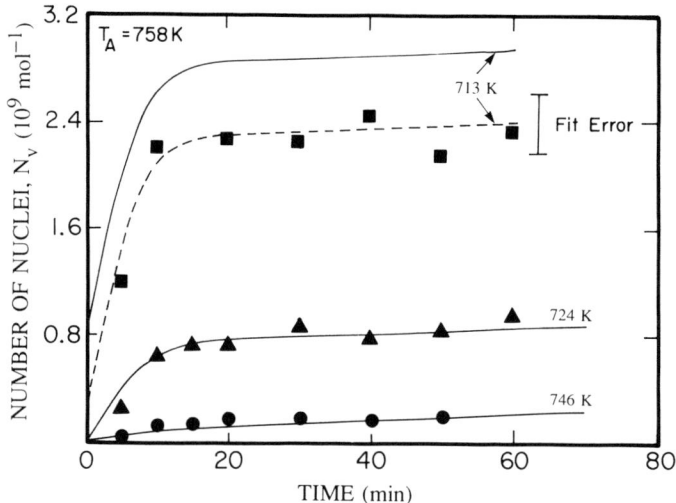

FIG. 31. The number of crystal nuclei for $Li_2O \cdot 2SiO_2$ as a function of annealing time at 758 K after annealing to steady state at 713 K, 724 K, and 746 K. The solid lines are from a numerical solution of the master equation of classical nucleation theory. The dashed line shows a shift of the numerical solution to match the experimentally measured number of nuclei at the end of the 713 K anneal (see text). Taken from K. F. Kelton and A. L. Greer, Phys. Rev. B **38**, 10089 (1988).

and position of that maximum depend critically on the temperatures and times of the annealing treatments. Within the classical theory, this behavior must result from the response of the cluster distribution to the annealing treatments. It should therefore be predictable, given the appropriate thermodynamic and kinetic parameters, and the annealing schedule. Using the measured data for ΔG_v, Kelton and Greer[44,193] fit the magnitude and temperature dependence of the steady-state nucleation rates and transient times to obtain $\sigma(T)$ and $\gamma(T)$ [Eq. (3.18) and (3.19)]. Those derived values were then held fixed, and the annealing treatments were simulated numerically.

Figure 31 shows the number of nuclei, N_v, measured by Kalinina et al.,[191,192,259,273–275] obtained by nucleating at 758 K and growing at 899 K, following preanneals at 713 K, 724 K, and 756 K. The solid lines are the predictions from the numerical calculation. Excellent agreement is obtained with the 724 K and 746 K preanneals. The apparent disagreement with 713 K is due to an error in the steady-state value originally reported at that temperature; shifting the data to take that into account improves the agreement (dashed line). Similar agreement was obtained for other preanneal-

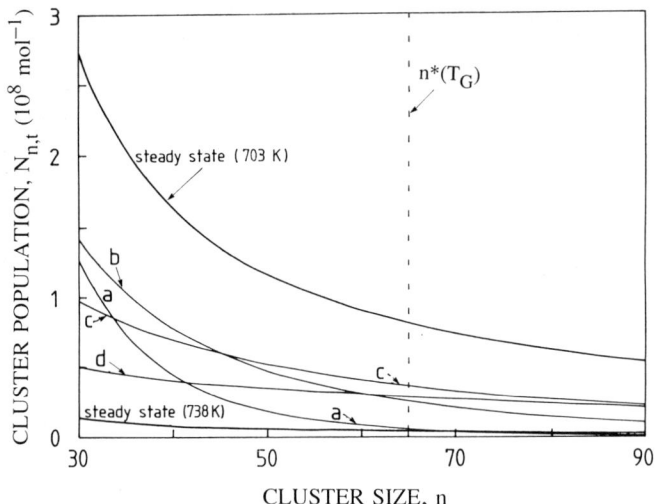

FIG. 32. The cluster size distribution computed from the numerical solution of the master equation showing the evolution with annealing. The steady-state distributions at 703 K and 738 K are shown. Curve a shows the distribution resulting from a 65-h anneal at 703 K. With annealing time at 738 K, the distribution evolves as shown by curves b (1000s), c (2000s) and d (3000s). Taken from A. L. Greer and K. F. Kelton, *J. Am. Cer. Soc.* (in press) by permission of the American Ceramic Society.

ing treatments, except the lowest one of 703 K. A list of possible causes for that anomalous behavior includes experimental error, a curvature dependence for σ or γ, or a breakdown in the classical theory at large departures from equilibrium.

The nucleation rate derived from the data in Fig. 31 shows a maximum with time as observed experimentally (Fig. 30). The cause can be observed directly from the calculated cluster distributions. As shown in Fig. 32, the cluster populations following a 65-h anneal at 703 K are significantly below the steady-state distribution at that temperature (curve a). Upon annealing at 738 K, the cluster distribution evolves toward the appropriate steady-state distribution (b, c, d). At larger sizes, however, the cluster density increases and subsequently decreases, giving rise to a "pulse" of clusters that sweep through the distribution to $n^*(T_G)$. These results demonstrate that nucleation in this glass proceeds as if there were an actual cluster distribution that responds to changes in the departure from equilibrium as predicted by the classical theory of nucleation.

VI. Computer Simulations of Steady-State Nucleation

Experiments to test homogeneous nucleation theories are difficult to design due to heterogeneities and compositional and strain effects. Even when data are available, their interpretation is complicated by an incomplete knowledge of the atomic interactions and mobilities appropriate to the material studied. Computer simulations, based on known atomic interactions and dynamics, can be used to "experimentally test" current nucleation theories and to identify relevant parameters that must be incorporated into more complete theories. Only since the early 1970s has an adequate level of computing power become available to realistically treat phase transitions. The ensembles considered are still relatively small and the potentials are often idealized, yielding only qualitative conclusions. Those results, however, demonstrate unambiguously that under appropriate conditions, undercooled liquids and supersaturated vapors phase transform by a fluctuation-based mechanism of nucleation and growth. They suggest that the capillarity assumption of a sharp interface between the cluster and the original phase is incorrect. If the correct atomic interactions for the initial and final phases were known, it should be possible to predict the structure of the nucleating solid. This has yet to be realized, however.

The time evolution of an initial system configuration is generally simulated using one of two methods: Monte Carlo or molecular dynamics. Because of their simplicity, Ising models, with artificial dynamics, have received the most attention. They have been used to probe the basic assumptions of nucleation theories and have guided the development of new models.[288,289] Numerical simulations of liquid clusters nucleating in a supersaturated vapor have focused primarily on thermodynamic questions. Crystal growth from the liquid has also received much attention,[290] with less work on the nucleation step. A brief discussion of the computer methods used (an abbreviated version of a discussion by Abraham[291]) and the main results for nucleation in a liquid or vapor are presented here.

15. Monte Carlo Calculations

The Monte Carlo technique is based on a stochastic process that produces a Markovian chain of configurations that are Boltzmann weighted. Unlike the molecular dynamics approach, there is no guarantee that the trajectory to

---

[288] M. E. Fisher, *Physics* 3, 255 (1967); K. Binder, *Ann. Phys.* 98, 390 (1976).
[289] D. Stauffer, A. Coniglio, and D. W. Heermann, *Phys. Rev. Lett.* 49, 1299 (1982); D. W. Heermann and W. Klein, Phys. Rev. B 27, 1732 (1983).
[290] K. A. Jackson, *J. Cryst. Growth* 24/25, 130 (1974); H. J. Leamy and G. H. Gilmer, *J. Cryst. Growth* 24/25, 499 (1974).
[291] F. F. Abraham, *Rep. Prog. Phys.* 45, 1113 (1982).

equilibrium is physically realizable. Monte Carlo methods are therefore best suited for investigations of the equilibrium thermodynamic properties. Assuming a canonical ensemble, the average potential energy can be defined,

$$\langle U \rangle = \int U(\xi)\rho(\xi)\,d\xi, \qquad (15.1)$$

where ξ represents the particle configurations occurring with probability $\rho(\xi)$. The straightforward method of obtaining the potential energy by averaging numerically over all possible system configurations, weighting each single average by the Boltzmann factor, is prohibitive due to the large number of configurations required. For faster calculations, Metropolis et al.[292] developed the Monte Carlo technique in which configurations are chosen with a frequency determined by the Boltzmann factor, and the average is computed by adding the energy from each selected configuration equally.

One of the most complete Monte Carlo calculations was made for the case of argon nucleating from the vapor, using a 6-12 Lennard-Jones potential.[293-296] These studies are mentioned since they provide information that bears on the general validity of the theory of nucleation.

Typical configurations in the canonical ensemble are shown in Fig. 33 for 100-atom cluster as a function of temperature. Below 30–40 K, the clusters appear to be more crystalline; at higher temperatures (particularly above the triple point, 84 K), they become more liquidlike. The diffuse interface is evident. Computed radial density distributions demonstrate that the density at the center of the cluster for the smaller cluster sizes is less than that of the bulk liquid. The computed free energies for cluster formation were in much better agreement with predictions from the Lothe–Pound theory[32] than with those from the Becker–Döring expression.[30]

Garcia and Torroja[297] found Monte Carlo predictions for supersaturation to be in better agreement with experimental measurements for argon than were those from the classical theory. One possible explanation is the neglect of rotational and translational contributions to the free energy; this is supported by the better agreement between the Lothe–Pound theory and the Monte Carlo calculation found by Lee et al.[293] Since both the classical and Lothe–Pound theories make additional approximations, however, that good agreement may be fortuitous.

[292] N. Metropolis, A. W. Rosenbluth, M. N. Rosenbluth, A. H. Teller, and E. Teller, J. Chem. Phys. **21**, 1087 (1953).
[293] J. K. Lee, J. A. Barker, and F. F. Abraham, J. Chem. Phys. **58**, 3166 (1973).
[294] F. F. Abraham, J. K. Lee, and J. A. Barker, J. Chem. Phys. **60**, 246 (1974).
[295] J. Miyazaki, J. A. Barker, and G. M. Pound, J. Chem. Phys. **64**, 3364 (1976).
[296] J. Miyazaki, G. M. Pound, F. F. Abraham, and J. A. Barker, J. Chem. Phys. **67**, 3851 (1977).
[297] N. J. Garcia and J. M. S. Torroja, Phys. Rev. Lett. **47**, 186 (1981).

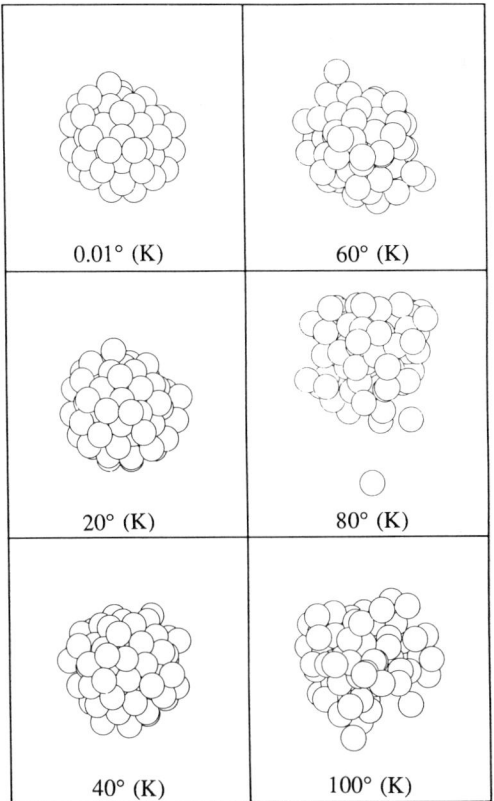

FIG. 33. Computed images of a 100-atom cluster during a Monte Carlo simulation. Taken from J. K. Lee, J. A. Barker, and F. F. Abraham, *J. Chem. Phys.* **58**, 3166 (1973).

16. Molecular Dynamics

Simulations of crystal nucleation from the liquid are generally based on molecular dynamics (MD) calculations with model potentials. These studies address such questions as these: (1) Does homogeneous nucleation occur? (2) What are the crystal structures of the nuclei? (3) How do the results compare with analytical model predictions? Technical issues such as the effects of the size of the atom distribution, the quench conditions from the stable to metastable regions, the boundary effects, and the influence of the chosen potential are also of interest. Currently molecular dynamics modeling is limited by available computational power. It is difficult to simulate a sufficiently large system for long enough times at significant undercoolings to make a direct comparison with experimental data or analytical model predictions.

The molecular dynamics technique follows the time evolution of a microcanonical ensemble of particles interacting via a given set of potentials, $u(r)$, by integrating Newton's equations of motion for all particles in the system:

$$m\frac{d^2\mathbf{r}_i}{dt^2} = -\sum_{j=1}^{N} \nabla u(|\mathbf{r}_i - \mathbf{r}_j|) \qquad \text{for } i = 1,\ldots,N. \tag{16.1}$$

By this method, the system trajectory can be observed directly, and the approach to equilibrium can be measured as a function of time, giving the dynamics of the phase transformation.

Three-dimensional studies of nucleation with MD are difficult since the fluctuations are extremely rare; undercooling and glass formation occur most frequently. Mandell et al.[298] first reported the observation of homogeneous nucleation and growth using a Lennard-Jones 6-12 potential for a 108-particle system. A well-defined limit to the degree of undercooling, indicated by an abrupt change in the system pressure, temperature, and mean displacement that increased with the density, was observed. General agreement with the classical theory of nucleation was noted. Interestingly, some evidence for subcritical nuclei was also observed. Later studies[299] on 500-atom systems separated nucleation, revealed by changes in $S(\mathbf{q})$, from crystal growth, causing the abrupt changes in the temperature, pressure, and mean square displacement. It was also possible to identify a nuclear region ($n^* \sim 40$–70 particles) in which a crystalline structure was established initially.

The unequivocal identification of a crystal nucleus in a molecular dynamics calculation is difficult. It will be revealed in $S(\mathbf{q})$ only when it is rather large. Tanemura et al.,[300] studying nucleation in a soft potential, suggested that Voronoi polyhedra might probe the local structure better. The Voronoi polyhedron is a generalized Wigner–Seitz cell, for which the planes are constructed along the perpendicular bisector between the atom and each of its nearest neighbors. The local configuration is specified by giving the number of edges on each face of the resulting polyhedra; bcc and fcc local structures are clearly identified. The Voronoi polyhedra construction has

---

[298] M. J. Mandell, J. P. McTague, and A. Rahman, *J. Chem. Phys.* **64**, 3699 (1976).

[299] M. J. Mandell, J. P. McTague, and A. Rahman, *J. Chem. Phys.* **66**, 3070 (1977).

[300] M. Tanemura, Y. Hiwatari, H. Matsuda, T. Ogawa, N. Ogita, and A. Ueda, *Prog. Theor. Phys.* **58**, 1079 (1977).

been used to describe the liquid and glass structures.[219,301-304] Using this method, Hsu and Rahman studied nucleation by MD in liquid rubidium, assuming a 500-particle system.[305] Again, nucleation occurred at a well-defined location ($n^* \sim 40$–70 atoms) and grew quickly to crystallize the system. They point out that these critical nuclei are so large that 500-particle systems are the smallest ensembles where reflection problems from the boundary are not significant. Cape *et al.* attempted to observe more than one nucleation event by studying a system of 4000 soft spheres.[306] Using the Voronoi construction, they gave an unequivocal demonstration of nucleation and growth in a system sufficiently large that boundary effects could be ignored. The barrier to nucleation was small at the observed undercoolings; the nucleation was limited by the relaxation time for cooperative translational diffusion.

Although most simulations support the hypothesis of fluctuations-induced phase transitions under appropriate conditions and agree, at least qualitatively, with the predictions of the classical theory, there is considerable disagreement on the structure of the nuclei. Mandell *et al.* observed fcc nucleation, using a 6-12 potential for a 108-atom distribution; they found bcc nuclei when they considered a 500-atom cluster.[298,299] Hsu and Rahman[305] found that the nuclei structure depends strongly on the potential assumed. Assuming rubidium, rubidium truncated, Lennard-Jones, and Lennard-Jones truncated potentials, they found that bcc nucleated only in the rubidium potential. Cape *et al.* observed both fcc and bcc nuclei in ensemble of 4000 soft spheres.[306] Mountain and Basu[307] studied a system of 432 rubidium atoms and performed 54 simulations. The nucleus was bcc in 23 cases and fcc in six cases, although fcc is the more stable structure.

VII. Beyond Classical Nucleation Theory

17. ANALYSIS OF THE CLASSICAL THEORY

The classical theory then provides a reasonably good description of both steady-state and time-dependent nucleation in undercooled liquids and glasses. It has several important virtues: (1) it is simple to use; (2) it provides a concrete physical model for nucleation in terms of the dynamics of microscopic clusters; and (3) it is adapted readily to describe a wide range of nucleation phenomena.

[301] J. D. Bernal, *Nature* **183**, 141 (1959).
[302] J. D. Bernal and J. L. Finney, *Disc. Farad. Soc.* **43**, 60 (1967).
[303] A. Rahman, *J. Chem. Phys.* **45**, 2585 (1966).
[304] J. L. Finney, *Proc. Roy. Soc. A* **319**, 479 (1970).

a. *Experimental Successes*

The classical theory of nucleation can describe successfully several features of nucleation in undercooled liquids and glasses:

1. The observed critical undercooling is predicted.
2. It allows an evaluation of the interfacial energy, σ, from measured nucleation rates. These data scale with the enthalpy of fusion and agree qualitatively with the few direct measurements of σ.
3. The predicted linear dependence between $\log(I\eta)$ and $(T \Delta G_v^2)^{-1}$ is observed over a wide temperature range in the glasses and liquids investigated.
4. It quantitatively explains the time and temperature dependence of the nucleation rate for arbitrary initial cluster distributions.

b. *Experimental Problems*

Several significant problems were also noticed, however.

1. The nucleation prefactor [A^* in Eq. (3.22)] is generally orders of magnitude larger than is predicted theoretically. This can often be corrected by assuming a temperature-dependent interfacial energy, which has not been tested directly.
2. The nucleation data for silicate glasses show a departure from linearity between $\log(I\eta)$ and $(T \Delta G_v^2)^{-1}$ at large undercoolings. A similar disagreement was also noticed in computing the number of nuclei produced following a preanneal in a similar temperature range. These may signal a failure of the classical theory at large departures from equilibrium.

c. *Theoretical Issues*

There are also significant questions of the theoretical assumptions of the theory.

1. On the basis of computer simulations, the capillarity approximation is probably erroneous for small clusters. The assumption that the free energy can be separated into a bulk and an isotropic interfacial component becomes meaningless when the cluster size is of the same order as the interfacial thickness.
2. The use of bulk thermodynamic quantities for clusters of 10 to 10^4 atoms is certainly questionable. The cluster radius of curvature is large, and the density in the cluster is probably less than that of the bulk.
3. Classical theory, based on Gibbs' approximation of a homogeneous nucleus and a sharp boundary between the nucleus and the metastable phase, predicts a nonzero barrier to transformation at all undercoolings. It fails, therefore, to predict spinodal transformation in the unstable region.

4. The dynamical model has not been tested adequately. Further, although the transient data support the assumption that the interfacial mobility scales with the bulk mobility, that needs much more investigation. Avramov et al.,[308] for example, offer data that question that scaling.

18. Curvature and Anisotropy Effects on σ

Predictions of the curvature dependence of the interfacial energy are often in disagreement, and there are few experimental results that can resolve the conflict. On thermodynamic grounds, Gibbs[21] and Tolman et al.[309] showed that the interfacial energy depends on the curvature of the surface. On the basis of quasithermodynamic arguments, considering a single-component, fluid–fluid system to lowest order, Tolman obtained the following expression for the curvature-dependent interfacial energy:

$$\frac{\sigma}{\sigma_\infty} = \frac{1}{1 + 2\delta/r}, \tag{18.1}$$

where

$$\delta = \Gamma/(\rho_1 - \rho_2). \tag{18.2}$$

Here, Γ is the excess interfacial density (defined as the difference between the density of the interface and the density it would have if the interface were perfectly sharp), σ_∞ is the energy for a planar interface, r is the cluster radius, and ρ_1 and ρ_2 are the densities of the respective phases. Similar conclusions were obtained by Kirkwood and Buff from a statistical theory of surface tension.[310] Generally, δ is taken to have a constant value appropriate for the superficial density of a planar liquid surface. As the size of the clusters decreases, however, this ceases to be true, and Tolman's expression becomes invalid. For example using a second-nearest-neighbor calculation, Benson and Shuttleworth[311] found a 15% reduction in the surface energy of a small crystal containing 13 atoms, a much smaller reduction than that predicted from Eq. (18.1). Assuming, however, a broken bond model, Oriani and Sundquist[312] estimated a 20–25% increase in surface energy for droplets ranging from 10 to 116 atoms. Abraham[75] suggests that the Tolman approximation is valid as long as $2\delta/r \ll 1$; for greater curvatures, results

---

[305] C. S. Hsu and A. Rahman, *J. Chem. Phys.* **71**, 4974 (1979).
[306] J. N. Cape, J. L. Finney, and L. V. Woodcock, *J. Chem. Phys.* **75**, 2366 (1981).
[307] R. D. Mountain and P. K. Basu, *J. Chem. Phys.* **78**, 7318 (1983).
[308] I. Avramov, I. Gutzow, and E. Grantscharova, *J. Cryst. Growth* **87**, 305 (1988).
[309] R. C. Tolman, *J. Chem. Phys.* **17**, 333 (1949).
[310] J. G. Kirkwood and F. P. Buff, *J. Chem. Phys.* **17**, 338 (1949).
[311] G. C. Benson and R. Shuttleworth, *J. Chem. Phys.* **19**, 130 (1951).
[312] R. A. Oriani and B. E. Sundquist, *J. Chem. Phys.* **38**, 2082 (1963).

must be obtained from more detailed considerations based on, for example, computer simulations or density expansions of the interface.

The nature of the crystal–melt interface is an important one that would require a separate review. Because of the problem complexity and the limited experimental data, molecular dynamics and Monte Carlo are important and particularly illuminating.[313–315] They show an increasing interfacial tension with temperature, supporting the negentropic argument proposed to reconcile the classical theory with experiment.

19. Field-Theoretic Models of Steady-State Nucleation

Gibbs assumed that the nucleus was homogeneous up to a sharp boundary although he recognized that this might be incorrect when the cluster size was very small. Cahn and Hilliard[33,316,317] relaxed this assumption by applying the Landau formalism to nucleation to avoid the division of the free energy into a surface and volume terms. Like the classical theory, this field-theoretic approach assumes that the system is in a metastable state that decays to the stable state by a localized fluctuation. The critical cluster can be viewed as a particular saddle-point configuration of the free energy. Field-theoretic approaches to phase transformations have the advantage that they can be formulated for any system for which a mean field theory exists. They only yield the free-energy barrier, however. No information is obtained about kinetic factors that control the width of the saddle region and, hence, the time that the "cluster" spends there. Hydrodynamic theory is used to graft kinetics onto the field theory. This has the advantage over the cluster dynamical approaches that accurate expressions for the cluster attachment and detachment kinetics are not required.

This development parallels that of Sarkies and Frankel.[318] Assuming that the Gibbs free-energy density functional, $g[\rho(\mathbf{r})]$, is analytic in the density and its gradients, $\rho, \nabla\rho, \nabla^2\rho, \ldots$, it can be expanded in a Taylor series about $g_0(\rho)$, the free-energy functional for the average density. Assuming an isotropic system with small density variations over some characteristic length scale, such as the coherence length, terms higher than second order can be ignored.

---

[313] A. J. C. Ladd and L. V. Woodcock, *Chem. Phys. Lett.* **51**, 155 (1977); *Mol. Phys.* **36**, 611 (1978); J. Hiwatari, E. Stoll, and T. Schneider, *J. Chem. Phys.* **68**, 3401 (1978); A. Bonissent and F. Abraham, *J. Chem. Phys.* **74**, 1306 (1981); E. Burke, J. Q. Broughton, and G. H. Gilmer, *J. Chem. Phys.* **89**, 1030 (1988); B. B. Laird and A. D. J. Haymet, *J. Chem. Phys.* **91**, 3638 (1989).

[314] I. Snook and W. van Megen, *J. Chem. Phys.* **70**, 3099 (1979); G. Bushnell-Wye, J. L. Finney, and A. Bonissent, *Phil. Mag. A* **44**, 1053 (1981).

[315] W. A. Curtin, *Phys. Rev. Lett.* **59**, 1228 (1987); W. E. McMullen and D. W. Oxtoby, *J. Chem. Phys.* **88**, 1967 (1988).

[316] J. W. Cahn and J. E. Hilliard, *J. Chem. Phys.* **28**, 258 (1958).

[317] J. W. Cahn, *J. Chem. Phys.* **30**, 1121 (1959).

[318] K. W. Sarkies and N. E. Frankel, *Phys. Rev. A* **11**, 1724 (1975).

Integrating over the volume to obtain the total free-energy functional, G, and using the divergence theorem,

$$G = \int_V [g_0(\rho) + \kappa(\nabla\rho)^2 + \cdots] \, d^3\mathbf{r} = \int_v g(\rho, \nabla\rho) \, d^3\mathbf{r}. \quad (19.1)$$

κ is taken to be a function of T only, independent of $\rho(\mathbf{r})$ and $\mathbf{r}$; this is discussed in detail by Rice.[319] The free energy is then composed of two parts: $g_0(\rho)$, which is a coarse-grained free-energy density for the homogeneous system that is well defined in the metastable and the unstable regions, and a gradient term that takes account of effects due to variations in the density. Care must be exercised in choosing the appropriate spatial average to construct the coarse-grained free-energy density.[320] The material within each uniform cell must reach local thermal equilibrium on a time scale that is short with respect to the time required for nucleation. The cells must be larger than the molecular diameter so that the continuum approximation is sensible. They cannot be so large, however, that nucleation can occur within a single cell. In general, the diameter cannot be larger than the coherence length. Thus these theories should be invalid when the critical size becomes less than that length.

The stable and metastable states lie near the density configurations that minimize G; they are given by the solution of the Euler equation

$$\nabla\left(\frac{\partial g}{\partial \nabla\rho}\right) - \frac{\partial g}{\partial \rho} = \lambda, \quad (19.2)$$

where λ is the Lagrangian multiplier introduced by the boundary condition that the average density remain constant. Substituting $g(\rho, \nabla\rho)$ into Eq. (19.2) gives

$$\lambda = (-\partial g_0(\rho)/\partial \rho)_{\rho_0}, \quad (19.3)$$

since $\nabla^2\rho \to 0$ as $\rho \to \rho_0$.

A prediction for the energy barrier to nucleation, W^*, is central to any nucleation theory. If the system is sufficiently large that there is no significant change in the density of the exterior phase during nucleation, W^* is determined readily from Eq. (19.1) by subtracting the free-energy functional of the initial phase $g_0(\rho_0)$. Assuming an isotropic system, W^* can be expressed in terms of the radius r for clusters of the nucleating phase:

$$W^* = 4\pi \int_0^\infty \left[\Delta g_0 + \kappa\left(\frac{d\rho}{dr}\right)^2\right] r^2 \, dr, \quad (19.4)$$

[319] O. K. Rice, *J. Phys. Chem.* **64**, 976 (1960).
[320] J. S. Langer, *Physica* **73**, 61 (1974).

where
$$\Delta g_0 = g_0(\rho) - g_0(\rho_0). \qquad (19.5)$$

The density profile is obtained by solving

$$\frac{\partial^2 \rho}{\partial r^2} + \frac{1}{r}\frac{\partial \rho}{\partial r} = \left[\left(\frac{\partial g_0(\rho)}{\partial \rho}\right)_T - \lambda\right]\frac{1}{2\kappa}, \qquad (19.6)$$

subject to the boundary conditions that

$$\rho \to \rho_0 \quad \text{as} \quad r \to \infty,$$

$$\frac{\partial \rho}{\partial r} \to 0 \quad \text{as } r \to 0.$$

Considering the stability of an incompressible homogeneous fluid, for small departures from equilibrium Cahn and Hilliard showed that the free-energy approaches that of a flat interface, and the radius of the critical nucleus approaches infinity. In this regime, then, the thermodynamic aspects of the classical theory are approximately valid. At large departures from equilibrium, W^* approaches zero, predicting spinodal transformation as the favored mechanism.

Cahn and Hilliard focused their attention on the correct form of the free energy for critical cluster formation. Recently, considering phase separation, the dynamics of nucleation theory were reformulated in the standard way for nonequilibrium processes so that the current density is related linearly by the atomic mobility to the gradient of the local free-energy functional.[321,322] Following this approach, a Fokker–Plank equation similar to that formulated from the continuum approximation to the classical theory [Eq. (3.17)] was obtained.

Langer extended the field theoretic method to account properly for the fundamental fluctuations leading to nucleation, and to consider fluctuations in cluster size and shape.[34,35,320] He introduced a Langevin force that fluctuates on a time scale that is fast with respect to the time scale for the molecular rearrangements accompanying nucleation. Following Cahn and Hilliard, the metastable state is taken to decay via a locally unstable fluctuation (critical fluctuation) that need not correspond to a physical droplet, but to a critical saddle-point configuration of the free energy of the system. This is illustrated in Fig. 34, showing the free-energy surface. The

---

[321] J. W. Cahn, "Metastability, Instability, and the Dynamics of Unmixing in Binary Critical Systems, in Critical Phenomena in Alloys, Magnets, and Superconductors" (R. I. Jaffe, R. Mills, and E. Ascher, eds.). McGraw-Hill, New York, 1971.

[322] J. E. Hilliard, "Spinodal Decomposition, in Phase Transformations" (H. I. Aronson, ed.). Am. Soc. for Metals, Metals Park, OH, 1970.

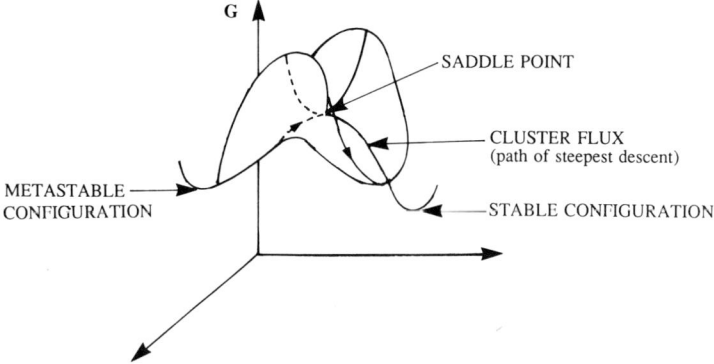

FIG. 34. A schematic illustration of the free-energy functional (as a function of two variables) saddle transition region between the metastable and stable configurations.

Langevin equation is generalized to describe the time dependence of the probability current rather than the dynamics of actual clusters. A set of variables η_i, $i = 1, 2, \ldots, M$, is introduced to describe the M degrees of freedom of the metastable system, and a distribution function $\rho(\eta, t)$ is introduced to describe the time-dependent configuration. A continuity equation for ρ is assumed:

$$\frac{\partial \rho}{\partial t} = -\sum_{i=1}^{M} \frac{\partial J_i}{\partial \eta_i}. \tag{19.7}$$

The probability current J_i is given by

$$J_i = -\sum_{j=1}^{M} M_{ij} \left(\frac{\partial G\{\eta\}}{\partial \eta_j} \rho + k_B T \frac{\partial \rho}{\partial \eta_j} \right), \tag{19.8}$$

where M_{ij} is a generalized mobility matrix and $G\{\eta\}$ is a coarse-grained free-energy functional. The equilibrium probability density has the form

$$\rho_{eq}\{\eta\} \propto \exp(-G\{\eta\}/k_B T). \tag{19.9}$$

The metastable and stable states have configurations $\{\eta\}$ that minimize G and, hence, maximize ρ_{eq}. A phase transformation proceeds from the metastable configuration $\{\eta_0\}$ to a configuration of lower free energy. During the transformation, the system passes through the saddle-point configuration $\{\bar{\eta}\}$. Langer argued that the saddle-point configuration is the same as that in the initial metastable state except for the presence of a single critical fluctuation. Near equilibrium, the nucleation rate is given by the rate of probability flow across the saddle point and can be expressed as

$$I = (\kappa \Omega_0 \exp[-W^*/k_B T]), \tag{19.10}$$

which is identical to the form predicted by the classical theory [Eq. (3.22)]. Here, κ is the dynamical prefactor, Ω_0 is the statistical prefactor, and $W^* = G\{\bar{\eta}\} - G\{\eta_0\}$. The statistical prefactor is a measure of the volume of the saddle-point region in configurational space, and is therefore a generalization of the Zeldovich factor [Eq. (3.15)]. The dynamical prefactor describes the exponential growth rate of the unstable mode of deformation, $\{\bar{\eta}\}$.

The description of nucleation in particular situations must be derived from this formal model by a proper choice of statistical variables $\{\eta\}$ and the corresponding coarse-grained free energy $G\{\eta\}$. In applying their model to condensation from a vapor, Langer and Turski[36] pointed out that a hydrodynamic description is most appropriate, since condensation (and nucleation from a liquid or solid) is characterized by a semimacroscopic density fluctuation involving large numbers of molecules. Making a comparison of the predictions from their theory with those from Becker and Döring theory for condensation of Xe and CO_2, they found that, despite all of these corrections, the calculated nucleation rates are almost identical. This good agreement between the hydrodynamic model and the classical theory supports the argument that the classical theory compensates correctly for all available degrees of freedom.

Field-theoretic and hydrodynamic models have been applied only recently to nucleation of a solid from a liquid. To describe nucleation, Haymet and Oxtoby[323,324] and Harrowell and Oxtoby[277] generalized an earlier density functional treatment of the liquid growing in a solid due to Ramakrishnan and Yussouff,[325] to include inhomogeneities (such as the solid-melt interface). The density of the solid is expanded about the liquid density, ρ_l,

$$\rho(\mathbf{r}) = \rho_l \left\{ [1 + \eta(\mathbf{r})] + \sum_Q \mu_Q(\mathbf{r}) e^{i\mathbf{Q}\cdot\mathbf{r}} \right\}, \qquad (19.11)$$

where, $\eta(\mathbf{r})$ is the fractional density change upon melting, $\mu_Q(\mathbf{r})$ are the Fourier coefficients of the solid, and $\mathbf{Q}$ are the reciprocal lattice vectors of the solid. Since η and μ_Q are nonzero in the solid, yet vanish in the liquid, they are taken as order parameters to construct a free-energy difference between the liquid and the solid, using a density functional expansion and the pair, triplet, and higher direct correlation functions of the liquid from the structure factor $S(\mathbf{q})$. The uniqueness and the completeness of these two parameters, however, is open to question. Also, assuming that η and μ_Q depend only on the radial

---

[323] A. D. J. Haymet and D. W. Oxtoby, *J. Chem. Phys.* **74**, 2559 (1981).
[324] D. W. Oxtoby and A. D. J. Haymet, *J. Chem. Phys.* **76**, 6262 (1982).
[325] T. V. Ramakrishnan and M. Yussouff, *Solid State Commun.* **21**, 389 (1977); *Phys. Rev. B* **19**, 2775 (1979).

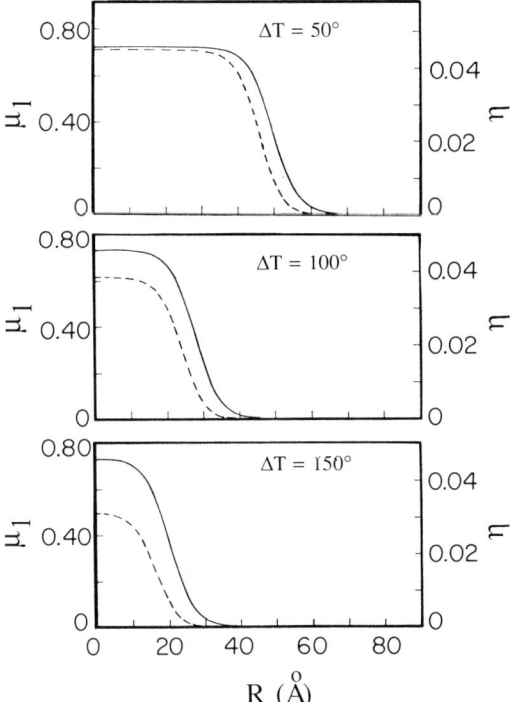

FIG. 35. The profiles of the fractional density change upon melting η (dashed curve), and the lowest-order Fourier coefficient of the solid density expansion, μ_1 (solid curve), from a field-theoretic calculation of the nucleation of a solid in an elemental liquid, demonstrating that the cluster interface becomes increasingly diffuse with undercooling. Taken from P. Harrowell and D. W. Oxtoby, *J. Chem. Phys.* **80**, 1639 (1984).

distance, this treatment ignores faceting. Calculations of the radial dependence of the order parameters, based on the extrapolation of temperature measurements of $S(\mathbf{q})$ into the undercooled regime, are shown in Fig. 35 for three undercooling values. The interface is broad, of the order of 30 Å or 6–7 atomic diameters, and is relatively independent of the degree of undercooling. This agrees qualitatively with the Monte Carlo calculations for liquid argon (Fig. 33). The fractional density change $\eta(\mathbf{r})$ falls off faster than $\mu_1(\mathbf{r})$, suggesting an ordered shell of liquid density around the nucleus. Finally, $\mu_1(0)$ remains approximately constant with undercooling, while $\eta(\mathbf{r})$ falls dramatically and approaches zero as the temperature decreases to 0 K, an unphysical prediction. This apparent failure in the calculation for such large undercoolings, however, probably does not affect the results for more reasonable undercoolings in the region where nucleation actually occurs. The

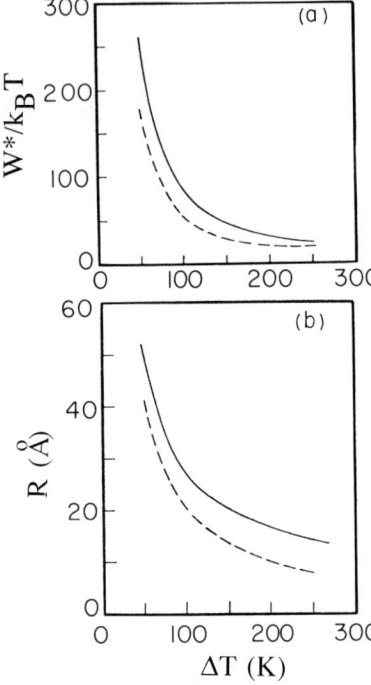

FIG. 36. (a) Free energy of cluster formation, W^* (——), calculated from a field-theoretic calculation, compared with that calculated using the capillarity approximation of the classical nucleation theory (- - -). (b) The critical nucleus radius as a function of undercooling, calculated from a field-theoretic calculation (——), and using the capillarity approximation (- - -). Taken from P. Harrowell and D. W. Oxtoby, *J. Chem. Phys.* **80**, 1639 (1984).

constant value of $\mu_1(\mathbf{r})$ disagrees with the results of Cahn and Hilliard but agrees with classical theory, suggesting that the undercoolings considered are far from the unstable (spinodal) region. This is also supported by the failure to observe a vanishing of the local free-energy minimum for the liquid and no observed divergence in the liquid structure factor. [A condition for the spinodal is $S(\mathbf{q}) \to \infty$.]

Figure 36a shows the free energy as a function of undercooling, computed from the density functional model and from the capillarity approximation. Figure 36b shows the radius dependence of the critical nucleus for the two calculations. In both cases, the functional forms are similar. The capillarity approximation, however, is lower than the density functional calculation, particularly at intermediate undercoolings.

Following Langer's method, Grant and Gunton[326] have grafted kinetics onto these thermodynamic treatments. Assuming that as the droplet grows, its kinetics principally involve thermal diffusion from the surface of the solid

---

[326]M. Grant and J. D. Gunton, *Phys. Rev. B* **32**, 7299 (1985); *Mat. Res. Soc. Symp. Proc.* (G. S. Cargill, F. Spaepen, and K. N. Tu, eds.), Vol. 57, p. 79. 1987.

droplet into the liquid, a hydrodynamical treatment gives the time-dependent radius of a droplet as

$$\frac{\partial R}{\partial t} \approx \frac{2\lambda\sigma T}{l^2\rho_1^2(1+\eta)^2}\frac{1}{R}\left[\frac{1}{R^*}-\frac{1}{R}\right], \qquad (19.12)$$

where l is the latent heat, given by Ramakrishnan and Yussouff, σ is the surface tension defined by Oxtoby et al.,[277,323,324] and λ is the thermal conductivity. As in previous treatments discussed in this review, an isotropic surface energy was assumed. The dynamical factor for the nucleation rate is

$$\kappa \approx \frac{2\lambda\sigma T}{l^2\rho_1^2(1+\eta)^2(R^*)^3}, \qquad (19.13)$$

and the statistical factor is

$$\Omega_0 \approx \left[\frac{L}{\xi}\right]^3\left[\frac{2}{3\sqrt{3}}\right]\left[\frac{\sigma\xi^2}{k_BT}\right]^{3/2}\left[\frac{R^*}{\xi}\right]^4, \qquad (19.14)$$

where L^3 is the volume of the system and ξ is a correlation length.

This calculation assumes that one order parameter is sufficient to provide a description of nucleation; this should be reasonable in simple bcc solids, such as the alkali metals. Provided that nucleation is limited by conduction of heat from the interface, the results of the calculation are reasonable. The nucleation rate is inversely proportional to the enthalpy change on transformation and directly proportional to the thermal conductivity in the original phase. Unfortunately, there currently exist no tests that can substantiate the validity of this approach over that of the classical theory. Further, nucleation for polymorphic devitrification of glasses is limited by interfacial kinetics; no field-theoretical calculations exist for that case.

20. FURTHER TREATMENTS OF NON–STEADY-STATE NUCLEATION

Nucleation can be viewed as one example of a class of processes with discrete states that evolve with time. Many such processes are best described by a master equation that relates the quantities of interest to transition probabilities between the different states. The coupled differential equations describing cluster evolution [Eq. (3.7)] are an example of such a master equation. Since it is generally difficult to obtain solutions to the master equation, various truncations and assumptions are made to generate approximate analytical solutions. One approach is to expand the master equation into an infinite-order differential equation, called the forward Kramers–Moyal equation.[327,328] Unfortunately, while this equation is equivalent to

[327]H. A. Kramers, *Physica* **7**, 284 (1940).
[328]J. E. Moyal, *J. R. Stat. Soc. London Ser. B* **11**, 150 (1949).

the original master equation, it is also difficult to solve. Instead it is often truncated after two terms to give a class of Fokker–Planck equations that approximately describe the process. The Zeldovich–Frenkel equation [Eq. (3.17)] is one such equation, founded on particular choices for the drift and diffusion terms. It is not unique, however. There have been several recent calculations aimed at obtaining the solution of alternative Fokker–Planck equations in the time-dependent regime.

The Zeldovich–Frenkel equation was first criticized by Goodrich[329] on the grounds that it incorrectly forced agreement with the solution of the discrete set of equations on the cluster boundary, n, rather than at the midpoints. Based on this, Goodrich derived a new Fokker–Planck equation that predicted a much larger nucleation rate. Numerical solutions of the discrete equations give results that are in good agreement with the solutions of the Zeldovich–Frenkel equation; a recent analysis by Shizgal and Barrett points to several errors in Goodrich's analysis that contribute to this discrepancy.[330]

Following a presentation by Shizgal and Barrett, the continuum form of the cluster distribution, $N(n, t)$, is given by the Kramers–Moyal expansion:

$$\frac{\partial N_{n,t}}{\partial t} = \sum_{i=1}^{\infty} \frac{1}{j!} \frac{\partial^j}{\partial n^j} [[k_n^- + (-1)^j k_n^+] N_{n,t}], \qquad (20.1)$$

where k_n^+ and k_n^- are the rate constants defined in Eq. (3.18). Retaining only the first two terms, we obtain

$$\frac{\partial N_{n,t}}{\partial t} = \frac{\partial}{\partial n} [A_n N_{n,t}] + \frac{\partial^2}{\partial n^2} [B_n N_{n,t}], \qquad (20.2)$$

where n is now taken to be continuous and A and B are given by

$$A_n = k_n^- - k_n^+ \qquad (20.3)$$

and

$$B_n = \tfrac{1}{2}[k_n^- + k_n^+]. \qquad (20.4)$$

A_n can be viewed as a drift coefficient, and B_n as a diffusion coefficient. In general, both are functions of the cluster size. The continuous form of the flux, equivalent to the nucleation rate at a particular cluster size, is given by

$$J_{n,t} = -\left[A_n N_{n,t} + \frac{\partial}{\partial n} B_n N_{n,t} \right]. \qquad (20.5)$$

---
[329] F. C. Goodrich, *Proc. R. Soc. London A* **277**, 155; 167 (1964).
[330] B. Shizgal and J. C. Barrett, *J. Chem. Phys.* **91**, 6505 (1989).

Since the flux vanishes in equilibrium, Eq. (20.4) can be used to determine the equilibrium density of clusters, N_n^e. Defining $F = N_{n,t}/N_n^e$, we can write the Fokker–Planck equation as

$$\frac{\partial F}{\partial t} = \frac{1}{N_n^e} \frac{\partial}{\partial n}\left[B_n N_n^e \frac{\partial F}{\partial n}\right]. \qquad (20.6)$$

Due to the truncation, the equilibrium distribution N_n^e calculated for this formalism differs from the equilibrium solution obtained from the original discrete equations. The very large steady-state nucleation rates found by Goodrich arose from a failure to fit the correct equilibrium distribution; they can be rectified by using the correct equilibrium distribution in Eq. (20.6). Then Eqs. (20.2) and (20.6) retain the same mathematical form, but A_n and B_n are no longer given by Eqs. (20.3) and (20.4). The central problem in formulating nucleation theories becomes choosing the proper values for A_n and B_n.

Shizgal and Barrett considered two possible sets of values for A_n and B_n and contrasted the solutions of the resulting Fokker–Planck equations with results from the Zeldovich–Frenkel equation and from a numerical solution of the discrete coupled differential equations. Both choices gave better agreement with the numerical solution than does the Zeldovich–Frenkel equation.

Rabin and Gitterman[331] argued that A_n and B_n are completely determined near critical points by the critical dynamics. They emphasized that it is essential to distinguish between slow and fast critical quenches, since a larger fraction of the clusters maintains the steady-state distribution during a slower quench. The calculated transient time can vary by orders of magnitude, depending on the quench rate; this was verified with a numerical simulation of nucleation in an Ising model.[332]

As discussed in Sections 8, 13, and 14, the importance of the quenched-in distribution on subsequent transient behavior was realized previously by several authors. The quench rate dependence has been modeled within the classical theory by determining the effective time lag as a function of the initial distribution of clusters. The differences, observed there are less than those reported by Rabin and Gitterman due to smaller critical sizes; they estimate that critical sizes of approximately 10^4 atoms are required to see such significant quench rate effects. It therefore appears that the effects noted by those authors can be described within the framework of the classical theory.

[331] Y. Rabin and M. Gitterman, *Phys. Rev. A* **29**, 1496 (1984).
[332] I. Edrei and M. Gitterman, *Phys. Rev. A* **33**, 2821 (1986).

VIII. Concluding Remarks

This review has focused primarily on measurements and theories pertaining to homogeneous nucleation in undercooled liquids and glasses that transform to phases of the same composition. Remarkably, within those constraints, close to equilibrium, and assuming a temperature-dependent interfacial energy, the classical theory predicts accurately time-dependent and time-independent nucleation phenomena. In particular, the dynamical model describing the cluster evolution appears to be quantitatively correct. As discussed in Section VI, although there are some experimental and theoretical difficulties with the classical theory, there are also many approaches to rectify these problems. At present, there are no experiments that can demonstrate or can refute conclusively the validity of the classical theory in the regime where it is expected to be valid. This is partially due to a lack of knowledge of the correct values for the parameters that are typically fit by the theory. Better measurements of the steady-state and time-dependent nucleation rates and the free-energy difference as a function of temperature, and the design of new experiments such as the multistep annealing treatments discussed in this review, can be used to test more rigorously nucleation theories. Time-dependent nucleation and the interface kinetics must be incorporated into the field-theoretic approaches to give theories that are as widely applicable and, hence, testable as the classical theory.

Unfortunately, there was insufficient time or space to discuss a wider range of topics in nucleation. For example, other theoretical approaches include alternative formulations of the Fokker–Planck equation[333] and more advanced atomistic models.[334] Studies of nucleation in fluids near the critical point were also not discussed. These are important since there the critical size is large and the classical theory is expected to work best. That the maximum undercooling in a variety of binary liquids near the critical point far exceeds that predicted by the classical theory[335] was originally taken as a signal of the failure of the classical theory. It is now understood to arise from a critical slowing down of droplet growth. Once again, the classical theory appears to be consistent with the experimental results. These topics and others are reviewed more extensively in Refs. 4 and 336.

---

[333] R. Lovett, *J. Chem. Phys.* **81**, 6191 (1984).
[334] K. Binder and D. Stauffer, *Adv. Phys.* **25**, 343 (1976); O. Penrose, J. L. Lebowitz, J. Marro, M. H. Kalos, and A. Sur, *J. Stat. Phys.* **19**, 243 (1978); H. Muller-Krumbhaar, in "Monte Carlo Methods in Statistical Physics" (K. Binder, ed.). Springer-Verlag, Berlin, 1979; O. Penrose, *Commun. Math. Phys.* **124**, 515 (1989); R. McGraw, *J. Chem. Phys.* **91**, 5655 (1989).
[335] B. E. Sundquist and R. A. Oriani, *J. Chem. Phys.* **36**, 2604 (1962); R. B. Heady and J. W. Cahn, *J. Chem. Phys.* **58**, 896 (1973).
[336] K. Binder, *Rep. Prog. Phys.* **50**, 783 (1987).

Appendix A
Approximations to ΔG

The free-energy difference between the liquid and crystal is computed by using entropy (ΔS) and enthalpy (ΔH) differences:

$$\Delta G = \Delta H - T \Delta S. \tag{A1}$$

If the heat capacities for the two phases are known as a function of temperature, the temperature-dependent free energy is computed readily, from the measured enthalpy. Given ΔH_f, the enthalpy of fusion at this melting point,

$$\Delta H = \Delta H_f - \int_T^{T_m} \Delta C_p \, dT', \tag{A2a}$$

and

$$\Delta S = \Delta S_f - \int_T^{T_m} \frac{\Delta C_p}{T'} \, dT', \tag{A2b}$$

where T_m is the melting temperature, ΔS_f is the entropy of fusion given by $\Delta H_f T_m$, and ΔC_p, is the difference between the specific heat capacities of the liquid and crystal phase. The free energy can then be written as

$$\Delta G = \frac{\Delta H_f \Delta T}{T_m} - \int_T^{T_m} \Delta C_p \, dT' + T \int_T^{T_m} \frac{\Delta C_p}{T'} \, dT', \tag{A3}$$

where ΔT is the undercooling, $T_m - T$. Measurements of ΔC_p in the highly undercooled liquid are difficult to make. Several expressions for ΔG have therefore been developed, based on an assumed temperature dependence for ΔC_p.

Turnbull[1] argued that if no data are available, the best approximation is to set ΔC_p equal to zero, giving the linear relation

$$\Delta G = \Delta H_f \Delta T / T_m. \tag{A4}$$

This should be a reasonable approximation for metallic systems since $C_p^l \approx C_p^x$, but it should be a poor approximation for polymers where the specific heats of the liquid and the solid are very different.

Assuming ΔC_p to be a nonzero constant and integrating Eq. (A3) gives

$$\Delta G = \frac{\Delta H_f \Delta T}{T_m} - \Delta C_p \left[\Delta T - T \ln\left(\frac{T_m}{T}\right) \right]. \tag{A5}$$

Making the approximation

$$\ln(T_m/T) \approx 2 \Delta T/(T_m + T) \tag{A6}$$

gives

$$\Delta G = \frac{\Delta H_f \Delta T}{T_m} - \frac{\Delta C_p \Delta T^2}{T_m + T}. \tag{A7}$$

This expression was first proposed by Jones and Chadwick,[337] although they did not indicate how ΔC_p should be determined. It is often approximated by the difference in specific heat at the melting point ΔC_p^m.

Hoffman[338] assumed that the difference in enthalpy between the two phases vanishes at a temperature T_∞ slightly below the glass transition temperature T_g of the liquid. For constant ΔC_p, using Hoffman's assumption, Eq. (A2) gives

$$\Delta C_p = \Delta H_f/(T_m - T_\infty). \tag{A8}$$

Using Eq. (A7), we have

$$\Delta G = \frac{\Delta H_f \Delta T}{T_m} \left[\frac{T}{T_m} + \left(\frac{\Delta T}{T_m + T} \right) \times \left(\frac{T}{T_m} - \frac{T_\infty}{T_m - T_\infty} \right) \right]. \tag{A9}$$

Hoffman argued that the last term in Eq. (A9) is negligible and obtained

$$\Delta G = \frac{\Delta H_f \Delta T}{T_m} \left(\frac{T}{T_m} \right). \tag{A10}$$

Thompson and Spaepen[339] and Battezzati and Garrone[340] pointed out that T_∞ is much less than zero in systems with a small ΔC_p^m, such as metals. This demonstrates that Hoffman's critical assumption, that T_∞ is close to T_g, is not generally valid. Thompson and Spaepen offer an adaptation of Hoffman's approach, noting that while the enthalpy difference changes rather slowly with undercooling, the entropy changes dramatically and goes to zero at some temperature T_0 slightly below T_g. Setting the specific heat difference equal to the entropy of fusion, ΔS_f, they predict

$$\Delta G = \frac{\Delta H_f \Delta T}{T_m} \frac{2T}{T_m + T}. \tag{A11}$$

Considering the cases of glass forming melts, Battezzati and Garrone also suggest that the specific heat difference is proportional to ΔS_f:

$$\Delta G = \frac{\Delta H_f \Delta T}{T_m} - \gamma \Delta S_f \left[\Delta T - T \ln \left(\frac{T_m}{T} \right) \right]. \tag{A12}$$

[337] D. R. H. Jones and G. A. Chadwick, *Phil. Mag.* **24**, 995 (1971).
[338] J. D. Hoffman, *J. Chem. Phys.* **29**, 1192 (1958).
[339] C. V. Thompson and F. Spaepen, *Acta Metall.* **27**, 1855 (1979).
[340] L. Battezzati and E. Garrone, *Z. Metallk.* **75**, 305 (1984).

The proportionality constant, γ, is taken to be a constant function of the heats of crystallization and fusion; it is approximately 0.8 for metallic glass-forming liquids.

Assuming the hole theory for the liquid state, in which the liquid is viewed as a quasi-lattice with many vacant holes (considerably smaller than atoms), Dubey and Ramachandrarao, computed the minimum free energy accompanying the introduction of the holes into the matrix, to approximate ΔG.[341] They obtained an expression that requires a knowledge of the energy necessary for the creation of a hole, the relative volume of holes and atoms, and T_0. These quantities occur in the theory as two constants, which can be determined from a knowledge of ΔC_p at two temperatures. Making several assumptions, they obtain the less accurate, but more readily usable, form

$$\Delta G = \frac{\Delta H_f \Delta T}{T_m} - \frac{\Delta C_p^m \Delta T^2}{2T}\left(1 - \frac{\Delta T}{6T}\right). \quad (A13)$$

The use of the hole theory for liquids has been criticized severely.[342] Lele et al. have derived a relation similar to Eq. (A13) by expanding ΔG about T_m.[343] Keeping the terms to first order and substituting the appropriate thermodynamic functions for the derivatives, they obtain

$$\Delta G = \frac{\Delta H_f \Delta T}{T_m} - \frac{\Delta C_p^m \Delta T^2}{T_m + T} + \left[\frac{\partial \Delta C_p}{\partial T}\right]_{T_m} \frac{\Delta T^3}{2(T_m + T)} + \cdots. \quad (A14)$$

Lele et al. argue that these first three terms should be adequate for calculating the free energy of most materials. The functional form will depend on the temperature dependence of ΔC_p. Keeping only the first three terms and assuming a linear dependence ($\Delta C_p = A - BT$), we obtain the free energy

$$\Delta T = \frac{\Delta H_f \Delta T}{T_m} - \frac{\Delta C_p^m \Delta T^2}{T_m + T} - \frac{B \Delta T^3}{2(T_m + T)}. \quad (A15)$$

Assuming a hyperbolic dependence ($\Delta C_p = A + B/T$), we recover Eq. (A7).

Taking a linear temperature dependence for ΔC_p, Singh and Holz[165] present an ansatz that allows ΔG to be expressed as

$$\Delta G = \frac{\Delta H_f \Delta T}{T_m}\left[\frac{7T}{T_m + 6T}\right]. \quad (A16)$$

Specific heat data exist for some undercooled liquid metals. Using droplet emulsions, Perepezko and Paik[123] have measured the difference in heat

---

[341] K. S. Dubey and P. Ramachandrarao, Acta Metall. 32, 91 (1984).
[342] M. Hillert, "Mat. Res. Soc. Symp. Proc." (B. H. Kear, B. C. Giessen, and M. H. Cohen, eds.), Vol. 8, p. 3. 1982.
[343] S. Lele, K. S. Dubey, and P. Ramachandrarao, Curr. Sci. 54, 994 (1985).

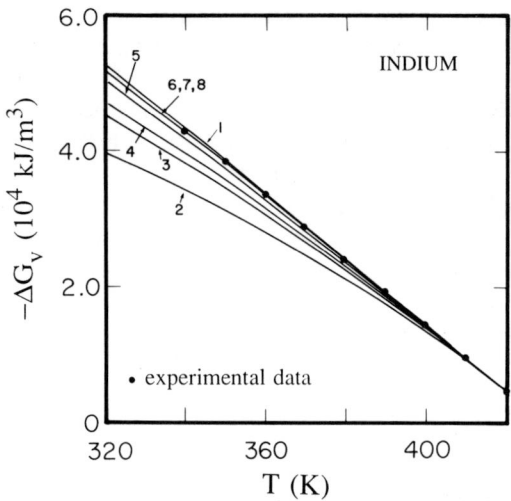

FIG. A1. A comparison between the measured volume free energy in liquid indium as a function of temperature and the predictions from analytical approximations. 1: Eq. (A4); 2: Eq. (A8); 3: Eq. (A11); 4: Eq. (A12); 5: Eq. (A16); 6: Eq. (A7); 7: Eq. (A13); 8: Eq. (A15). The measured data are from Ref. 123.

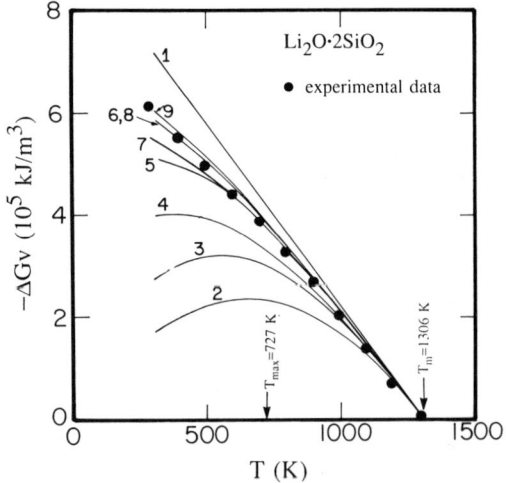

FIG. A2. A comparison between the measured volume free energy in $Li_2O \cdot 2SiO_2$ glass and the predictions from analytical approximations. 1: Eq. (A4); 2: Eq. (A8); 3: Eq. (A11); 4: Eq. (A12); 5: Eq. (A16); 6: Eq. (A7); 7: Eq. (A13); 8: Eq. (A15); 9: Eq. (A5) using ΔC_p at T_m. The measured data are from Ref. 253. The melting temperature T_m and temperature of maximum nucleation rate T_{max} are indicated.

capacity for undercooled Bi, Sn, and In and calculated ΔG_v, the free-energy difference per unit volume, as a function of undercooling. Figure A1 compares their measured data for indium with predictions from the analytical expressions discussed here. As they noted, the best agreement is with the Turnbull expression [Eq. (A4)] and the Singh and Holz expression [Eq. (A16)]. Evans found excellent agreement between measured values for ΔG_v in $Pd_{0.4}Ni_{0.4}P_{0.2}$ and those calculated from the Dubey and Ramachandrarao expression [Eq. (A13)];[96] data obtained by Kui and Turnbull,[344] however, are in worse agreement. Dubey and Ramachandrarao and Lele et al. also report good agreement between experimental data and their respective model predictions.

Figure A2 shows ΔG_v computed from data for $\Delta C_p(T)$ for lithium disilicate, a glass that crystallizes homogeneously to a crystal of the same composition. These are compared with model predictions. Here, the Turnbull approximation describes the data poorly. The Singh and Holz expression, however, is in good agreement with the data.

Acknowledgments

I am grateful to David Turnbull, who introduced me to the problem of nucleation, and to Lindsay Greer, who continues to be an invaluable collaborator in the investigation of time-dependent nucleation. I thank both of them for useful discussions during the preparation of this manuscript. I thank Anders Carlsson, Patrick Gibbons, Michael Ogilvie, Robert Phillips, and Frans Spaepen for a critical reading of the manuscript in various stages of its preparation and for many useful suggestions. I also thank Pranoat Suntharothok-Priesmeyer for her excellent job of preparing the manuscript and the figures. This work was partially supported by the National Science Foundation under Grant Number DMR 89-03081.

---

[344]H. W. Kui and D. Turnbull, *J. Non-Cryst. Solids* **94**, 62 (1987).

Glass Transition and Relaxation of Disordered Structures

FUMIKO YONEZAWA

Department of Physics
Keio University
Yokohama, Japan

I.	Introduction	179
II.	General Discussions	181
	1. What Is a Glass Anyway?	181
	2. Models of Amorphous Structures	185
III.	Glass Transition	186
	3. Experimental Results	186
	4. Theoretical Approaches	192
IV.	Computer Simulations of Glass Transition	197
	5. Merits and Demerits	197
	6. Molecular Dynamics Methods	197
	7. Simulations under Constant Pressure	201
V.	Simulations of Glass Transition at Constant Pressure	202
	8. Quenching of a Lennard-Jones System	202
	9. Physical Properties through Glass Transition	205
	10. Glass Transition versus Crystallization	219
	11. Microscopic Structure Parameters	224
VI.	Fluctuation and Relaxation in Disordered Systems	234
	12. Relaxation of a Supercooled Liquid and a Glass	234
	13. Relaxation of a Glass Prepared with Lower Quench Rate	244
	14. Visual Presentation of Atomic Configurations	249
VII.	Summary and Future Work	254

I. Introduction

Glasses appeared in the history of the mankind as early as at the Mesopotamian age more than 4000 years ago, and it is over a century since glasses entered on the stage of science. And yet, some essential natures of glasses are not fully revealed. The properties that need further investigation include the mechanism of the glass transition, the characteristic features of glassy or amorphous structures, and the key factors that make glasses stable.

Research on glasses or amorphous solids has become one of the fashionable subjects in the field of physics for the last three or four decades, partly because of the outstanding progress in the preparation techniques of these materials, and partly because of a promising prospect for these materials as devices. The more widely the importance of these systems is recognized, the more urgent it becomes to clarify their properties.

In order to understand the physical properties of a material most adequately, the atomic structure therein should be characterized in the first place. As for crystals, almost complete information was obtained already at the beginning of the present century, owing to the structure analyses by means of x-ray diffractions. The structure analyses of this kind are remarkably powerful for crystals in which the atomic configurations are specified by the periodicity or, in other works, by the long-range order (LRO). On the other hand, when the materials to be studied are noncrystalline or disordered, as are glasses or amorphous solids as well as liquids, the same method of x-ray diffraction could give at best the pair distribution functions (PDF) $g(r)$, from which it has been ascertained that all disordered structures have the common aspects—the absence of the LRO and the existence of the short-range order (SRO). Here, the term SRO is used to indicate that the number and positions of nearest-neighbour atoms around a given atom in a disordered system are almost identical to those in a corresponding crystal.

Accordingly, the intermediate-range order (IRO) is expected to be a deciding factor in distinguishing one disordered system from another. Here, the term IRO is used to represent the order, if any, in the atomic distributions beyond the immediate nearest neighbours. The determination of the IRO by experiments has so far been performed only by some indirect methods such as the measurements of densities or other physical quantities. These indirect methods, however, have limitations, especially in the sense that they cannot give the structures uniquely. From a theoretical point of view, it is very difficult to predict from first principles the existence or absence of the IRO and to propose the characteristics of the IRO if it exists at all. This is the reason why the approaches through structure modelings have played a significant role in the investigations of disordered structures. Models of disordered structures have been constructed first by hand and recently by computers.

With this situation in mind, it is the purpose of this chapter to pursue the possibilities and extent of computer studies for analyzing disordered structures. Since space is limited, a comprehensive survey of the literature is by no means intended. The background of recent computer studies is briefly touched upon, and discussions of concrete structure analyses are given mainly with reference to work of the author and her co-workers.

II. General Discussions

1. WHAT IS A GLASS ANYWAY?

a. *A Glass versus an Amorphous Solid*

In science and technology, the term *glass* is used to describe the special group of amorphous solids that have been prepared by the melt-quench method via glass transition. Common window glasses naturally belong to this group. In recent years, this term is often mentioned as synonymous with *amorphous solid*. The situation will be grasped when the processes for preparing amorphous solids are studied.

An amorphous solid is generally prepared by the sequence of the following two processes:

1. A process of bringing the original system into a state of very high energy.
2. A process of taking from the system its kinetic energy very rapidly. This process is sometimes called *quenching*. The quench rate required to realize an amorphous state varies from material to material. Any material can transform into an amorphous state when the requirement for the quench rate appropriate to that material is fulfilled.

Various efficient techniques for realizing the aforementioned two processes have been developed, and some of them are listed in Table I. As described in the preceding subsection, the roles of the two fundamental processes are the same in all preparation techniques of amorphous solids.

The method presented in the top row in Table I is called the *melt-quench* method because in this case a high-energy state reached as a result of process 1 is a liquid, which is quenched either in a gas, or in a cold liquid, or by contact with a cold solid. The ranges of quench rates attained through these three quench methods are, respectively, $1-10$ K/s, 10^2-10^5 K/s, and 10^6-10^8 K/s.

It is difficult to estimate the quench rates achieved in the methods listed in the second and third rows, but there are some indications that they are higher than 10^9 K/s.

The characteristics of the aforementioned two fundamental processes are reflected in the characteristics of the amorphous solids thereby obtained. In the high-energy state attained as a result of process 1, the atomic distribution in the system is generally disordered. In process 2, the kinetic energy of the system is taken away so rapidly that the atoms in the system do not have enough time to search the whole phase space in order for the system to achieve a state of minimum free energy, which is normally a crystalline state. Instead, the atoms can adjust their configurations only locally in real space: this means that only some local portion of the phase space is surveyed, and,

TABLE I. VARIOUS PREPARATION METHODS OF AMORPHOUS SOLIDS

Process 1		Process 2		
Technique for Bringing the System into a High-Energy State	A High-Energy State Reached as a Result	Technique for Quenching	Examples of Amorphous Materials Thus Prepared	Quench Rate (K/s)
Heating	Liquid	Quenching (a) in a gas (b) in a liquid (c) by contact with a cold solid	Silica Glass Alloy Wire Alloy Ribbon	$1 \sim 10$ $10^2 \sim 10^5$ $10^6 \sim 10^8$
Physical Vapour Deposition (solid → gas) (a) Vapor Deposition (b) Sputtering (c) Ion Plating Chemical Vapour Deposition (CVD) (a) Photo CVD (b) Thermal CVD	Gas	• Deposition as a Thin Film on a Cold Substrate • Projection onto a Substrate • Deposition on a Substrate • Growth of a Film through Chemical Reaction on a Substrate	Thin-Film Alloy	10^9?
Plasma CVD	Plasma	Growth of a Film in a Gas or on a Substrate		

accordingly, the system is forced to have an atomic configuration corresponding to the local minimum of free energy within that local portion. Unless this local minimum happens to be the lowest minimum of the whole phase space, the state is not in thermal equilibrium but is metastable, because there is a finite probability with which the system transforms into the lowest minimum.

The free energies in phase space and the amorphous states are shown schematically in Fig. 1, in which different types of amorphous states are related to the different shapes of the free energy. Amorphous state A_1 is *relatively stable* because there exists a barrier with high activation energy at P_1 between A_1 and a crystalline state denoted by C. Therefore, it is rather difficult for the system to hurdle the barrier. On the other hand, amorphous state A_2 is *relatively unstable* because the activation energy at P_2 between A_2 and C is not very high, and consequently, there is a finite probability of crystallization through P_2. The terms *relatively stable* and *relatively unstable* are used since the duration of time spent before the relaxation of the system into a crystal is defined only *in relation to* some reference time scale such as the lifetime of a human being or of a universe. In fact, the actual time scale for the relaxation is not important, but the relative magnitude with respect to the reference time scale has a significant meaning.

As a natural outcome of rapid quenching, the atomic structures of amorphous systems are noncrystalline.

b. *Classification of Condensed Matter*

The whole class of condensed matter can be divided into categories as listed in Table II: (1) liquid, which is structurally disordered and thermally stable, realized at temperatures higher than the melting temperature T_m; (2) crystal, which is structurally ordered and thermally stable, realized at temperatures

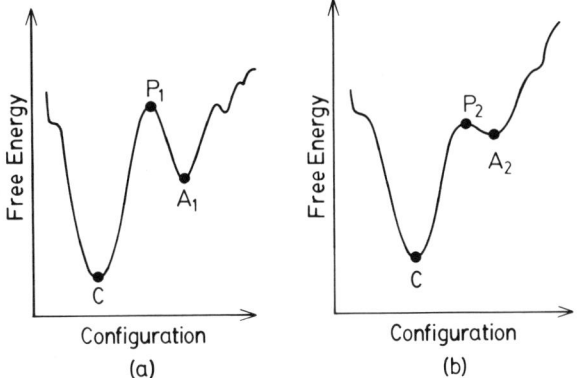

FIG. 1. Free energy versus schematic variable in phase space that defines the atomic configurations in real space.

TABLE II. CLASSIFICATION OF CONDENSED MATTER

Condensed Matter		Temperature Range	Thermal Condition	Atomic Configuration
Liquid		$T > T_m$	Stable	Disordered
Solid	Crystal	$T < T_f$	Stable	Ordered
	Amorphous Solid (glass)	$T < T_g$ or T'_c	Metastable	Disordered

lower than the freezing temperature T_f; and (3) an amorphous solid, which is structurally disordered and thermally metastable, realized at temperatures lower than the glass transition temperature T_g or than the crystallization temperature T'_c. Here, T'_c is the temperature at which an amorphous solid, when heated, transforms to a crystal.

As mentioned at the beginning of this section, the conventional definition limits glasses to those amorphous solids prepared by the melt-quench techniques demonstrated in the top row of Table I and treats them as different from other amorphous solids. The idea for this definition is based upon the assertion that amorphous solids prepared by the melt-quench methods through the glass transition tend to be relatively stable, corresponding to the condition as shown by A_1 in Fig. 1a, and accordingly are expected to show different behaviours from the other amorphous solids.

This assertion applies to easy-glass-formers, such as polymers and silicates, which can be prepared into glasses even when the quench rates are as low as 1–10 K/s, the kinds of quench rates realized by quenching in gases. However, the remarkable progress of quench techniques in recent years has raised the quench rates in laboratories, and, consequently, it has become possible to construct glasses of materials that were previously regarded as non-glass-formers or difficult-glass-formers. Among these difficult-glass-formers are metals in which atoms are mostly isotropic. Then, the situation changed in the sense that even the amorphous solids prepared by the melt-quench methods could be relatively unstable because, in the systems of isotropic atoms, the relaxations toward crystalline structures are highly probable. This means that all possible behaviours and properties are expected to take place even in the amorphous solids of isotropic atoms prepared by the melt-quench methods. Besides, although the melt-quench techniques are simply one class out of various methods of preparing amorphous structures, they embody the essential aspects characteristic of any preparation method. On the basis of all these considerations, the study in this chapter is confined to the glass transition and to the glassy or amorphous structures in the systems composed of isotropic atoms.

2. MODELS OF AMORPHOUS STRUCTURES

Amorphous solids or glasses can be roughly categorized according to the types of constituents into the following three groups: amorphous solids made of (1) polymers, (2) covalent atoms, and (3) isotopic atoms. Polymeric glasses of course belong to the first group, popular silicate glasses and amorphous semiconductors to the second group, and metallic glasses and amorphous solids of the rare gases to the third group. Note, however, that so far the rare gases have never been prepared in the forms of amorphous solids. As can be easily predicted, structure modellings of polymeric glasses are least tractable, and no systematic approaches have been proposed yet. Modellings of amorphous structures in the second and third groups have attained considerable success, and some of them are presented in what follows.

a. *Continuous Random Network (CRN)*

The first really outstanding model of amorphous structures for covalent materials was due to Zachariasen,[1] who some 60 years ago proposed the fundamental concept of a continuous random network (CRN). A CRN is a network of covalent bonds that is random in the sense that there exists no LRO, and is continuous in the sense that no broken or dangling bonds are included. The central rule in the construction of a CRN is that the number of the covalent bonds around each atom must be the same as that in the corresponding crystal. An important point of Zachariasen's model lies in the recognition that, even with this rule, a nonperiodic arrangement of covalent atoms can be constructed by introducing variations in bond lengths, bond angles, and dihedral angles. Another important point is that this model is consistent with the experimental indication, as stated in the introduction for the existence of the SRO and the absence of the LRO in amorphous materials.

This model of CRN has been applied to tetrahedrally bonded amorphous semiconductors[2] and studied extensively. Some detailed explanations and relevant references are given in the books by Eliott[3] and Zallen.[4]

b. *Dense Random Packing of Spheres (DRPHS)*

A dense random packing of hard spheres (DRPHS) can be constructed by mixing balls in such a way as to avoid a periodic arrangement. The concept of the DRPHS was first introduced by Bernal[5,6] and elaborated by Finney.[7]

[1] W. H. Zachariasen, *J. Am. Chem. Soc.* **54**, 3841 (1932).
[2] D. E. Polk, *J. Non-Cryst. Solids* **5**, 365 (1971).
[3] S. R. Eliott, "Physics of Amorphous Materials." Longman Group, 1983.
[4] R. Zallen, "The Physics of Amorphous Solids." Wiley, New York, 1983.
[5] J. D. Bernal, *Nature* **183**, 141 (1959).
[6] J. D. Bernal, *Proc. Roy. Soc.* **A 280**, 299 (1964).
[7] J. L. Finney, *Proc. Roy. Soc.* **A 319**, 479 (1970).

This model has greatly aided our understanding of the atomic configurations of metallic glasses. Discussions in this connection are found in a review paper by Cargill[8] and in the aforementioned books.[1,3]

III. Glass Transition

3. EXPERIMENTAL RESULTS

In a well-known book by Wong and Angell,[9] the definition of "the glass transition," given by the National Research Council Ad Hoc Committed on Infrared Transmitting Materials,[10] is cited as "that phenomenon in which a solid amorphous phase exhibits with changing temperature a more or less sudden change in the derivative thermodynamic properties, such as heat capacity and expansion coefficient, from crystal-like to liquid-like values." The temperature of the transition is called T_g.

The embodiment of this definition is presented in Fig. 2, where the schematic illustrations of experimental results are shown for (a) the volume-versus-temperature relation and for (b) the isobaric specific heat-versus-temperature relation. Experiments are performed by cooling a liquid. When the cooling rate is low, the first-order phase transition from the melt to a crystal takes place at the freezing temperature T_f. When the cooling rate is high, on the other hand, the volume changes continuously, as shown by the solid curve in Fig. 2a. The experimental determination of the glass transition temperature T_g is based upon the extrapolation method, as indicated by broken lines, the extrapolation being achieved from the linear part of a high-temperature liquid and from the linear part of a low-temperature glass.

There exist great differences in the characteristic features of the manifestation of the glass transition in specific heat C_p at constant pressure, as schematically displayed in Fig. 2b. Roughly speaking, there are two categories of materials: those showing a monotonic change (solid curve) and those showing a maximum at T_g (broken curve). In the former category are included "strong" network liquids, such as SiO_2, GeO_2, and BeF_2, while some non-network and ostensibly ionic substances belong to the latter category. The substances whose structures are intermediate between the strong network type and the ionic nature exhibit the C_p-versus T relations of the intermediate behaviour between two typical curves presented in Fig. 2b.

---

[8]G. S. Cargill, III, "Solid State Physics," (F. Seitz, D. Turnbull, and H. Ehrenreich, eds.), Vol. 30. Academic Press, New York, 1975.
[9]J. Wong and C. A. Angell, "Glass-Structure by Spectroscopy." Marcel Dekker, New York, 1976.
[10]Material Advisory Board, Nat. Aead. Sci. Res. Council MAB-243, 1968.

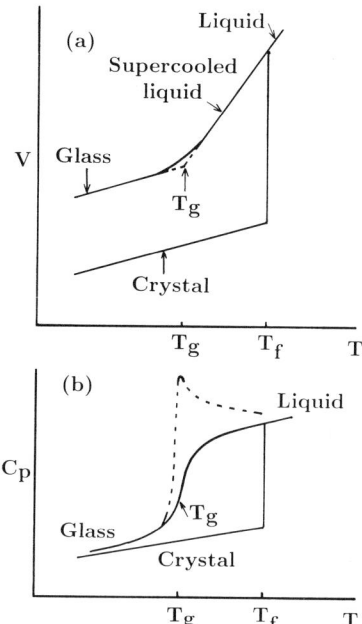

FIG. 2(a). Schematic illustrations of the change in volume (V) with temperature (T) as a liquid is cooled. The solid curves schematically display experimental results, while the broken curves represent the extrapolations in order to determine T_g. (b) Schematic illustrations of the change in specific heat at constant pressure (C_p) as a liquid is cooled. The solid and broken curves respectively express the behaviours of C_p for two typical classes of substances.

The flow properties such as viscosity and diffusion constant are continuous through the glass transition. It has been empirically attested that the shear viscosity η is of the order of 10^{13} poise at T_g in a rather wide variety of materials. By taking this point into account, we show the schematic illustrations of viscosities η obtained from experiments in Fig. 3 as a function of reduced reciprocal temperature T_g/T where T_g is chosen as the temperature at which $\eta = 10^{13}$ poise. The behaviours of measured viscosities are divided into two main classes; one class obeys an Arrhenius rate law over wide ranges of temperature as given by the solid curve in the figure, while the other class is well-described by the Vogel–Fulcher law

$$\eta_i = \eta_0 \exp\left[-\frac{\text{const}}{T - T_0}\right], \tag{3.1}$$

as expressed by the broken curve in the figure, where the temperature dependence of η_0 is small and $T_0 > T_g$.

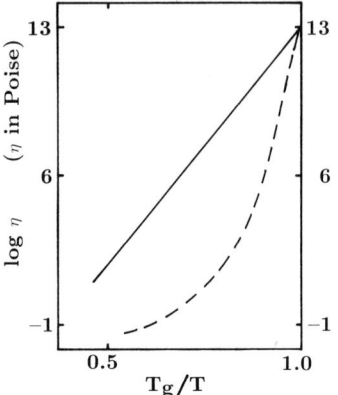

FIG. 3. Schematic illustrations of logarithm of viscosity (η) versus reduced reciprocal temperature (T_g/T). There are two classes of materials: those obeying the Arrhenius rate law as given by the solid curve, and those obeying the Vogel–Fulcher law as given by the broken curve.

Experimental results indicate that there exists a correlation between the temperature-dependent characteristics of specific heat C_p and those of viscosity η. As stated earlier, substances such as "strong" network liquids have C_p, which changes monotonically with temperature while the liquid states of those substances exhibit viscosities that follow the Arrhenius behaviour. On the other hand, the specific heats C_p of substances such as non-network materials display the distinct features at T_g, as shown in Fig. 2b, while the viscosities of these materials deviate considerably from the Arrhenius behaviour, as demonstrated by the broken curve in Fig. 3, and they are better fitted to the Vogel–Fulcher relation given by Eq. (3.1).

These differences in the features of C_p and η are regarded as attributable to the higher degree of conformational freedom near T_g in the substances of the second class, such as non-network materials, when compared with the degree of freedom in network materials whose atomic configurations are largely restricted by the network structures.

A remarkable aspect of the glass transition is that the transition temperature T_g depends on the preparation conditions such as the quench rate. The volume-versus-temperature curves in cooling experiments are presented in Fig. 4 for the organic glass polyvinylacetate ($CH_2CHOOCCH_3$). The two curves fitted to solid circles are respectively labeled by experimental time scales 0.02 h and 100 h. These times are those elapsed in cooling an initial liquid at a fixed temperature down to the glass transition region. The difference between the quench rates of the fast cooling and of the relatively slow cooling is a factor of 5000. The quench rate dependence does appear in the glass transition temperature T_g; the shift, however, is only 8 K and $\Delta T_g/T_g$

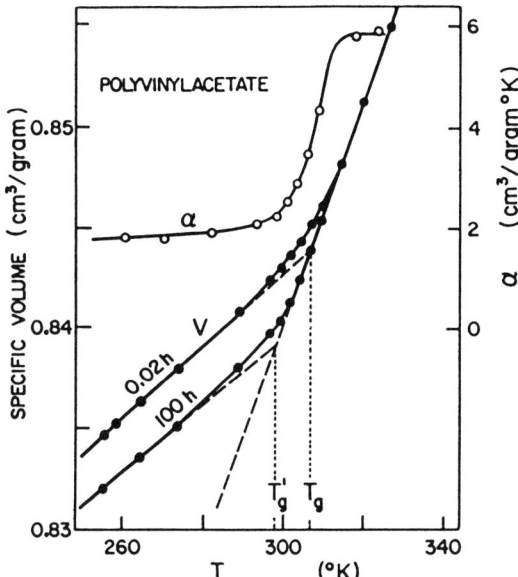

FIG. 4. Volume versus temperature on cooling for an organic material near the glass transition range, presented to show the quench rate dependence of T_g. The curve denoted by 0.02 h demonstrates the results of fast cooling, while the curve denoted by 100 h represents the result of relatively slow cooling. The coefficient $\alpha(T)$ of thermal expansion for fast cooling is also given. The break in $V(T)$ and the corresponding step in $\alpha(T)$ monitor the occurrence of the transition from liquid to glass—that is, the occurrence of the glass transition. Taken from A. J. Kovacs, J. M. Mutchinson, and J. J. Aklonis, in "The Structure of Noncrystalline Materials" (P. H. Gaskell, ed.), p. 153. Taylor and Francis, London, 1977.

is less than 3%, which is comparatively small, considering that the quench rates differ by three or four orders of magnitude.

The definite existence of the quench rate dependence, mild though it is, has a significant meaning for a better understanding of glassy states because the existence of the quench rate dependence indicates that the glassy states are metastable and that the glass transition has a kinetic aspect. In this context, it would be rather appropriate to say that the "glass transition" is simply the reflection of the appearance in the measured physical properties of the contributions from the metastable atomic structures. This viewpoint explains the aforementioned correlations between the thermodynamic properties such as C_p and the flow properties such as η. It also accounts for the fact that the observed physical quantities such as T_g undoubtedly depend on the preparation conditions and on the time scales of measurements.

Once this viewpoint is accepted, it is not difficult to imagine that the higher the degree of conformational freedom of a system, the wider is the variety of behavior in the glass transition region and in the glassy state. The degree of structural freedom in the atomic configurations is expected to be highest in non-network systems composed of spherical atoms. Examples of such systems are metals, metallic alloys, and rare gases. Since a wider variety of information would provide more help for a deep understanding of glassy structures, a promising first stage is to carry out detailed analyses of the glass transition and the glassy state of relatively simple systems consisting of atoms with spherical interactions.

In this connection, it is interesting to touch upon the characteristic features of the pair distribution functions, denoted by $g(r)$, of metallic glasses in comparison with the PDF of liquid metals. A typical shape of $g(r)$ for a liquid metal is given in Fig. 5, which has (1) a reasonably broad first peak, (2) a smooth second peak, and (3) a third peak with a considerably diminished intensity. On the other hand, metallic glasses and metal–metalloid glasses are specified by $g(r)$ with (1) a considerably narrow and sharp first peak, (2) a split second peak, and (3) a distinct third peak. These features manifest themselves in the reduced radial distribution function $G(r)$ for amorphous $Ni_{76}P_{24}$, as demonstrated by the solid curve in Fig. 6.[12,13]

In particular, the splitting of the second peak of $g(r)$ or $G(r)$ seems to be common to most metallic and metal–metalloid glasses. The split second peak has never been observed in the PDF of liquids, and accordingly it is highly probable that this splitting of the second peak is due to a denser packing of atoms in a glass. Quite a number of structural models[12] have been proposed and analyzed to clarify the origin of the split second peak. For instance,

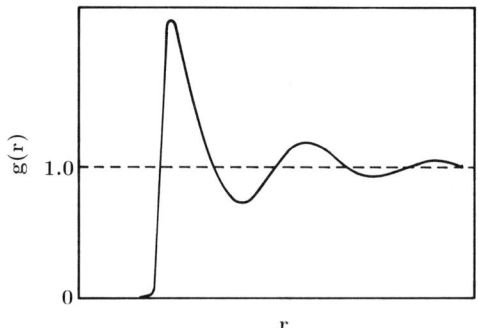

FIG. 5. Typical behaviour of the PDF $g(r)$ for liquids.

[11] A. J. Kovacs, J. M. Hutchinsen, and J. J. Aklonis, in "The Structure of Noncrystalline Materials" (P. H. Gaskell, ed.). Taylor and Francis, London 1977.
[12] G. S. Cargill III, J. Appl. Phys. 41, 12 (1970).
[13] G. S. Cargill, III, "Solid State Physics" (F. Seitz, D. Turnbull, and H. Ehrenreich, eds.), Vol. 30, p. 227. Academic Press, New York, 1975.

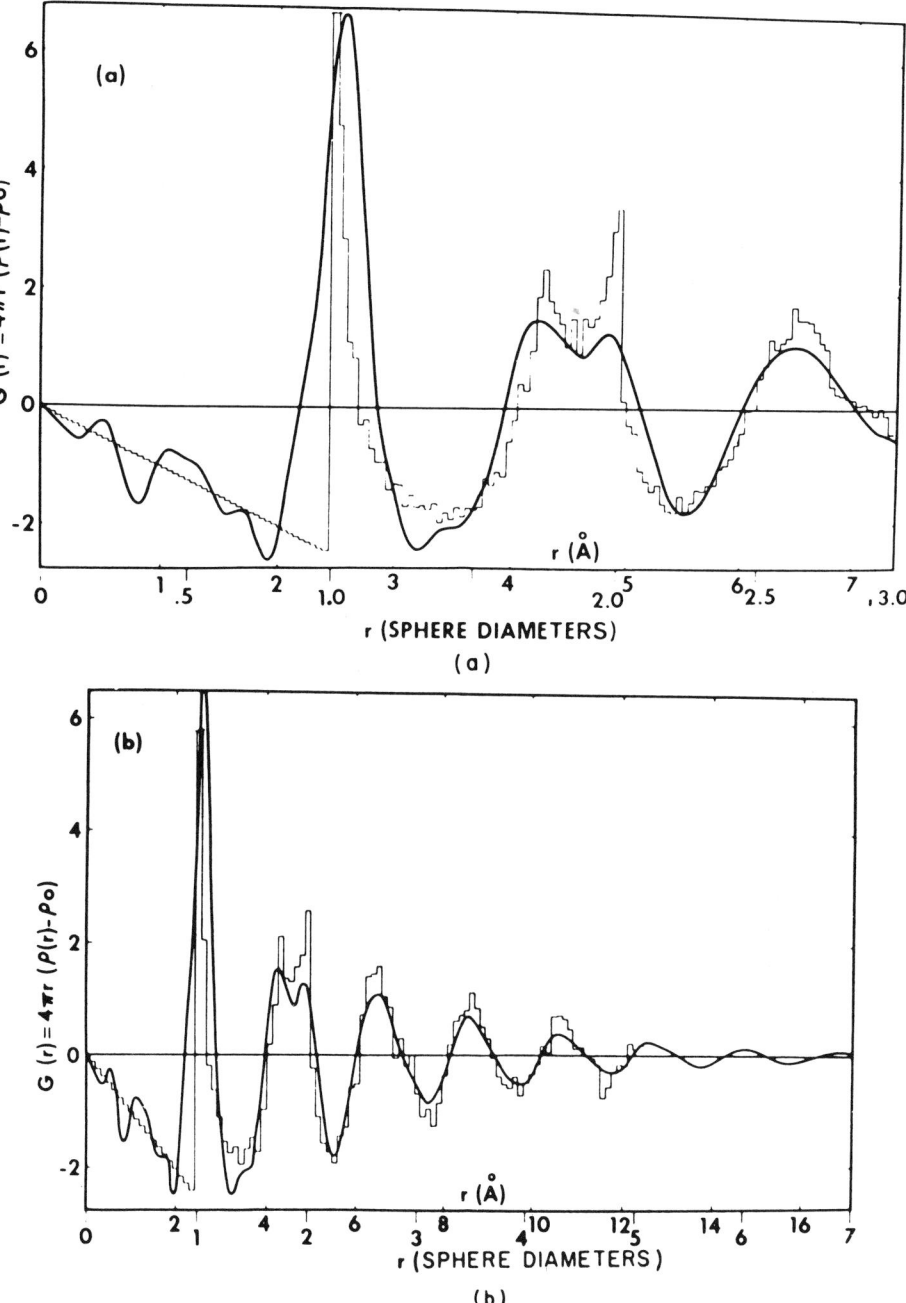

FIG. 6. Comparison of reduced radial distribution functions $G(r)$ for Finney's DRPHS structure[7] and for $Ni_{76}P_{24}$ glass after Cargill III[12,13], in which $\rho(r)$ and ρ_0 are, respectively, the number density of atoms at r and the average number density.

Finney's DRPHS distribution function $G(r)$ is illustrated by the histogram [7] in Fig. 6, which also exhibits a clear splitting of the second peak. Finney[7] and Bennett[14] have put forward the kinds of local atomic configurations that could account for the split second peak. These local configurations hardly appear in a sparse packing of atoms in a liquid, and therefore, are typical of a dense packing in a glass.

It has been pointed out by Cargill [12] that $G(r)$ of many metal–metalloid glasses are very similar to the $G(r)$ obtained by Finney[7] and others for the DRPHS models as is obvious from Fig. 6. This fact implies that some essential aspects of local atomic configurations in metal–metalloid glasses are simulated by the DRPHS model. A detailed discussion and analysis of the single-size spheres and binary DRPHS models are given in a review by Cargill.[13]

4. THEORETICAL APPROACHES

Quite a number of theoretical attempts have been made to account for the conditions for glass formation, the relations between glass formation and crystallization, the nature of the glass transition, the relaxation mechanism through the glass transition as well as in glasses, the microscopic structures in supercooled liquids as well as in glasses, and all related problems. In this section, some of these attempts are mentioned to sketch a rough picture of the present status of the field.

a. *Kinetic Theory of Crystal Nucleation and Glass Transition*

As is clear from Section 1, an amorphous or glassy state is less stable than a corresponding crystalline state, and consequently the former can be realized only when some appropriate conditions are satisfied in the process of preparation. The problem of clarifying these conditions for the formation of a glass has been intensively studied by Turnbull and his co-workers.[15–20] In particular, Turnbull has formulated the kinetics of crystal nucleation and concluded that the conditions for glass formation are determined by a set of factors such as the quench rate $q = -dT/dt$, the liquid volume v_l, and the density ρ_s of crystal seeds under each given set of material constants, such as the reduced crystal–liquid interfacial tension σ, the fraction f of acceptor

---

[14] C. H. Bennett, *J. Appl. Phys.* **43**, 2727 (1972).
[15] D. Turnbull and J. C. Fisher, *J. Chem. Phys.* **17**, 71 (1949).
[16] D. Turnbull, in "Solid State Physics—Advances in Research and Applications" (F. Seitz and D. Turnbull, eds.), Vol. 3, p. 225. Academic Press, New York, 1956.
[17] D. Turnbull, *J. Chem. Phys.* **18**(2), 198 (1950).
[18] D. Turnbull, *Trans. AIME* **221**, 422 (1961).
[19] D. Turnbull, *J. Phys. Chem.* **66**, 609 (1962).
[20] D. Turnbull, *Contemp. Phys.* **10**, 473 (1969).

sites in the crystal surface, and the reduced glass transition temperature T_g/T_m. It has been found that the glass-forming tendency is greater for larger $q = -dT/dt$, σ, and T_g/T_m and for smaller v_l, ρ_s, and f. All these results are consistent with simple physical predictions.

The work of Turnbull and co-workers has played an important role in attaining a better understanding of glassy structures and in stimulating the development of metallic–glass devices.

b. Thermodynamic Considerations

As observed in Fig. 2, the glass transition is generally characterized by a continuous change in volume V and by rather discontinuous changes in the coefficient α_T of thermal expansion and in C_p. Actually, besides V, some other extensive thermodynamic properties, such as entropy S and enthalpy H, are continuous at T_g, while the coefficient κ_T, isothermal compressibility, is discontinuous at T_g. In this way, the glass transition shows the appearance of a *second-order* phase transition in the Ehrenfest sense. If the glass transition is a genuine second-order phase transition, the continuity of entropy at the glass transition requires the relation [21,22]

$$\frac{dT_g}{dp} = TV\frac{\Delta\alpha_T}{\Delta C_p}, \tag{4.1}$$

and the continuity of volume at the glass transition requires the relation

$$\frac{dT_g}{dp} = \frac{\Delta\kappa_T}{\Delta c_T}, \tag{4.2}$$

in which a quantity expressed with Δ, such as $\Delta\alpha$, denotes the magnitude of the discontinuity of the corresponding physical quantity across the glass transition. Experimental data attest that the first relation is almost always satisfied, while the right-hand side of the second equation is considerably higher than the left-hand side. This proves that the glass transition is not the ordinary second-order phase transition.

It was shown later by Prigogine and Defay[23] that, without using the conditions of the second-order phase transition, the relation

$$\mathscr{R} \equiv \frac{\Delta\kappa_T \Delta C_p}{TV(\Delta\alpha_T)^2} = 1 \tag{4.3}$$

---

[21] P. Ehrenfest, *Commun. Kammerlingh Onnes Lab., Univ. London,* 756 (1933).
[22] H. B. Callen, "Thermodynamics." Wiley, New York, 1960.
[23] I. Prigogine and R. Defay, in "Chemical Thermodynamics." Longman Greens, New York, 1954.

can be obtained if the thermodynamic state of the glass at T_g is completely determined by a single-order parameter. The ratio $\mathscr{R}$, named the Prigogine–Defay ratio, is larger than unity,

$$\mathscr{R} > 1, \qquad (4.4)$$

if more than one ordering parameter is involved. Experimental data show that $\mathscr{R} > 1$, thus suggesting that the thermodynamic properties of a glass cannot be specified by a single ordering parameter. This result reinforces our understanding that a glassy state is thermally metastable and, accordingly, dependent on preparation parameters.

c. *Entropy Considerations*

It has been experimentally established, as demonstrated in Fig. 4, that the lower the quench rate, the lower the glass transition temperature T_g. From theoretical considerations,[24] the lowest limit T_0 of the glass transition temperature is defined by the relation

$$S_f = \int_{T_0}^{T_m} [C_p(\text{supercooled liquid}) - C_p(\text{crystal})] \, d \log T, \qquad (4.5)$$

where S_f is the entropy of fusion observed at the melting point T_m of the first-order phase transition. It seems that this theoretical limit has never been reached experimentally because both the nucleation rate and growth rate are expected to increase sharply as the temperature is lowered below T_m, and as a consequence the crystallization instead of the glass transition takes place rather easily.

In this connection, Gibbs and DiMarzio[25,26] have studied the entropy of linear polymer chains through the glass transition. Although their models and treatments are fairly limited, the application of their idea to more realistic models is possible.

Since the specific heat C_p of a glass is almost the same as that of a crystal at a given temperature $T < T_g$, a system transformed into a glassy state at $T_g < T_m$ has an entropy larger than that of a crystal when compared at the same temperature. This excess entropy is entirely due to the higher degree of configurational freedom in a glass than that in a crystal.

d. *Free-Volume Theory*

It has been experimentally shown that the concept of *free volume* makes sense in explaining molecular transport. The same concept has also played a

[24] W. Kauzmann, *Chem. Rev.* **43**, 219 (1948).
[25] J. H. Gibbs and E. A. DiMarzio, *J. Chem. Phys.* **28**(3), 373 (1958).
[26] E. A. DiMarzio, *J. Appl. Phys.* **45**(10), 4143 (1974).

central role in several theoretical attempts to investigate the nature of the glass transition. Actually, the free-volume theory is the most popular among theories of glass transition because the idea is very simple and intuitive and the mathematical treatment is relatively easy. While the free-volume theory is enjoying a considerable success in the study of liquids and glasses, it has some obscure aspects. In spite of the fact that the concept of free volume itself has quite a long history, no clear-cut and unanimous definition of free volume has ever been given. Besides, some arguments assert that, although free volume is useful as a guide for discussion, it has no physical reality. Therefore, it is extremely interesting and important to examine the scope of the free volume theory.

In 1913, Batschinski[27] suggested a linear relation between the fluidity $\phi \equiv \eta^{-1}$ of a simple liquid and the free volume v_f, where v_f is defined as the difference between the specific volume v and the van der Waals volume v_0. The specific volume v is identical to the atomic volume determined by $v \equiv V/N$, V and N being, respectively, the volume of the system and the number of atoms therein. The van der Waals volume, on the other hand, is equal to the constant volume that appears in the van der Waals potential, and corresponds to the situation in which the atoms or molecules in the liquid are packed so tightly that self-diffusion hardly occurs.

Concerning early theoretical attempts to elucidate the mechanism of the glass transition in terms of free volume, we should mention Frenkel,[28] Eyring,[29] and Fox and Flory[30] The concept of free volume was also used in the analysis of experimental data: Doolittle[31] related the fluidity ϕ to v_f by the expression

$$\phi \propto \exp\left[-\frac{bv_0}{v_f}\right], \qquad (4.6)$$

where b is a constant of order unity and v_0 is again the van der Waals volume. Later, Williams, Landel, and Ferry[32] developed a theory leading to the relation between v_f and temperature T,

$$v_f = v_g[0.025 + \Delta\alpha(T - T_g)], \qquad (4.7)$$

in which v_g and T_g are, respectively, the glass transition volume and temperature, and $\Delta\alpha$ is defined by $\Delta\alpha = \alpha(\text{liquid}) - \alpha(\text{glass})$, α being the coefficient of thermal expansion. Substitution of Eq. (4.7) into Eq. (4.6) yields

---

[27] A. J. Batschinski, *Z. Phys. Chem.* **84**, 643 (1913).
[28] J. Frenkel, in "Kinetic Theory of Liquids." Clarendon Press, Oxford, 1946.
[29] H. Eyring, *J. Chem. Phys.* **4**, 283 (1936).
[30] T. G. Fox and P. J. Flory, *J. Appl. Phys.* **21**, 581 (1950).
[31] A. K. Doolittle, *J. Appl. Phys.* **22**, 147 (1951).
[32] M. Williams, R. F. Landel, and J. D. Ferry, *J. Am. Chem. Soc.* **77**, 3701 (1955).

the Vogel–Fulcher relation described by Eq. (3.1).

From the viewpoint of molecular theory, Turnbull and Cohen[33,34] laid the foundation of Eqs. (4.7), (4.6), and (3.1). They also introduced the idea of a *cage structure* in relation to the concept of free volume, as follows. An atom in a supercooled liquid or in a glass vibrates within a cage determined by its surrounding atoms. From time to time, a hole is generated in its neighborhood as a result of the cooperative movements of the surrounding atoms, where the hole could be so large that it can contain one atom. If the original atom under consideration gains an activation energy high enough to exceed the barrier between the cage, to which the atom belongs, and the hole, which has been created in the neighbourhood of the atom, then the atom can escape from the cage to the hole where the atom will be caught in a new cage. Repeated occurrences of this process allows the atom to diffuse from its original position.

Cohen and Grest[35,36] assumed that it is possible to associate a free volume with each atom. An additional assumption is that atoms can be classified into two major categories: liquidlike and solidlike. The criterion used for the classification is the probability that the atom escapes from the original cage, where the probability is defined by the free volume of the atom. When liquidlike and solidlike atoms are respectively regarded as percolating and nonpercolating atoms, the problem can be discussed in the framework of percolation theory. The work of Cohen and Grest has extended the vision of the free-volume theory.

e. Other Attempts

Other interesting approaches to the elucidation of glass transition include theories based upon local fluctuations [37–40] and analyses in terms of mode-coupling theory.[41–48]

---

[33] D. Turnbull and M. H. Cohen, *J. Chem. Phys.* **34**(1), 120 (1961).
[34] D. Turnbull and M. H. Cohen, *J. Chem. Phys.* **52**(6), 3038 (1970).
[35] M. H. Cohen and G. S. Grest, *Phys. Rev. B* **20**(3), 1077 (1979).
[36] M. H. Cohen and G. S. Grest, *J. Non-Cryst. Solids* **61, 62**, 749 (1984).
[37] A. R. Eastwood, *J. Chem. Soc., Faraday Trans.* **2**, 1411 (1981).
[38] T. Egami and S. Aur, *J. Non-Cryst. Solids* **89**, 60 (1987).
[39] T. Egami, in "Glassy Metals I" (H.-J. Guntherodt and H. Beck, eds.), p. 25. Springer-Verlag, New York, 1981.
[40] T. Egami, K. Maeda, and V. Vitek, *Phil. Mag. A* **41**, 883 (1980).
[41] E. Leutheusser, *Z. Phys. B—Condensed Matter* **55**, 235 (1984).
[42] E. Leutheusser, *Phys. Rev. A* **29**(5), (1984).
[43] E. Leutheusser, *J. Phys. C: Solid State Phys.* **15**, 2801 (1982).
[44] W. Götze and L. Sjögren, *Z. Phys. B—Condensed Matter* **65**, 415 (1987).
[45] T. Munakata, *Prog. Theor. Phys.* **68**, 1900 (1982).
[46] U. Bengtzelius and L. Sjogren, *J. Chem. Phys.* **84**(3), 1744 (1986).
[47] S. P. Das and G. F. Mazenko, *Phys. Rev. A* **34**, 2265 (1986).
[48] U. Bengtzelius, W. Gotze, and A. Sjolander, *J. Phys. C: Solid State Phys.* **17**, 5915 (1984).

IV. Computer Simulations of Glass Transition

5. Merits and Demerits

As pointed out in the brief survey of the preceding section, no definite theory has ever succeeded in accounting for all phenomena related to the glass transition. In this situation, the most promising approach would probably be *computer experiments* or *computer simulations*. In fact, computer-assisted approaches are quite new to all natural sciences, which have long been supported by two major classes of investigations, experimental and theoretical. It goes without saying that this third possibility of exploiting the power of modern computers has appeared as a consequence of remarkable progress in computer technology.

The advantages of computer-assisted science go beyond the ability to perform calculations very fast. Not only quantitative but also essentially qualitative changes are expected to take place, which might open a new era in the history of science. This is especially true in a field where no systematic methods are analytically established, and research on glass transition is such a field.

The scope of computer simulations cannot be conveyed simply by words because it contains much wider possibilities than our imagination could ever reach. However, it is useful to name a few advantages, specifically in connection with the study of the glass transition:

1. Complete information is available concerning the positions (and velocities in the case of molecular dynamics methods) of all atoms in the system, which can never be hoped for from experiments.

2. Extreme limits can be realized in computers, such as exceedingly high pressures and extraordinarily rapid quench rates, which could never be achieved in laboratories.

3. The relatiohships between cause and effect are clarified because the input data can be controlled at will and the resulting output data can be analyzed in detail. For instance, it is possible to examine the dependencies of the glass properties on quench rates or on interatomic potentials.

No method is without limitations or disadvantages, and the application of computer simulations to the glass transition is no exception. One problem is the size effect in the sense that the physical system to be simulated is necessarily limited in size. The size problem naturally leads to the necessity of choosing appropriate boundary conditions under which the computer simulations are carried out.

6. Molecular Dynamics Methods

For the purpose of studying the microscopic situation across the glass transition and the relaxation of atomic configurations in glasses, the most appropriate simulation method is molecular dynamics (MD), which concerns itself with the time evolution of the classical motion of atoms (molecules) in a condensed state. In practice, N coupled Newtonian equations of motion, N being the number of atoms in the system, are solved as time proceeds under a given set of thermodynamic variables, such as volume, pressure, and temperature. Moreover, it is possible to design the whole scheme so that it can cope with instantaneous changes of these thermodynamic variables, and it is this ability that makes the MD methods most appropriate for the study of the glass transition and the relaxation processes in glasses. A complete knowledge of the atomic positions and velocities provided from MD methods permits the evaluation of almost all physical properties of the system.

The first systematic study of disordered systems, such as liquids and glasses, by means of MD techniques is due to Rahman and his co-workers.[49] Some of the early MD applications to the investigations of the glass transition were reviewed by Angell *et al.*[50] In the MD simulations of the glass transition reported so far, the interactions among atoms are approximated by the sum of pairwise and nondirectional potentials, some of which are described in the rest of this section, but the references are not comprehensive. In what follows, a pairwise and nondirectional potential between atoms at $\mathbf{r}_i$ and $\mathbf{r}_j$ is written $\phi(r_{ij}) = |\mathbf{r}_i - \mathbf{r}_j|$.

a. *Hard Core Potential*

A hard core system is characterized by pair potentials of the form

$$\phi_{\text{HC}}(r_{ij}) = \begin{cases} \infty; & r < \sigma, \\ 0, & r \geq \sigma. \end{cases} \tag{6.1}$$

MD simulations of the glass transition for hard core systems have been performed by Alder and Wainwright,[51-55] Woodcock,[56,57] and Gordon *et al.*[58]

The investigation of hard core systems started with a series of works by Alder and Wainwright[51-55] They asserted that even hard core systems, in which attractive interactions are absent, can crystallize when the temperature

---

[49] A. Rahman, *Phys. Rev.* **136**(2A), 405 (1964).
[50] C. A. Angell, J. H. R. Clarke, and L. V. Woodcock, *Adv. Chem. Phys.* **48**, 397 (1981).
[51] B. J. Alder and T. E. Wainwright, *J. Chem. Phys.* **27**, 1208 (1957).
[52] B. J. Alder and T. E. Wainwright, *J. Chem. Phys.* **31**, 459 (1959).
[53] B. J. Alder and T. E. Wainwright, *J. Chem. Phys.* **33**, 1439 (1960).
[54] B. J. Alder and T. E. Wainwright, *Phys. Rev. Lett.* **18**, 988 (1967).
[55] B. J. Alder and T. E. Wainwright, *Phys. Rev. A* **1**, 1 (1970).
[56] L. V. Woodcock, *J. Chem. Soc. Faraday Trans II* **72**, 1677 (1976).
[57] L. V. Woodcock, *J. Chem. Soc. Faraday Trans II* **74**, 11 (1978).

is reduced or the pressure is applied. The transition from liquid to crystal in a hard core system is named the *Alder transition* after the inventor.

b. *Soft Core Potential*

A soft core system is represented by the pair potentials

$$\phi_{SC}(r_{ij}) = \varepsilon(\sigma/r_{ij})^n. \tag{6.2}$$

This potential has an advantage in that the equations of state for a soft core system are described by a single reduced variable defined by[59,60]

$$\rho^* = \frac{N\sigma^3}{V}\left(\frac{\varepsilon}{kT}\right)^{3/n}. \tag{6.3}$$

MD methods were applied to soft core systems for the study of the glass transition by Clarke et al.,[61] Cape et al.,[62,63] Hiwatari et al.,[60,64] and Wolynes.[65] Nucleation in supercooled liquids of soft core systems has also been studied by MD methods.[63,66-69]

c. *Lennard–Jones Potential*

A Lennard–Jones system is defined by the pair potentials

$$\phi_{LJ}^{mn}(r_{ij}) = 4\varepsilon[(\sigma/r_{ij})^m - (\sigma/r_{ij})^n], \tag{6.4}$$

and the glass transition of Lennard-Jones systems has been studied by MD methods by Verlet,[70] Rahman et al.[71] Damgaard Kristensen,[72] Clarke,[73]

---

[58] J. M. Gordon, J. H. Gibbs, and P. D. Fleming, *J. Chem. Phys.* **65**, 2771 (1976).
[59] W. G. Hoover and M. Ross, *Contemp. Phys.* **12**, 239 (1971).
[60] Y. Hiwatari, *J. Phys. C: Solid State Phys.* **13**, 5899 (1980).
[61] J. H. R. Clarke, J. F. Maguire, and L. V. Woodcock, *Discuss. Faraday Soc.* No. **69** (Exeter), 273 (1980).
[62] J. N. Cape and L. V. Woodcock, *J. Chem. Phys.* **72**, 976 (1980).
[63] J. N. Cape and L. V. Woodcock, *J. Chem. Phys.* **75**, 2336 (1981).
[64] Y. Hiwatari, H. Matsuda, T. Ogawa, N. Ogita, and A. Ueda, *Prog. Theor. Phys.* **52**, 1105 (1974).
[65] P. G. Wolynes, *J. Non-Cryst. Solids* **75**, 443 (1985).
[66] M. J. Mandell, J. P. McTague, and A. Rahman, *J. Chem. Phys.* **64**, 3699 (1976).
[67] M. J. Mandell, J. P. McTague, and A. Rahman, *J. Chem. Phys.* **66**, 3070 (1977).
[68] M. Tanemura, Y. Hiwatari, H. Matsuda, T. Ogawa, N. Ogita, and A. Ueda, *Prog. Theor. Phys.* **58**, 1079 (1977).
[69] R. D. Mountain and A. C. Brown, *J. Chem. Phys.* **80**, 2730 (1984).
[70] L. Verlet, *Phys. Rev.* **159**(1), 98 (1967).
[71] A. Rahman, M. J. Mandell, and J. P. McTague, *J. Chem. Phys.* **64**(4), 1564 (1976).
[72] W. Damgaard Kristensen, *J. Non-Cryst. Solids* **21**, 303 (1976).
[73] J. H. R. Clarke, *J. Chem. Soc. Faraday Trans. II* **75**, 1371 (1979).

Hansen and Verlet,[74] Fox and Andersen,[75] Yonezawa et al.,[76-80] and Steinhardt et al.[81] Nucleation and crystallization in supercooled liquids of Lennard–Jones systems are also studied by MD techniques.[66,67,69,82-86]

d. *Gaussian Core Potential*

The Gaussian core potentials of the form

$$\phi_{GC}(r_{ij}) = \varepsilon \exp[-(r_{ij}/\sigma)^2] \quad (6.5)$$

were introduced and applied to the MD study by Stillinger and Weber.[87,88]

e. *Potential Proposed for Simple Ionic and Covalent Systems*

The model potentials for simple ionic and covalent systems are introduced in the form

$$\phi(r_{ij}) = \frac{z_i z_j e^2}{r_{ij}} + \left(1 + \frac{z_i}{n_i} + \frac{z_j}{n_j}\right) b \exp\left[-\frac{\sigma_i + \sigma_j + r_{ij}}{\rho}\right] \quad (6.6)$$

where z is the electronic charge, n is the number of outer shell electrons, σ is a distance parameter characteristic of the ionic radius, and b and ρ are constants. When b is chosen to be unity, the potentials are simple Coulomb potentials. These potentials were used for the MD study of supercooled liquids, the glass transition, and glasses by Woodcock et al.,[89] Angell et al.,[90] and Mitra et al.[91,92]

Stillinger and Weber[87,88] have proposed the potentials for Si in terms of seven parameters in the form

$$\phi(r_{ij}) \equiv \varepsilon u(r_{ij}/\sigma), \quad (6.7)$$

$$u_{SW}(x) = \begin{cases} A(Bx^{-p} - x^{-q})\exp[(x-a)^{-1}], & x < a \\ 0, & x \geqslant a, \end{cases} \quad (6.8)$$

[74]J.-P. Hansen and L. Verlet, *Phys. Rev.* **184**(1), 151 (1969).
[75]J. R. Fox and H. C. Andersen, *Ann. N.Y. Acad. Sci.* **371**, 123 (1981).
[76]M. Kimura and F. Yonezawa, in "Topological Disorder in Condensed Matter" (F. Yonezawa and T. Ninomiya, eds.). Springer, 1983. Computer Glass Transition.
[77]S. Nosé and F. Yonezawa, *Solid State Commun.* **56**, 1005 (1985).
[78]F. Yonezawa, S. Nosé, and S. Sakamoto, *J. Non-Cryst. Solids* **95, 96**, 83 (1987).
[79]F. Yonezawa, S. Nosé, and S. Sakamoto, *J. Non-Cryst. Solids* **95, 96**, 373 (1987).
[80]F. Yonezawa, S. Nosé, and S. Sakamoto, *Neue Folge* **156**, 77 (1988).
[81]P. Steinhardt, D. R. Nelson, and M. Ronchetti, *Phys. Rev.* **28**, 784 (1983).
[82]M. J. Mandell and J. P. McTague, *J. Chem. Phys.* **66**(7), 3070 (1977).
[83]C. S. Hsu and A. Rahman, *J. Chem. Phys.* **71**, 4974 (1979).
[84]J. D. Honeycull and H. C. Andersen, *Chem. Phys. Lett.* **108**, 533 (1984).
[85]S. Nosé and F. Yonezawa, *Solid State Commun.* **56**, 1009 (1985).
[86]S. Nosé, S. Sakamoto, and F. Yonezawa, *Neue Folge* **156**, 91 (1988).
[87]F. H. Stillinger and T. A. Weber, *J. Chem. Phys.* **70**(11), 4879 (1979).
[88]F. H. Stillinger and T. A. Weber, *J. Chem. Phys.* **68**, 3837 (1978).

where q could be either positive or negative, but all of the other six parameters must be positive. Stillinger and Weber[88,93] have used these potentials in the MD study.

f. *Potential Proposed for Alkali Metals*

The density-dependent effective pair potentials for alkali metals, especially for rubidium, have been introduced by Price et al.[94,95] and used in MD studies of the glass transition[96,97] and the nucleation.[68,83,98–101]

Comparisons of the effects from different pair potentials have been examined by Rahman and his coworkers,[98,99] who have shown that different pair potentials lead to different crystal structures.

7. SIMULATIONS UNDER CONSTANT PRESSURE

Most of the MD simulations mentioned in the preceding section are carried out under the condition of constant volume, because this is easier to handle than the condition of variable volume. In order for the outcomes of the simulations to stand comparisons with experimental results, the simulations must be carried out under the same conditions as realized in laboratories. For instance, simulations under constant pressure are desired. Some isobaric simulations of the glass transition have been performed by Monte Carlo methods,[102,103] and some isoenthalpic–isobaric simulations by MD methods have also been presented.[104,105]

Andersen[106] has put forward a technique of the constant-pressure MD, in which the volume of a simulation cell is allowed to vary. This technique was extended by Parrinello and Rahman[99] to the case in which the shape of a simulation cell is allowed to vary.

[89] L. V. Woodcock, C. A. Angell, and P. Cheeseman, *J. Chem. Phys.* **65**, 1565 (1976).
[90] C. A. Angell, P. A. Cheeseman, J. H. R. Clarke, and L. V. Woodcock, in "The Structure of Non-Crystalline Materials" (P. H. Gaskel, ed.), p. 191. Taylor and Francis, London, 1977.
[91] S. K. Mitra, M. Amini, D. Fincham, and R. W. Hockney, *Phil. Mag. B* **43**(2), 365 (1981).
[92] S. K. Mitra, *Phil. Mag. B* **45**(5), 529 (1982).
[93] S. Sakamoto, Ph.D. Thesis, Keio University, 1990.
[94] D. L. Price, *Phys. Rev. A* **4**, 358 (1971).
[95] D. L. Price, K. S. Singwi, and M. P. Tosé, *Phys. Rev. B* **2**, 2983 (1970).
[96] D. H. Li, R. A. Moore, and S. Wang, *J. Chem. Phys.* **88**(4), 2700 (1988).
[97] D. H. Li, R. A. Moore, and S. Wang, *J. Chem. Phys.* **89**(7), 4309 (1988).
[98] C. S. Hsu and A. Rahman, *J. Chem. Phys.* **70**, 5234 (1979).
[99] M. Parrinello and A. Rahman, *Phys. Rev. Lett.* **45**, 1196 (1980).
[100] R. D. Mountain, *Phys. Rev. A* **26**, 2859 (1982).
[101] R. D. Mountain and P. K. Basu, *J. Chem. Phys.* **78**, 7318 (1983).
[102] H. R. Wendt and F. F. Abraham, *Phys. Rev. Lett* **41**(18), 1244 (1978).
[103] F. F. Abraham, *J. Chem. Phys.* **72**(1), 359 (1980).
[104] J. M. Haile and H. W. Graben, *J. Chem. Phys.* **73**(5), 2412 (1980).
[105] J. M. Haile and H. W. Graben, *Mol. Phys.* **40**(6), 1443 (1980).
[106] H. C. Andersen, *J. Chem. Phys.* **72**(4), 2384 (1980).

V. Simulations of Glass Transition at Constant Pressure

8. QUENCHING OF A LENNARD–JONES SYSTEM

a. *Model*

In order that the outcomes from a computer simulation can be compared seriously to the experimental results on the glass transition, the simulation must be performed under a constant-pressure condition, preferably by the MD method. The reason is that almost all experiments concerning the glass transition have so far been carried out under atmospheric pressure.

One of the distinguished pioneer studies along this line is the constant-pressure MD simulations of the glass transition due to Damgaard Kristensen.[72] Since his work provides the fundamental ideas of essential importance, we frequently refer to it in this section.

His motivation was to establish whether an amorphous phase of a one-component system can be achieved at all by quenching a liquid of spherical particles, to clarify, from the theoretical point of view, the quench rate dependence of the glass transition temperature T_g that had been observed experimentally and to test the stability of the amorphous structures thus obtained.

The examination of these problems by computer simulations is significant because it is expected that an extremely high quench rate is necessary to bring about the glass transition in a system composed of uniform spherical particles. Since quench rates as high as this cannot be attained in laboratories, investigations by experiment are not possible.

In order to study a system of spherical particles, the model system consisting of Lennard–Jones atoms is often adopted. The paper of Damgaard Kristensen (hereafter called the DK paper) also treats the model system composed of 336 atoms interacting via the 12-6 Lennard-Jones potentials. These atoms are placed in a simulation cell with periodic boundary conditions. The calculations are conveniently performed in dimensionless variables by scaling the energy in ε, length in σ, mass in M, temperature in ε/k_B, pressure in ε/σ^3, and time in $(M\sigma^2/\varepsilon)^{1/2}$, where ε and σ are the potential parameters of the Lennard–Jones potential, M is the atomic mass, and k_B is the Boltzmann constant. It is widely accepted that the Lennard–Jones potential applies very well to the rare gases. The magnitudes of the units relevant to the rare gases are given in Table III.

b. *Method*

The MD technique is used to simulate the atomic motion at finite temperature. The calculated time series of atomic configurations may be regarded as a microcanonical ensemble of the thermodynamic properties of

TABLE III. UNITS FOR ENERGY, LENGTH, MASS, AND TIME OBTAINED BY FITTING THE PARAMETERS OF THE LENNARD-JONES POTENTIAL TO THE EXPERIMENTAL VALUES OF THE SUBLIMATION ENERGY AND LATTICE PARAMETER AT ZERO TEMPERATURE FOR THE NOBLE GASES ARGON, KRYPTON, AND XENON

	ε (K)	σ (Å)	M (10^{-22} g)	$t_0 = (\sigma^2 M/\varepsilon)^{1/2}$ (ps)
Ar	118	3.84	0.66	2.44
Kr	170	4.08	1.39	3.14
Xe	244	4.43	2.18	3.56

W. Damgaard Kristensen, J. Non-Cryst. Solids 21, 303 (1976).

the system, and thermodynamic averages can be taken accordingly.

The classical equations of motion are approximated by finite difference equations with time step Δt of the order of 10^{-14} s. The temperature is defined classically in terms of the velocities $\{v_i\}$ by

$$T(t) = \sum_i \frac{[v_i(t)]^2}{3N}, \qquad (8.1)$$

where the summation runs over the N atoms in the simulation cell. The temperature can be changed at will by introducing a scaling of the atomic velocities. The pressure is calculated via the Virial theorem to be

$$P(t) = \rho T + \frac{4\rho}{3N} \sum_i \sum_j \left[\frac{m}{(r_{ij})^m} - \frac{n}{(r_{ij})^n} \right], \qquad (8.2)$$

where ρ is the number density, defined by $\rho = N/V$, V being the volume of the simulation cell. It is clear from this equation that the pressure can be controlled by expanding or contracting the simulation cell. In constant-pressure simulations, the pressure is kept constant by changing in Eq. (8.2) the number density ρ (or the volume V).

In the DK paper, the quenching of LJ liquids is realized through three different quench rates corresponding to the range 10^{12}–10^{13} K/s for argon.

c. *Results and Discussion*

The equation-of-state data ρ *versus* T are shown in Fig. 7. The figure definitely demonstrates that, when the Lennard-Jones liquid is quenched rapidly, the system is densified continuously without any discontinuous transition. Moreover, the system below $T = 0.5$ has a density a few percent lower than that of an fcc crystal at the same temperature. The structures in

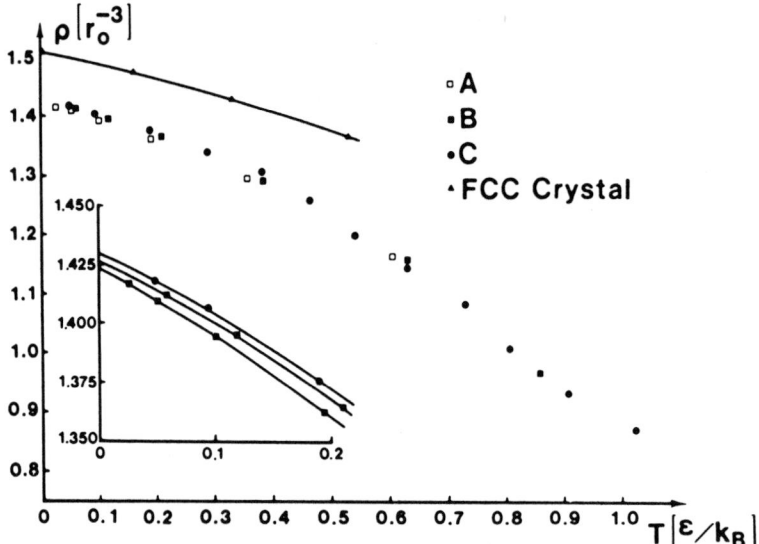

FIG. 7. The average number density ρ (or the inverse of the specific volume, $1/v_s$) versus temperature T (due to Damgaard Kristensen[72]). Notations A, B, and C represent the systems prepared with three different quench rates, while the filled triangles denote the results for fcc crystals. The excess volume for the disordered systems relative to the perfect crystal is approximately 6%.

this temperature range are ascertained to be disordered by analyzing closely the pair distribution function $g(r)$. All these facts suggest that the amorphous solid appears at lower temperatures.

The smoothing out of phase transitions is a well-known effect of the finite size, and it is readily seen that this is the case for the results in Fig. 7. As a consequence, the glass transition temperature T_g cannot be identified.

Notations A, B, and C in Fig. 7 denote the states achieved in three different quench rates. A slight dependence on the quench rate is observed in the ρ-versus-T curves at low temperatures, but the quench rate dependence is ascertained not to be serious in $g(r)$ as well as in the diffusion constant.

Figure 8 illustrates the comparison between the quench simulation of the Lennard–Jones liquids (solid curve) and the result from the dense random packing of hard spheres (DRPHS).[7] It has been pointed out in the DK paper that, in spite of the large difference in packing density, the DRPHS model and the Lennard-Jones model show striking similarities with respect to the derived pair distribution functions.

It is also argued in the DK paper that the splitting of the second peak in $g(r)$ as shown in Fig. 8, which is considered as characteristic of amorphous structures, can be explained by the dislocation model, the DRPHS model, and even by the microcrystal model. From this argument and from other

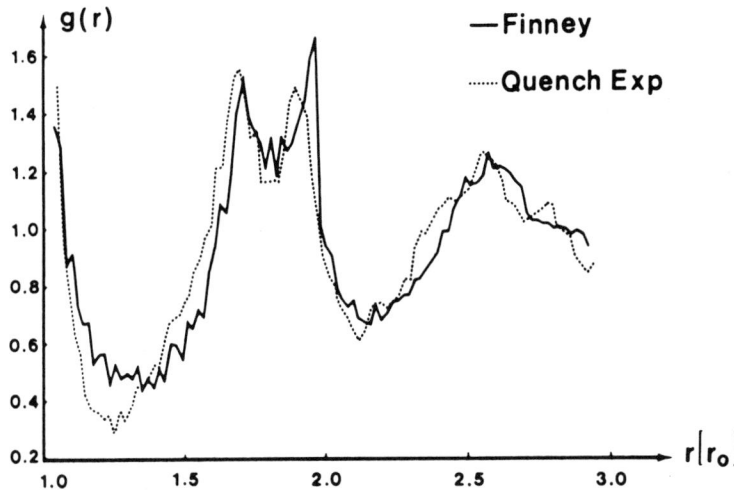

FIG. 8. Pair distribution functions for the dense random packing of hard spheres due to Finney[7] and for the MD simulation at $T = 0.094(\varepsilon/\kappa_B)$ due to Damgaard Kristensen.[72] The hard sphere result has been rescaled to match the equivalent hard sphere diameter of the fcc crystal phase at zero temperature and pressure.

considerations, it is generally concluded that the information given from the pair distribution function $g(r)$ alone is not sufficient to understand the detailed configurations of atoms in disordered systems.

9. PHYSICAL PROPERTIES THROUGH GLASS TRANSITION

A desire to derive more profound and comprehensive information about the glass transition from the simulation study is the chief motive of the work of Kimura and Yonezawa,[76] who followed the idea of Damgaard Kristensen,[72] as described in the preceding section. Morrel H. Cohen[107] has emphasized the value and importance of simulation studies of glasses and the glass transition under constant-pressure conditions as in the Kimura–Yonezawa[76] study.

In this chapter, the outline of the work[76] is explained with a view to establishing the fundamental concepts about the microscopic mechanism of the glass transition. We emphasize the investigation of the possibilities and limitations of simulation studies on amorphous structures.

The work [76] of Kimura and Yonezawa uses the MD technique in which the Newtonian equations of motion are integrated on the basis of the Verlet algorithm[70] with a time step Δt for the difference equations, Δt being chosen

[107]M. H. Cohen, in "Topological Disorder in Condensed Matter" (F. Yonezawa and T. Ninomiya, eds.). Springer, 1983. The author is honored by professor Cohen's kind comments on the research of Kimura and herself.[76]

as $\Delta t = 0.01$, which in the case of argon corresponds roughly to 2×10^{-14} s. The initial system of the simulation is an fcc crystal consisting of 864 Lennard–Jones atoms, whose velocities are chosen so that they satisfy the Maxwellian velocity distribution at $T = 0.72$, this temperature being near the triple point. Note that the variables such as temperature, length, and time are expressed in the reduced units, unless otherwise stated, as defined in the preceding section.

In order to construct a liquid in an equilibrium state, the initial system is heated to $T = 1.0$, which is well above the melting temperature $T_m = 0.7$ at $P = 1$ bar. The heating simulation is carried out by allowing the expansion of the volume while the pressure is fixed at $P = 0.0024$, corresponding to the atmospheric pressure. The system achieves thermal equilibrium after 3000 time steps. Thermodynamic averages are taken by using the data over 1000 time steps after thermal equilibrium has been reached.

This equilibrium liquid is quenched in a stepwise manner. The MD technique is organized such that the pressure is always constant ($P = 1$ bar) at each step of the temperature reduction. The decrement of temperature is $\Delta T = -0.1$ during 500 time steps, which corresponds to 1.4×10^{12} K/s for argon. This quench rate is greater by a few orders of magnitude than the highest possible quench rates realized in laboratories. The limited capability in producing high quench rates in laboratories has hitherto prohibited the glass formation of simple monoatomic systems in which atoms interact through isotropic potentials. This is where the techniques of computer simulations are decisive, because in computers the quench rates can be raised, in principle, as high as desired.

The physical properties evaluated from the data of the simulations are listed in Table IV. The quantities in the first three categories are the kinds of physical properties that can be experimentally measured in laboratories, and thus it is possible to compare the results obtained from the computer simulations with the experimental results. This kind of comparison gives a clue to a better understanding of the nature of the glass transition and of amorphous structures.

In this section, the physical properties in the first three categories are discussed, while the quantities in the last category will be studied in Section 11.

a. *Structural Properties*

The representative quantities about the atomic structure include the (PDF) $g(r)$ and the structure factor $S(q)$. Experimentally, $S(q)$ is directly measured, while $g(r)$ is calculated by taking the Fourier transformation of $S(q)$. In computer simulations, $g(r)$ is obtained from the data about the atomic configurations, and thereafter $S(q)$ is enumerated through the Fourier

TABLE IV. PHYSICAL PROPERTIES DERIVED FROM THE RESULTS OF COMPUTER SIMULATIONS

I Structural Properties

(1) $g(r)$ Pair Distribution Function
(2) $S(q)$ Structure Factor
(3) R Wendt–Abraham Parameter

II Thermodynamic Properties

(1) V vs T Equation-of-State Data (isobars in a $V\text{-}T$ plane)
(2) α_P Isobaric Thermal Expansion Coefficient
(3) H Enthalpy
(4) C_P Specific Heat at Constant Pressure
(5) C_V Specific Heat at Constant Volume
(6) κ_T Isothermal Compressibility

III Dynamical and Vibrational Properties

(1) $M(t)$ Mean Square Displacement
(2) $\psi(t)$ Velocity Autocorrelation Function (VAF)
(3) D Diffusion Constant (both from VAF and MSD)
(4) η Shear Viscosity (from stress autocorrelation function)
(5) $f(\omega)$ Power Spectra

IV Microscopic Structure Parameters

a. Quantities to Reveal the Hidden Structures
(1) $g(\bar{r})$ Pair Distribution Function Defined by the Time Average of the Position Vectors
b. Quantities to Describe the Fluctuation in the Microscopic Structure
(2) $\langle v_i \rangle_i$ Average of the Volume of the Voronoi polyhedra
(3) δv Fluctuation Parameter Defined by the rms of the Volumes $\{v_i\}$ of the Voronoi Polyhedra
(4) $p(v_i)$ Distribution of the Volumes of the Voronoi Polyhedra
c. Quantities to Describe the Distortion in the Microscopic Structure
(5) ξ_i Shape Parameter
(6) $\langle \xi_i \rangle_i$ Distortion Parameter Defined by the Average of the Shape Parameter
(7) $p(\xi_i)$ Distribution of the Shape Parameters
d. Quantities to Describe the Local Symmetry of the Atomic Configuration
(8) $p(\text{Voronoi})$ Distribution of the Selected Types of Voronoi polyhedra
(9) n_α Voronoi Face Parameters
(10) $\hat{w}_6(i)$ Local Symmetry Parameter
(11) $\langle \hat{w}_6(i) \rangle_i$ Average of the Local symmetry Parameters $\hat{w}_6(i)$ over all Atoms $\{i\}$ in the System
(12) $p(\hat{w}_6(i))$ Distribution of the Local Symmetry Parameters

transformation of $g(r)$. The PDF $g(r)$ thus derived from the simulation data is presented in Fig. 9.

From Fig. 9, systematic changes in $g(r)$ are clearly observed to accompany the decrease of temperature. The characteristic features of $g(r)$ are

1. The lower the temperature, the higher and sharper the first peak.
2. The lower the temperature, the more distinct is the splitting of the second peak.
3. The lower the temperature, the lesser is the falling off of the further (third, fourth, etc.) peaks.
4. For all temperatures, there appear no peaks characteristic of the crystalline structures, thus indicating the absence of long-range order and no sign of crystallization even at low temperatures.

Actually, these features are experimentally shown to be typical for the PDF of more complex laboratory glasses. This fact would serve as circumstantial evidence that our model system has transformed into a glass on quenching.

The same argument applies to the structure factors $S(q)$ derived in the way stated earlier. The results are illustrated in Fig. 10. Here again, the aspects

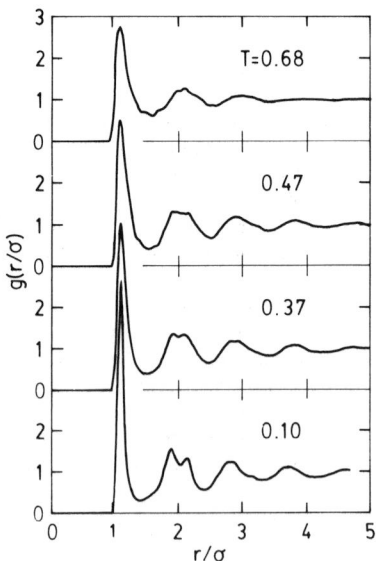

FIG. 9. Pair distribution functions $g(r)$ for LJ systems.[76] The real space distance r is normalized by the distance σ at which the 12-6 LJ potential becomes zero. The numbers in the figure denote temperatures.

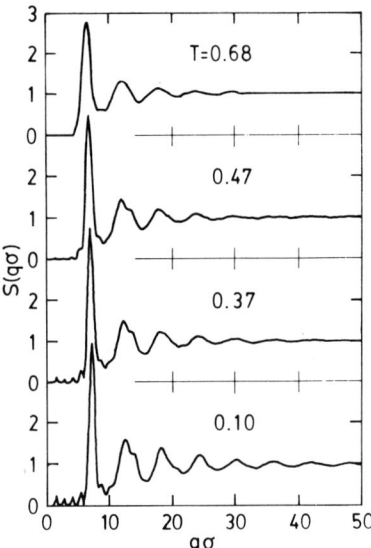

FIG. 10. Structure factors $S(q)$ derived from the corresponding PDF in the previous figure through a Fourier transformation.[76] The wave number q is normalized by $1/\sigma$.

characteristic of the structure factors of laboratory liquids and glasses are reproduced. The three characteristic features listed in connection with the PDF are also true for the structure factors, though the origins of the second-peak splitting are completely different for $g(r)$ and for $S(q)$.

In order to estimate the glass transition temperature T_g from the temperature dependence of $g(r)$ or $S(q)$, we need some key hidden in $g(r)$ or in $S(q)$. One example of such a key is the empirical criterion for identifying the glass transition proposed by Wendt and Abraham.[102] They introduced the parameter R defined by $R = g_{min}/g_{max}$, where g_{min} and g_{max} are, respectively, the value of $g(r)$ at its first minimum and at its first maximum. From the Monte Carlo calculation of the Lennard–Jones system, they have found that R changes linearly with temperature T both in the liquid region and in the glassy region, although the slope in the liquid region is larger than that in the glassy region. They have also found that the linear extrapolations from the liquid and glassy branches intersect at a value of $R = 0.14$. This empirical criterion has later been studied theoretically on the basis of a simple model.[108]

The Wendt–Abraham parameter R derived from the PDF of the simula-

---

[108] S. Basak, R. Clarke, and S. R. Nagel, *Phys. Rev. B* **20**(8), 3388 (1979).

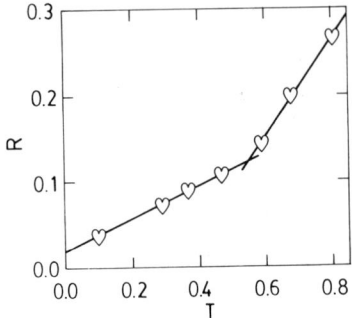

FIG. 11. Wendt–Abraham parameter R versus temperature T, where R is given by the ratio of the PDF's first minimum to the first maximum.[76]

tion under consideration is illustrated in Fig. 11, where the linear relations between R and T *clearly hold both in high- and low-temperature regions. The value of R at the intersection is slightly below 0.14*, while the temperature at the intersection is about 0.57. The assertion of Wendt and Abraham is that this temperature can be identified as the glass transition temperature T_g.

The behaviour of R as plotted in Fig. 11 indicates that the nature of the glass transition is related to the kinetic property of the system because the parameter R reflects some aspect of the glass structure in which the atoms are kinetically frozen.

As we see in the following, the described value of T_g is almost identical with T_g defined from thermodynamic properties, thus suggesting that the glass transition is also related to the thermodynamic properties of the system.

b. *Thermodynamic Properties*

The specific volume V/N, V being the volume of the simulation cell, is presented in Fig. 12 as a function of temperature T. Circles denote the results of quenching at $P \sim 1$ bar, triangles at $P \sim 200$ bar, while the squares represent the data for the Lennard–Jones crystals of the fcc structure. Both for $P \sim 1$ bar and $P \sim 200$ bar, the variations of the specific volumes in the high- and low-temperature regions are respectively linear with temperature. This feature is consistent with the results observed in laboratory liquids and glasses of more complex structures, thus implying that the glass transition has really occurred in the model system when it is quenched.

As mentioned in the preceding section, the phase transition is smoothed out in a system of a finite size. Therefore, the glass transition temperature T_g for a model system is defined by the value of temperature at the intersection of the extrapolations from the liquid and glassy branches. From the figure, we see that $T_g \sim 0.57$ at $P \sim 1$ bar, while $T_g \sim 0.68$ at $P \sim 200$ bar.

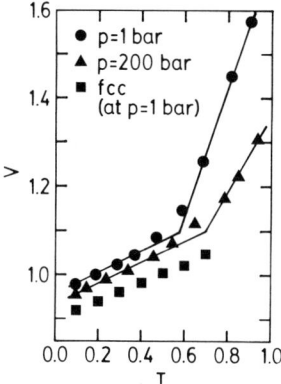

FIG. 12. Specific volume v versus temperature T. Circles and triangles respectively denote the results of rapid cooling under atmospheric pressure and under 200 bar, while squares denote the V-T relation for an fcc crystal.[76]

As we can see from the figure, the pressure effects may be summarized as follows:

1. The higher the pressure, the higher is the glass transition temperature.
2. The glass transition range is slightly narrower for the quenching under the higher pressure.
3. The higher the pressure, the smaller the isobaric thermal expansion coefficient α_P in the liquid phase. In the solid phase, on the other hand, α_P seems to be rather insensitive to pressure. Note that the coefficient α_P is calculated from the first derivative of the $V - T$ curve.
4. From the comparison of the $V - T$ curve for the Lennard–Jones fcc crystal with the curves for the Lennard–Jones glasses, it can be concluded that α_P for the solid state is not influenced very much by the atomic structure, in the sense that the coefficient α_P is almost the same for the amorphous and crystalline solids. This is interesting in connection with the fact stated in 3 that α_P in a solid phase is hardly affected by pressure. These results indicate that the atoms in the amorphous phase are packed as tightly as the atoms in the crystal.

Enthalpy per particle is defined by

$$H/N = (U + PV)/N, \tag{9.1}$$

where the internal energy U is calculated by

$$U = \sum_{i>j} \phi(r_{ij}), \tag{9.2}$$

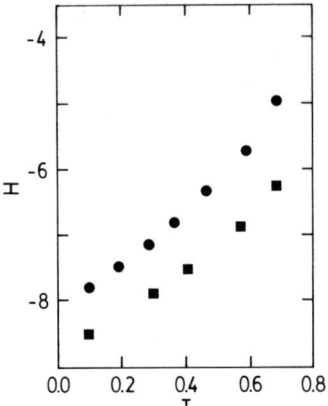

FIG. 13. Enthalpy H per particle versus temperature T. Circles represent rapidly cooled LJ systems, while squares correspond to fcc crystals.[76]

$\phi(r_{ij})$ being the interatomic potential. The temperature dependence of H/N is shown by circles with error bars in Fig. 13, in which the corresponding values for the Lennard–Jones fcc crystals are also presented by squares for the sake of comparison. The change of H/N in the glass transition range is rather gradual, as is the case for the specific volume in Fig. 12. This means that the specific heat C_P (per particle) at constant pressure also varies gradually in the glass transition range since C_P is defined as

$$C_P = \frac{1}{N}\left(\frac{\partial H}{\partial T}\right)_P. \tag{9.3}$$

The temperature dependence of C_P through the quench process is given by circles in Fig. 14. However, the data in the vicinity of T_g are missing because the estimation therein is difficult. The specific heat C_P of the Lennard–Jones fcc crystal is also presented in the figure. Note that, when T goes to zero, C_P approaches the expected value of 3.0 both for a glass and for a crystal.

The specific heat C_V at constant volume is derived from the well-known relation

$$C_V = \frac{1}{k_B T^2}\langle(\delta E)^2\rangle, \tag{9.4}$$

where δE is the fluctuation in the total energy, and $\langle \cdots \rangle$ denotes the thermodynamic average. The estimated value of C_V on the high-temperature side of the glass transition range is about 2.41, while that on the low-temperature side is about 2.93.

Once we have obtained C_P, C_V, α_P, and V at a given temperature, then the isothermal compressibility κ_T can be evaluated from the relation

$$\kappa_T = TV \frac{\alpha_P^2}{C_P - C_V}. \tag{9.5}$$

The value of κ_T estimated on the liquid side is about $\kappa_T(\text{liq}) = 0.250(\sigma^3/\varepsilon)$, while that on the glassy side is about $\kappa_T(\text{glass}) = 0.114(\sigma^3/\varepsilon)$. When these values are scaled in the units appropriate for argon, $\kappa_T(\text{liq}) = 5.9 \times 10^{10}\,\text{cm}^2/\text{dyne}$ and $\kappa_T(\text{glass}) = 2.6 \times 10^{10}\,\text{cm}^2/\text{dyne}$. These values are comparable with those for liquid and crystalline argon near the melting temperature.

c. *Dynamical Properties*

One of the advantages of the molecular dynamics technique over the Monte Carlo method is that the former method provides us with information about dynamical properties of model systems in computers.

The self-diffusion constant D is monitored first by calculating the mean square displacement function $M(t)$, defined by

$$M(t) = \left\langle \frac{1}{N} \sum_i [\mathbf{r}_i(t+s) - \mathbf{r}_i(t)]^2 \right\rangle_s, \tag{9.6}$$

where $\mathbf{r}_i(t)$ is the position vector of the ith atom at t, and second by performing a least-squares fit of the asymptotic part at large t to a straight line of the form

$$M(t) = 6Dt + \text{const.} \tag{9.7}$$

The brackets $\langle \cdots \rangle_s$ indicate averaging over the initial conditions $\{s\}$. The diffusion constant D is also calculated from the velocity autocorrelation

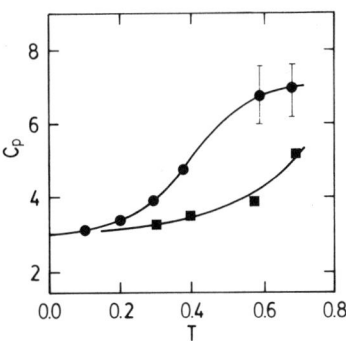

FIG. 14. Specific heat C_P at constant pressure versus temperature T. Circles denote rapidly cooled systems, while squares illustrate fcc crystals.[76]

function $\psi(t)$, defined by

$$\psi(t) = \frac{\langle (1/N)\Sigma_i \mathbf{v}_i(t+s)\cdot \mathbf{v}_i(s)\rangle_s}{\langle |\mathbf{v}(s)|^2\rangle_s} \tag{9.8}$$

through the relation

$$D = \tfrac{1}{3}\int_0^\infty \psi(t)\,dt. \tag{9.9}$$

The MSD and the VAF are given in Figs. 15 and 16, respectively, for several

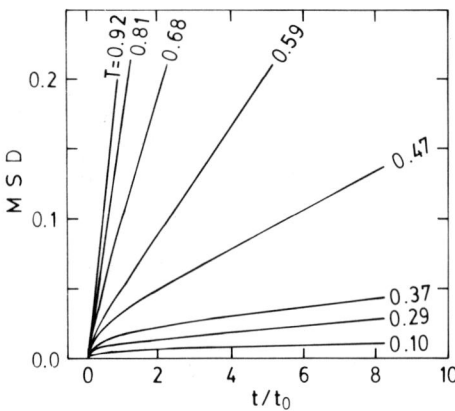

FIG. 15. Mean square displacement (MSD) versus time step t. The numbers denote temperatures.[76]

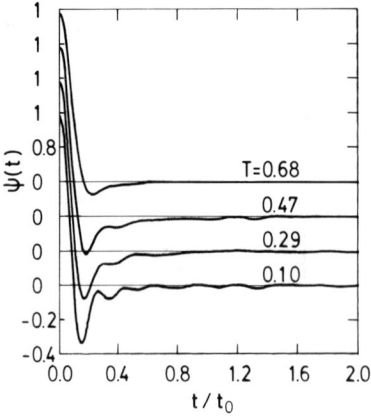

FIG. 16. Velocity autocorrelation function $\psi(t)$ versus time step t, where the time steps are scaled by the unit of time $t_0 = (M\sigma 2/\varepsilon)^{1/2}$. When the values of M, σ, and ε are chosen to be appropriate for argon, the unit of time t_0 is approximately 2 ps.[76]

temperatures in the quench process. The vibrational behaviour of the atoms is observed from the shapes of the VAF at low temperatures. From the figure, the characteristic time τ_{vib} for vibration is estimated to be 0.2–0.3 in the reduced units, which corresponds to 0.4–0.6 ps in the case of argon.

The calculated diffusion constant D is presented in Fig. 17, in which the solid circles denote the values evaluated from the VAF by using Eq. (9.9), while the open triangles represent the data from the MSD defined by Eq. (9.7). The agreement is remarkably good between these two sets of values derived through the two different methods. This proves the validity of the simulation technique. In the figure, the solid square denotes the experimentally obtained value of the diffusion constant at the temperature slightly above the melting temperature. The agreement between the theoretical and experimental result is also satisfactory, showing again the validity of the simulation. From Fig. 17, it is clearly seen that D obeys the Arrhenius relation at high temperatures, which is consistent with experimental results.

The possible interpretations for the nonvanishing values of D at lower temperatures are as follows:

1. The system could still be in the process of structure relaxation toward a state with lower free energy.
2. At low temperatures, the calculated value of D is smaller than

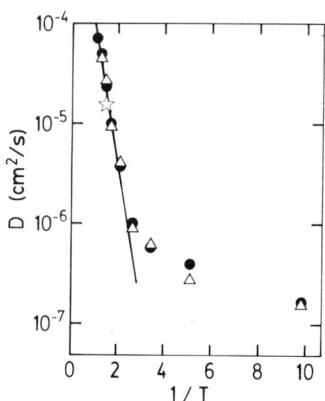

FIG. 17. Logarithmic plot of the diffusion constant D versus $1/T$, where D is scaled by the unit appropriate for argon.[76] The solid circles denote the results derived from the MSD, while the open triangles represent the results evaluated from the VAF. The open star indicates the experimentally observed diffusion constant of liquid argon at a temperature slightly above the melting temperature ($T_m \sim 0.7$ in the reduced unit) and under atmospheric pressure. The deviation from the Arrhenius behaviour (solid line) in the high-D region starts at about $T \sim 0.4$. The source of nonzero diffusion constants in the low-T region is discussed in the text.

10^{-6} cm^2/s, which is the same order of magnitude as the statistical fluctuations.

Therefore, the finite D values at low temperatures might simply be due to the fluctuations.

The shear viscosity η can be calculated from the integration of the stress autocorrelation function (SAF) $\chi(t)$ through the relation

$$\eta = \frac{\rho}{Nk_B T}\int_0^\infty \chi(t)\,dt, \tag{9.10}$$

where

$$\chi(t) = \langle J_{xy}(t+s)J_{xy}(s)\rangle_s \tag{9.11}$$

and

$$J_{xy} = \sum_i \left(\frac{p_{ix}p_{iy}}{M} + \tfrac{1}{2}\sum_{i\neq j} F^x_{ij}y_{ij}\right), \tag{9.12}$$

p_{ix} being the x component of the momentum of the ith atom, M the mass of the atom, F^x_{ij} the x component of the stress acting on the ith atom from the jth atom, and y_{ij} the y component of the vector $\mathbf{r}_{ij} = \mathbf{r}_i - \mathbf{r}_j$. The evaluation of η from Eq. (9.10) is far more difficult than the evaluation of D from Eq. (9.9), especially in the glass transition range. The reason is that the value of the SAF remains appreciable at large values of t, and as a consequence the numerical calculation of the integral in Eq. (9.9) becomes a very delicate problem. On the other hand, the VAF decays comparatively fast and, accordingly, the integration in Eq. (9.9) is easy to estimate numerically.

By means of some careful analysis, η is obtained and shown in Fig. 18. The

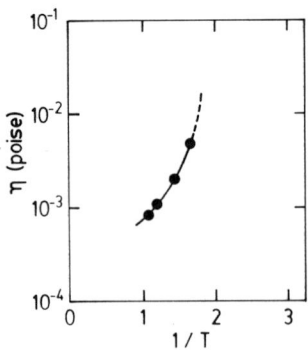

FIG. 18. Shear viscosity η in the logarithmic scale versus $1/T$, where η is measured in the unit appropriate for argon.[76]

temperature dependence of η in the figure is consistent with experimental results, but calculation over a wider range of temperature is required to draw meaningful conclusions.

In order to perform a more detailed investigation into the dynamics of atoms in a glass over a short time, it would help to study the typical features of the VAF for a crystal. For this purpose, the VAF for the Lennard–Jones fcc crystal at $T = 0.1$ is plotted in Fig. 19. The vibrational motion of an atom around each lattice point is obvious from this figure. The power spectrum $F(\omega)$ defined through the Fourier transformation of the VAF in the form[71]

$$F(\omega) = \frac{6}{\pi} \int_0^\infty \psi(t)\cos(\omega t)\, dt. \qquad (9.13)$$

The power spectra of the Lennard–Jones fcc crystal are calculated from Eq. (9.13) and illustrated in Fig. 20. The solid curve represents $F(\omega)$ at $T = 0.1$, while the broken curve denotes $F(\omega)$ at $T = 0.6$. A remarkable point about the solid curve is the appearance of two characteristic peaks at frequencies corresponding to the typical peaks in the phonon spectra.[109,110] Actually, it has been checked in detail that $F(\omega)$ derived from the VAF via Eq. (9.13) exhibits many features similar to those common to the ordinary phonon spectra. The distinct peaks observed in $F(\omega)$ at $T = 0.1$ are broadened at $T = 0.6$ owing to the larger amplitudes of vibrations, and the whole spectrum is systematically shifted to lower frequencies.

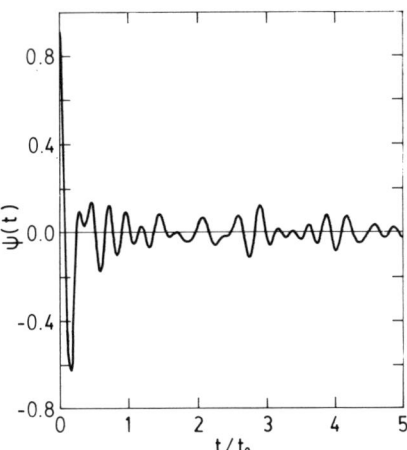

FIG. 19. The time dependence of the velocity autocorrelation function (VAF) of an LJ crystal of the fcc structure at $T = 0.1$.[76]

[109] J. M. Dickey and A. Paskin, *Phys. Rev.* **188**(3), (1969).
[110] G. S. Grest, S. R. Nagel, A. Rahman, and T. A. Witten, Jr., *J. Chem. Phys.* **74**(6), 3532 (1981).

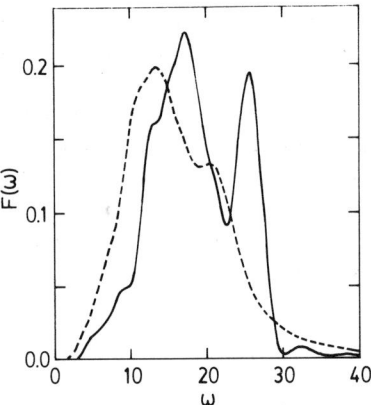

FIG. 20. Power spectra $F(\omega)$ of the fcc LJ crystal versus frequency ω, where $F(\omega)$ is defined by the Fourier transformation of the VAF. The solid curve denotes the $F(\omega)$ at $T = 0.1$, while the broken curve corresponds to $T = 0.6$.[76]

The power spectrum $F(\omega)$ for Lennard–Jones glass at $T = 0.1$ is illustrated by the solid curve in Fig. 21 in comparison with the corresponding data of the Lennard–Jones fcc crystal at $T = 0.1$, denoted by the broken curve. The power spectrum of the glass has the shape of a plateau with the edges located at the frequencies at which the two characteristic peaks appear in the power spectrum of the fcc crystal. This fact implies that the vibrational motion of atoms in a glass bears some resemblance to that of an atom around each lattice point of a crystal, thus lending support to the concept of the cage structure described in Section 5.

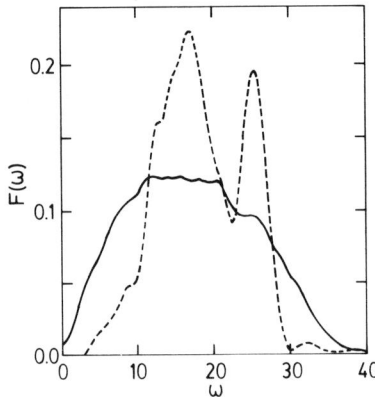

FIG. 21. Comparison of the power spectra for quenched LJ glass (solid curve) and for the fcc LJ crystal (broken curve), both at $T = 0.1$.[76]

10. Glass Transition versus Crystallization

a. Quench Rate Dependence

It has been known by experiment that glass transition takes place only when the quench rate is high enough, while the system crystallizes when it is cooled slowly. Now, it is interesting to see if computer simulations can reproduce this situation depending on the speed of quenching. To this end, a series of researches due to the author and her co-workers may be the first systematic investigation.[77-80,85,86,93,111,112] The discussion in this section will be presented by making particular reference to this research.

Constant-pressure simulations have become relatively tractable after a new technique for simulations at constant pressure was introduced by Andersen.[106] In addition to this, an accurate temperature control is required when the quench rate is a sensitive key parameter. In this sense, the temperature-control MD technique proposed by Nosé[113,114] is appropriate for the quench experiments by MD simulations. The combination of these two methods is used in the aforementioned work.

The model system here is again composed of 864 Lennard–Jones atoms. The equations of motion are solved by the fifth-order predictor–corrector algorithm, and periodic boundary conditions are assumed. A time step for integration is $\Delta t = 0.00234$ in the reduced unit, which corresponds to 0.5×10^{-14} s for argon. The pressure P is adjusted so that it substantially corresponds to the atmospheric pressure.

The results of the simulations are summarized in Fig. 22. Let us first concern ourselves with the results denoted by open and solid squares in this figure. The Lennard–Jones liquid at $T = 1.0$ in the reduced unit is quenched with several quench rates. The reduced temperature $T = 1.0$ corresponds to $T \cong 120$ K for argon. In what follows, temperatures will be mentioned mostly in the real scale for argon, but the rescaling for other rare gases is straightforward (see Table III).

The V-T data for $T \geqslant 60$ K do not depend on the quench rate. In other words, the curve as expressed by open squares is reproduced irrespective of the quench rate. This result indicates that the system is in equilibrium down to 60 K. The quench rate dependence appears below 60 K. The solid squares illustrate the result of quenching with $\Delta T = -2 \times 10^{13}$ K/s (decrease of 10 K over 100 time steps). The continuous change in volume is similar to those as

---

[111] S. Nosé and F. Yonezawa, *J. Chem. Phys.* **84**(3), 1803 (1986).
[112] F. Yonezawa, presented at the International Conference by Institute of Advanced Science, Kyoto, 1988.
[113] S. Nosé, *Mol. Phys.* **52**, 255 (1984).
[114] S. Nosé, *J. Chem. Phys.* **81**, 511 (1984).

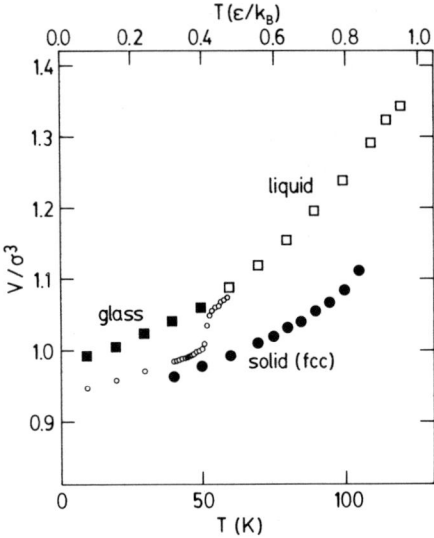

FIG. 22. Volume (V) versus temperature (T).[111] Squares represent the rapid quenching of a LJ system, open squares denoting the liquid phase and solid squares denoting the glassy phase. Small dots show the process of slow cooling and crystallization. Solid circles represent the heating of an LJ fcc crystal; melting takes place when the temperature is raised from 105 to 110 K.

shown in Fig. 7 and in Fig. 12. A detailed analysis of such physical properties as described in the preceding section proves that the glass transition occurs approximately at 50 K and the system is in a glassy state at lower temperatures.

When the quench rate is $\Delta T = -4 \times 10^{11}$ K/s (decrease of 10 K over 5000 time steps), which is lower by almost two orders of magnitude than the previous quench rate, the glass transition still takes place, and it is found that the lower the quench rate, the lower the glass transition temperature T_g and the smaller the volume at T_g.

Then, the question arises: What happens when the quench rate is much lower? Is it possible to reduce T_g at will by reducing the quench rate? This was actually the theme that led to the concept of the Kauzmann paradox.[24] From the experimental viewpoint, the answer to the second question seems to be negative, since the glass transition is achieved only when the quench rate is high enough, as mentioned at the beginning of this subsection.

The simulation under consideration also lends support to experiments; the crystallization instead of the glass transition takes place when the quench rate is as low as 4×10^{10} K/s (decrease of 1 K over 5000 time steps). The volume–temperature relation for this quench rate is given by the small dots in Fig. 22, which exhibit a rather drastic change in volume from 54 to 50 K. The system at lower temperatures is shown to be a crystal from the structural analyses as follows. The pair distribution function $g(r)$ is given in Fig. 23 for the five different temperatures between 54 and 50 K. Accompanying the decrease in temperature, various peaks become marked, suggesting that the system has some kind of long-range order. In particular, the subpeak between the first and second main peaks is known to be characteristic of a close-packed structure, such as an fcc crystal.

For the purpose of studying the atomic configuration more closely, the projections of atoms onto several planes are depicted. One example of such a projection is given in Fig. 24, in which the atomic configuration at 40 K (well below the drastic volume contraction between 54 and 50 K) is projected onto the yz plane. The atomic distribution in each of 10 layers in the simulation cell is demonstrated in Fig. 25. The radius of each circle is taken to be $\sigma/2$. Although there appear occasional point defects, the atoms form a regular close-packed triangular lattice without any dislocations. The difference in volume between the small dots (the crystal obtained through the crystallization) and the solid circles (the starting crystal) is accounted for by the volume increase due to these point defects.

The whole system consists of the close-packed stacking of these 10 layers. It turns out that the number of layers in the structure attained through the

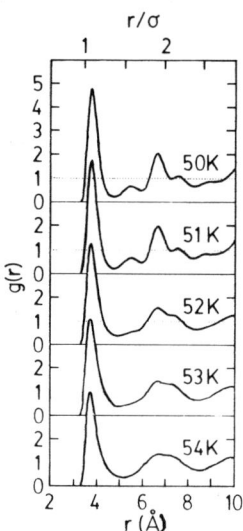

FIG. 23. Pair distribution function $g(r)$ at temperatures between 54 and 50 K.[111]

FIG. 24. Projection of all atoms onto the yz plane at 40 K.[111]

process just described, or through other processes, is not necessarily a multiple of 3. This indicates that the system does not form an fcc structure with the regular stacking of the ABCABC type, but rather takes a close-packed layer structure with stacking faults. Since the differences in the minimum potential energies are negligibly small among a hexagonal close-packed structure, a random stacking, and an fcc structure (the differences being less than 0.01%), it is not surprising that the system does not show as much preference for a particular stacking.

The results presented so far suggest the existence of a critical quench rate of the Lennard–Jones system, which separates the crystal-forming quench rates from the glass-forming quench rates. According to the foregoing simulations, the critical quench rate falls between 4×10^{10} and 4×10^{11} K/s.

b. *Potential Dependence*

So far, the discussions in this chapter have been concerned with the Lennard–Jones system. Although the essential aspects of the glass transition and amorphous structures are expected to be revealed from the study of the Lennard–Jones system, it would still be instructive to get an idea about the potential dependence of physical properties through the glass transition.

As stated in Section 8, the Lennard–Jones potential applies very well to the rare gases whose crystalline structure is fcc. On the other hand, alkali metals are known to form bcc crystals. In order to simulate the situations in alkali metals, the original Stillinger–Weber potential[81] is modified to the form[78]

$$\phi(r) = \varepsilon u(r/\sigma), \tag{10.1}$$

$$u(x) = 8.56438 \left(\frac{1}{x^8} - \frac{1}{x} \right) \exp\left(\frac{1}{x - 1.8} \right), \quad \text{for } x < 1.8,$$
$$= 0, \tag{10.2}$$

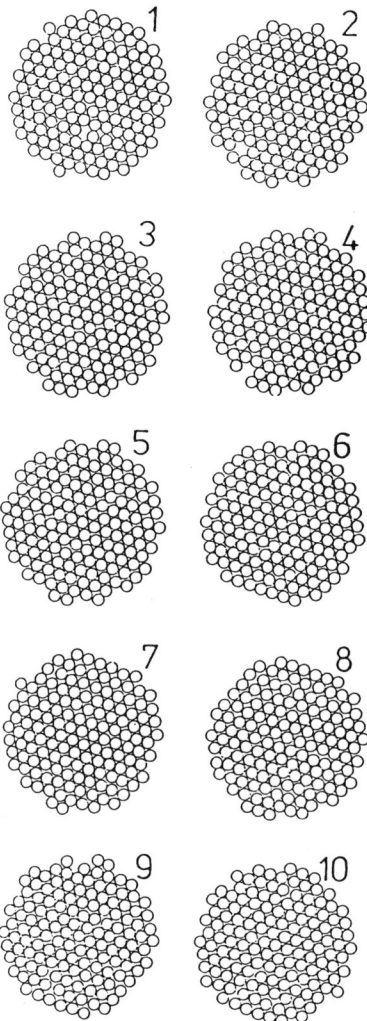

FIG. 25. Atomic configuration[111] in each of 10 layers described in Fig. 24.

where the parameters ε and σ are chosen so that $\phi(r)$ reproduces the pair potential of Rb.

By means of the constant-pressure MD simulations with accurate temperature control as explained in the preceding subsection, the system of atoms with the modified Stillinger–Weber (SW) potential of this form is studied, and the results obtained therefrom are as follows:[78]

1. When the quench rate is high enough, the glass transition occurs, and the physical properties across the glass transition are qualitatively similar to those for the Lennard–Jones system.
2. When the quench rate is appropriately low, the system crystallizes into the bcc structure.

The investigation of the modified SW system, as a whole, confirms the validity of the assumption that the important features about the glass transition and amorphous strucrures would be provided from studying the Lennard–Jones system.

c. *Heating and Melting*

It is of great interest to see if the same scheme of the MD technique can properly simulate the melting of the Lennard–Jones fcc crystal. For the heating process, an fcc crystal at 40 K is chosen, and the temperature is raised with each increment of 10 K up to 70 K and with each increment of 5 K up to 105 K. The volume changes continuously up to 105 K as shown in the figure.

When the temperature is switched from 105 to 110 K, the volume of the system starts expanding rather drastically until it reaches the liquid branch. The atomic configuration before this expansion is ascertained to be an fcc crystal [111] while, after the expansion, the system takes a disordered atomic structure typical of liquids. This fact leads to the conclusion that the volume expansion at this temperature corresponds to melting.

From the phase diagram expressed as Fig. 22, a rather large hysteresis is observed in the whole process of heating and melting followed by cooling and crystallization. Note that the experimentally measured melting temperature of argon is 84 K. The large hysteresis is considered to be due to the finite-size effect.

11. Microscopic Structure Parameters

As mentioned in Section 1, one advantage of computer simulations is that complete information about the positions of all the atoms in the system is available, which makes possible the microscopic analyses of atomic configurations. However, too much information often amounts to no information at all; the list of all atomic positions by itself hardly makes sense. It is therefore important to find some method for extracting useful knowledge from these data. For this purpose, microscopic structure parameters as described in the fourth category in Table IV are introduced.[76–80,85,86,93,111,112,115] The definitions and indications of these microscopic structure parameters are explained in this section along with the demonstrations of some actual applications.

a. *Hidden Structures*

The peaks in the PDF $g(r)$ are broadened as a result of thermal motions of atoms, and, consequently, the underlying atomic configurations are smeared out at high temperatures. This means that, even when the system concerned is a crystal, there appears no peak structure in $g(r)$ characteristic of that crystal if the temperature is high enough.

For instance, the bottom curve in Fig. 26a represents $g(r)$ of the Lennard–Jones system at 105 K, which is just below the melting, as described toward the end of the preceding section. The first peak at the interatomic distance a is not so sharp as it should be, and the peaks at $\sqrt{2}a$ and $2a$, typical of an fcc crystal, are not observed. Since the system at this temperature is in the overheated state and the thermal vibrations of the atoms therein are quite active, a snapshot of the atomic positions might not catch the backbone structure. What is relevant in this context is the PDF $g(\bar{r})$ defined by the time-averaged position vectors

$$\mathbf{r}_i(t) = \frac{1}{\tau} \int_{t-\tau/2}^{t+\tau/2} \mathbf{r}_i(s) \, ds. \tag{11.1}$$

This kind of PDF $g(\bar{r})$ for $\bar{r} = |r(t)|$ was first introduced by Kimura and Yonezawa.[76] The idea is to eliminate the effects of thermal vibrations by taking the time average, and, consequently, the hidden structure is expected to show up. It is clearly seen from the top curve in Fig. 26a that this actually is the case, where the top curve $g(\bar{r})$ illustrates the result of the time average over $\tau = 100$ time steps. The PDF for the time-averaged positions shows that the system is undoubtedly in an fcc structure even at 105 K, and this is strong evidence for the conclusion in the preceding section that the melting does occur above 105 K.

It is very interesting to study the relationship between $g(r)$ and $g(\bar{r})$ of a liquid in comparison with that for a crystal, as shown in Fig. 26a. Figure 26b presents $g(r)$ and $g(\bar{r})$ of a liquid at 100 K, which is almost the same temperature as treated in Fig. 26a. The time average for $g(\bar{r})$ in Fig. 26b is taken over $\tau = 120$ time steps, which again is almost the same as for the $g(\bar{r})$ in Fig. 26a. These figures show a remarkable difference between $g(\bar{r})$ in Fig. 26a and in Fig. 26b; in $g(\bar{r})$ of a liquid, the first peak is sharper than $g(r)$, but no peak structures are revealed, confirming that the system has no long-range order at all.

It is also instructive to examine the relationship between $g(r)$ and $g(\bar{r})$ of a glass as depicted in Fig. 26c. When compared with $g(r)$, $g(\bar{r})$ has a sharper first peak and the splitting of the second peak thereof is more distinct. This result

---

[115]F. Yonezawa, S. Sakamoto, and S. Nosé, *Int. J. Supercomputer Appl. Special Issue* (1990).

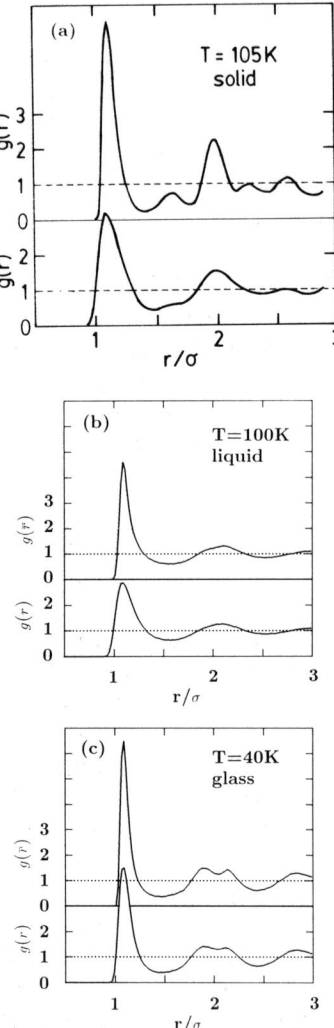

FIG. 26. The pair distribution function $g(r)$ defined by the instantaneous position vectors compared to the PDF $g(\bar{r})$ defined by the time average of position vectors. All figures are concerned with the LJ systems. (a) an fcc crystal at 105 K;[111] (b) a liquid at 100 K; and (c) a glass at 40 K.

indicates that the splitting of the second peak of the PDF is truly an essential aspect of the amorphous structures of isotropic atoms. The result also gives one support to the validity of the free-volume concept and the cage structure.

In calculating $g(\bar{r})$ especially for liquids and glasses, the time duration τ

over which the average is taken plays a key factor. When τ is too small, the underlying structure will not be disclosed. On the other hand, when τ is too large, a completely random distribution of a jellium type will be found. The most appropriate value for τ is the time span during which an atom stays in the cage structure, and may be estimated from the minimum position of the Wendt-Abraham parameter R for $g(\bar{r})$ expressed as a function of τ. The relation R-τ is given in Fig. 27, in which solid and open circles respectively denote the results for the liquid at 100 K and for the glass at 40 K. In both cases, the minima are observed around 100–150 K, which is of the order of the atomic vibration. The fact that the R-τ curve as shown in Fig. 27 has a minimum serves as supporting evidence for the cage structure concept.

b. *Fluctuations in the Microscopic Structure*

In order to study the microscopic structures of disordered systems such as liquids and glasses, it is convenient to introduce structure parameters beyond $g(\bar{r})$. The tessellation of the whole space according to the atomic configuration[7,14,107,116] has been regarded as one promising method of characterizing amorphous structures. Of several ways for the tessellation, the Voronoi tessellation (or equivalently tessellation by Wigner-Seitz cells) has turned out to be most useful. One advantage of the Voronoi tessellation is that it tessellates the whole space uniquely.

Each Voronoi polyhedron is specified by the volume, the number of the faces, the number of the edges, the number of the vertices, the types of the faces, the area of each face, the length of each edge, the area of the whole surface, and so on. The microscopic structure parameters discussed in this and following two subsections are defined on the basis of the Voronoi

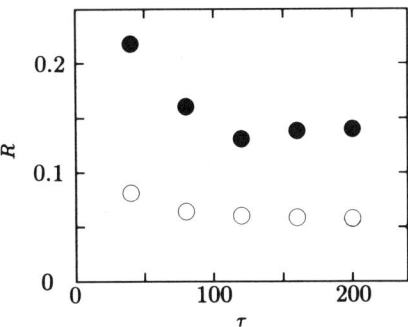

FIG. 27. The Wendt-Abraham parameter R as a function of τ. Solid circles denote a liquid at 100 K, while open circles denote a glass at 40 K.

[116] G. Kahl and J. Hafner, *Phys. Chem. Liq.* **17**, 267 (1988).

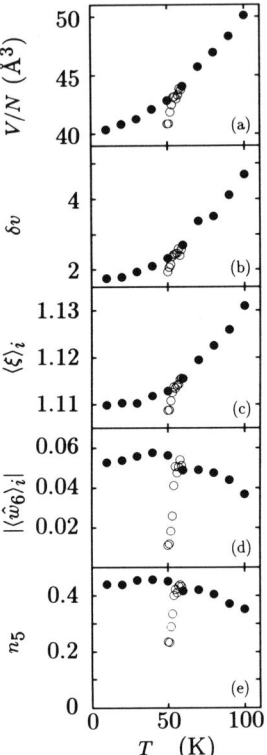

FIG. 28. Temperature dependencies of the microscopic parameters: (a) volume $V/N = \langle v_i \rangle_i$; (b) fluctuation parameter δv; (c) distortion parameter $\langle \xi \rangle_i$; (d) local symmetry parameter $\langle \hat{\omega}_6 \rangle_i$; and (e) Voronoi face parameters n_α. Solid circles denote the change from liquid to glass on rapid quenching, while open circles represent crystallization on slow cooling.

polyhedra. The average $\langle v_i \rangle_i$ of the volumes of all Voronoi polyhedra in the system is identical to the volume per atom V/N, and therefore the top figure $\langle v_i \rangle_i$ versus temperature in Fig. 28 is the same as shown in Fig. 22. In Fig. 28, the other parameters are also presented as functions of temperature, where solid circles denote the process of rapid quenching while open circles depict the process of the crystallization on slow cooling.

In the second column from the top in Fig. 28, a fluctuation parameter δv defined by the root mean square of the volumes of the Voronoi polyhedra is illustrated. The figure demonstrates that the fluctuation in the volumes of the Voronoi polyhedra decreases as the average volume itself decreases from a liquid to a glass or from a liquid to a crystal. This means that the more packed the distribution of the atoms becomes, the more evenly the volume is allotted to each atom. In other words, the free volume is expected to become smaller as V and δv decrease.

c. *Distortion in the Microscopic Structure*

The fact described at the end of the previous subsection also indicates that the local distortion in the atomic configuration will also be reduced when the packing of the atoms is closer as a result of quenching. One possible quantity to measure the local distortion is the shape parameter of each Voronoi polyhedron defined by

$$\xi_i = \frac{v_o^{2/3}/s_o}{v_i^{2/3}/s_i} = \frac{(4\pi/3)^{2/3}/4\pi}{v_i^{2/3}/s_i}, \qquad (11.2)$$

where v_i and s_i are, respectively, the volume and whole surface of the Voronoi polyhedron i, while v_o and s_o are, respectively, the corresponding values of a sphere. The minimum value of ξ is unity, which corresponds to a sphere, and the more distorted the Voronoi polyhedron, the larger the value ξ.

In the third column of Fig. 28, the average $\langle\xi_i\rangle_i$ is given as a function of temperature. It shows a behaviour very similar to that of the δv-T curve, which leads to the conclusion that, as far as the quench process is concerned, the fluctuation δv reflects the distortion in the microscopic structure.

Another interesting quantity to study in this connection is the distribution of the shape parameters, $p(\xi_i)$, as well as the distribution of the volumes of the Voronoi polyhedra, $p(v_i)$. The latter is shown in Fig. 29, the former in Fig. 30. In both cases, the distribution of a glass at 40 K is compared with that of a crystal at the same temperature in (a) and to that of a liquid at 100 K in (b).

In each distribution, the mean is the lowest for the crystal, slightly larger than that for the glass, and much larger for the liquid as indicated in Fig. 27. For the glass, $p(v_i)$ is roughly symmetric around its mean, while $p(v_i)$ for the

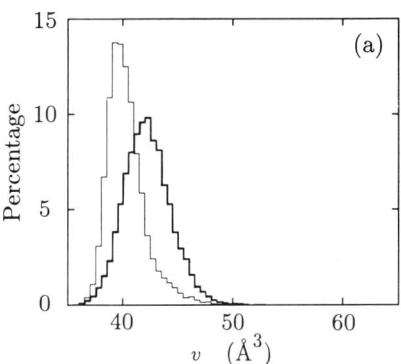
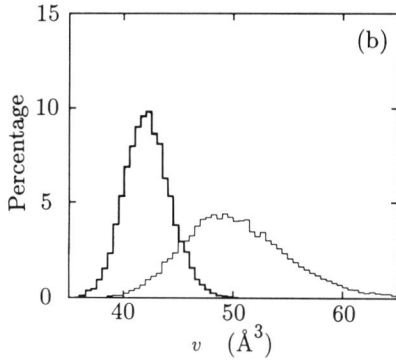

FIG. 29. Distribution of the volumes of the Voronoi polyhedra: (a) comparison of a glass at 40 K (thick solid curve) and a crystal at 40 K (thin solid curve); (b) comparison of a glass at 40 K (thick solid curve) and a liquid at 100 K (thin solid curve).

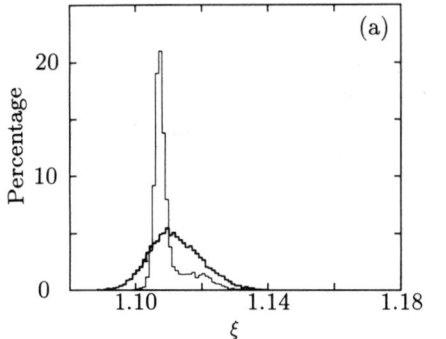 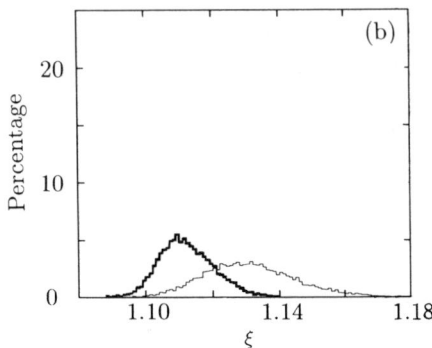

FIG. 30. Distribution of the shape parameters of the Voronoi polyhedra: (a) comparison of a glass at 40 K (thick solid curve) and a crystal at 40 K (thin solid curve); (b) comparison of a glass at 40 K (thick solid curve) and a liquid at 100 K (thin solid curve).

crystal has a tail at higher values of v_i, and the situation is the same for the liquid. The asymmetry of $p(v_i)$ for the crystal indicates that the packing of the atoms in the crystal is closest, allowing no smaller Voronoi polyhedra. On the other hand, the asymmetry of $p(v_i)$ for the liquid indicates that the free volume is large in the liquid. Similar behaviours are observed in $p(\xi_i)$ when careful comparisons are made among the crystal, the glass, and the liquid.

d. *Local Symmetry*

Another popular method of characterizing the Voronoi polyhedra is due to Finney,[7] in which a local configuration around an atom i is described by a list of the indices $(n_3(i), n_4(i), n_5(i),$ etc.), where $n_\alpha(i)$ denotes the number of α-edged faces of the Voronoi polyhedron i. One example of this classification method is given in Fig. 31 in the form of histograms for three different structures of the Lennard–Jones systems.[76] The histograms express the percentages of the Voronoi polyhedra with the corresponding indices. From left to right the results are presented for a supercooled liquid at $T = 0.59$ (just above T_g in the reduced unit), a low-temperature glass at $T = 0.29$ (in the reduced unit), and a crystal at $T = 0.1$ (in the reduced unit).

A perfect fcc structure contains only one kind of Voronoi polyhedron with indices $(0, 12, 0, 0, 0)$. But a Voronoi polyhedron of this type is degenerate in the sense that even a small degree of perturbation easily transforms it into other polyhedra, such as those described by $(0, 4, 4, 6, 0)$, $(0, 4, 4, 7, 0)$, or $(0, 3, 6, 4, 0)$. The figure shows that the set of indices $(0, 4, 4, 6, 0)$ appears most often, the percentage being 17%, the percentages of $(0, 4, 4, 7, 0)$ and $(0, 3, 6, 4, 0)$ are a little smaller. These three percentages amount to nearly 50%.

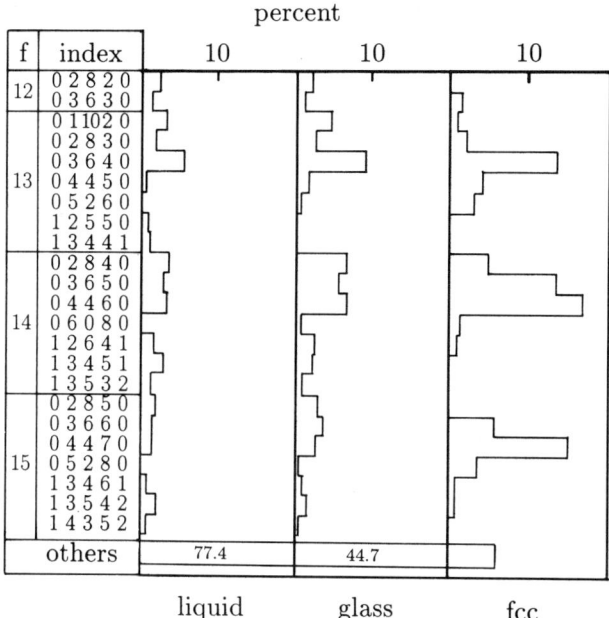

FIG. 31. Histograms of the distribution of the selected types of the Voronoi polyhedra, which are classified according to the sets of integers $(n_3(i), n_4(i), n_5(i),...)$, where $n_3(i), n_4(i), n_5(i),...$ represent the number of faces with 3, 4, 5,... edges, respectively. Histograms are for a supercooled liquid at $T = 0.59$, a glass at $T = 0.39$, and a fcc crystal at $T = 0.1$. Temperatures are scaled in the reduced unit. Taken from M. Kimura and F. Yonezawa, in "Topological Disorder in Condensed Matter" (F. Yonezawa and T. Ninomiya, eds.). Springer, 1983.

When the histogram for the glass is compared with that for the liquid, it is noticed that the variety is wider in the case of the liquid.

Although the list as displayed in Fig. 31 is helpful to a certain extent, it is nevertheless difficult to extract some definite conclusions therefrom. One problem is that a Voronoi index does not uniquely correspond to one kind of a Voronoi polyhedron.

A more useful quantity is the Voronoi face parameter $n\alpha$, which is defined by the ratio of the number of α-edged faces in the system to the total number of the Voronoi faces. Some examples of the Voronoi face parameters are shown in Fig. 32 both for the process of rapid quenching [in (a)] and for the process of crystallization on slow cooling [in (b)]. When the system transforms from liquid to glass, n_5 increases from 35% to 44%, indicating that the local fivefold symmetry is more distinct in the glass. On the other hand, n_5 decreases from 40% to nearly 20% as the system crystallizes below 56 K, thus implying that the degree of the local fivefold symmetry is reduced with

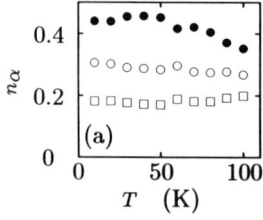
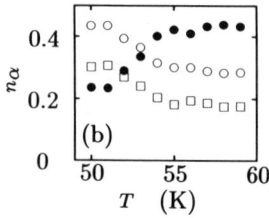

FIG. 32. Voronoi face parameters n_4, n_5, and n_6: (a) the process of rapid quenching; (b) the process of crystallization (note the scale of temperature).

increasing extent of crystallization. The temperature dependencies of n_5 for the processes of rapid quenching and of slow cooling are also shown in the fifth column in Fig. 28 for the sake of comparison with other parameters.

In the field of crystallography, it has been known since the 19th century that fivefold symmetry is incompatible with translational symmetry of a crystal. On the other hand, it was pointed out by Frank[117] in 1952 that small particles can take shapes having fivefold (or more precisely icosahedral) symmetry since the icosahedral structure gives the lowest local energy, owing to the large coordination numbers. For the same reason, the local icosahedral configurations are expected to appear in disordered systems, such as liquids and amorphous solids, in which there exist neither long-range order nor translational invariance. In fact, no odd-edge faces except for triangles appear in the Wigner–Seitz cells in any crystalline structure. As a consequence, it seems reasonable to suppose that the degree of fivefold symmetry could serve as a measure for the degree of amorphousness. The behaviour of n_5, as stated above, lends support to the importance of the local fivefold symmetry in amorphous structures. The significance of the fivefold or icosahedral symmetry in disordered systems was also mentioned in the paper of Damgaard Kristensen.[72]

Steinhardt et al.[81] have proposed bond orientational parameters to discuss the relation between the bond orientational order and the glass transition. Their parameters are defined as follows. With a bond whose midpoint is at $\mathbf{r}$, the set of numbers

$$Q_{lm} = Y_{lm}[\theta(\mathbf{r}), \phi(\mathbf{r})] \tag{11.3}$$

is associated where the $Y_{lm}[\theta(\mathbf{r}), \phi(\mathbf{r})]$ are the spherical harmonics, and $\theta(\mathbf{r})$ and $\phi(\mathbf{r})$ are the polar angles of the bond measured with respect to some reference coordinate system. The average

$$\bar{Q}_{lm} = \langle Q_{lm}(\mathbf{r}) \rangle \tag{11.4}$$

[117]F. C. Frank, Proc. Roy. Soc. London Ser. A **215**, 43 (1952).

is defined, where the average $\langle \cdots \rangle$ is taken over some suitable set of bonds in the system. Then rotationally invariant parameters of third order are introduced as

$$\hat{w}_6(\mathbf{r}_i) = \sum_{\substack{m_1, m_2, m_3 \\ m_1 + m_2 + m_3 = 0}} \begin{pmatrix} l & l & l \\ m_1 & m_2 & m_3 \end{pmatrix} \langle Q_{lm_1}(\mathbf{r}_i) \rangle \langle Q_{lm_2}(\mathbf{r}_i) \rangle \langle Q_{lm_3}(\mathbf{r}_i) \rangle$$

$$\times \left(\sum_m |\langle Q_{lm}(\mathbf{r}_i) \rangle|^2 \right)^{-3/2} \tag{11.5}$$

where the coefficients are the Wigner 3j symbols. Among the parameters in Eq. (11.5) for various values of l, $\hat{w}_6$ is very useful because it is about -0.17 for an icosahedron while $|\hat{w}_6|$ is about 0.13 for a cluster with crystalline symmetry, such as a simple cubic cluster, an fcc cluster, and a bcc cluster.

Steinhardt et al.[81] originally asserted that, when the average in Eq. (11.4) is taken over all bonds in the sample, $|\hat{w}_6|$ can be employed as the criterion for the existence of the local icosahedral symmetry. In research of the author's group,[76-80,85,86,93,111,112,115] it has been checked that $|\hat{w}_6|$ is more relevant when the average in Eq. (11.4) is taken *over all bonds associated with atom i* rather than over all bonds in the system. In addition, it is also shown that the weighted average rather than the ordinary average is more appropriate, where the weighted average means that the contribution from a given bond (connecting the atom i under consideration with the nearest-neighbour atom defined by the Voronoi polyhedron) is weighted by the area of the corresponding face of the Voronoi polyhedron. The purpose of taking the weighted average is to eliminate the spurious effect from the bonds connecting distant nearest neighbours.

The average of $|\hat{w}_6(i)|$ over all atoms $\{i\}$ therefore is expected to provide information about the abundance of the atoms with icosahedral symmetry in the system. This average is given in the fourth column of Fig. 28 versus temperature for the process of rapid quenching (solid circles) and for the crystallization process on slow cooling (open circles). A remarkable point is that the temperature dependence of $\langle |\hat{w}_6(i)| \rangle_i$ behaves almost in the same manner as that of the Voronoi face parameter n_5. From the definitions, the former measures the degree of the local icosahedral symmetry while the latter measures the degree of the local fivefold symmetry, and therefore the similarity between the temperature dependencies of these two parameters indicates that the degree of the local fivefold symmetry closely reflects the degree of the local icosahedral symmetry.

The distribution of $|\hat{w}_6(i)|$ is also worth examining. It is illustrated in Fig. 33, where comparisons are made between the glass and the crystal at 40 K [in (a)] and between the glass at 40 K and the liquid at 100 K [in (b)]. The figure shows that the atoms in the liquid are distributed more randomly than in the glass.

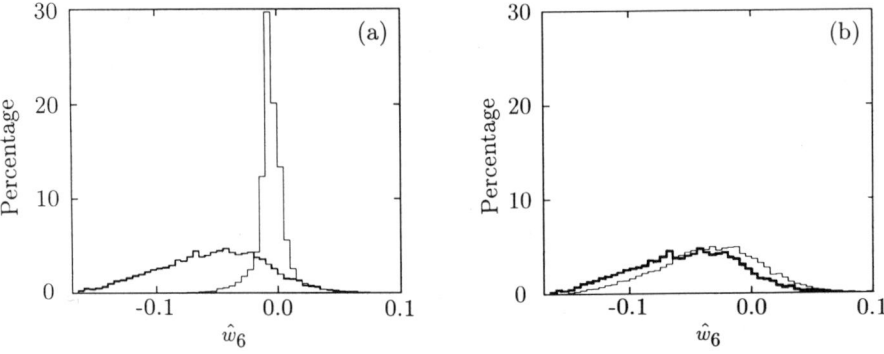

FIG. 33. Distribution of the local symmetry parameters of the Voronoi polyhedra: (a) comparison of a glass at 40 K (thick solid curve) and a crystal at 40 K (thin solid curve): (b) comparison of glass at 40 K (thick solid curve) and a liquid at 100 K (thin solid curve).

VI. Fluctuation and Relaxation in Disordered Systems

12. RELAXATION OF A SUPERCOOLED LIQUID AND A GLASS

a. *Annealing*

To get an idea about the mechanism of the glass transition and to understand the essential nature of amorphous structures, it is very helpful to study the ways in which various kinds of disordered systems in the metastable states are relaxed to the stable states across activation barriers. As examples of such disordered systems, a supercooled Lennard–Jones liquid at 54 K and a Lennard–Jones glass at 40 K are chosen for the detailed analyses in the present section. These systems are both in the metastable states because both temperatures 54 K and 40 K are well below the experimental melting temperature 84 K and also below the onset of the crystallization ~ 58 K, as suggested from Fig. 22. At 54 K and 40 K, the close-packed crystalline structure is known to have minimum energy.

To make these systems undergo relaxation, the temperatures of the systems are fixed at their respective original values (54 K and 40 K), referred to as *annealing*. The changes on annealing of several physical quantities are observed as functions of the annealing time.

The situation is explained in Fig. 34 for the volume at 40 K. To begin with, the V-T relation is one typical example of the problems that have been treated in this review—being the investigations of the temperature dependencies of physical quantities and the studies of their behaviors through the quenching processes with different cooling rates. When the time axis is added to the V-T plane as the third axis, it becomes possible to define the V-t plane at a desired temperature. The drama of annealing at that temperature is played in this V-t plane.

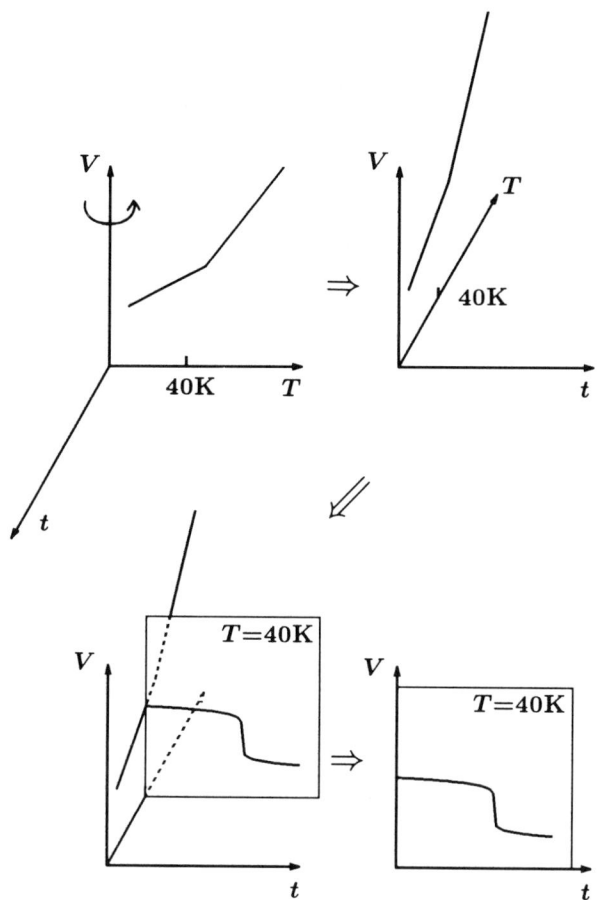

FIG. 34. Three axes, time t, temperature T, and volume V, which show the concept of annealing.

b. *PDF through the Annealing Process*

As stated in the preceding sections, the PDF $g(r)$ is a most convenient quantity for examining the atomic configurations, which is the theme of this subsection. In particular, a careful analysis of the time evolution of $g(r)$ during the annealing processes provides information about the arrangements of atoms and, among other things, about the short-range structures of atomic distributions.

Figure 35 demonstrates the change of the atomic distributions at several time steps through the process of annealing a supercooled liquid at 54 K.

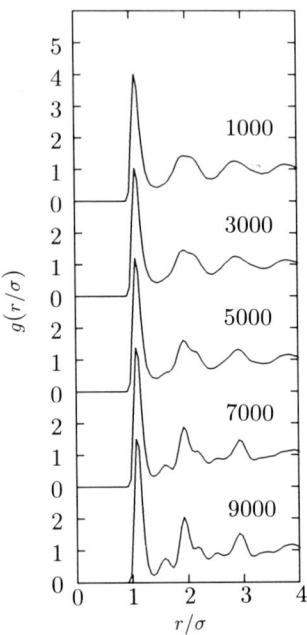

FIG. 35. The PDF $g(r)$ at 54 K in the process of annealing; from top to bottom, $t = 1000, 3000, 5000, 7000,$ and 9000.

At $t = 1000$, which is 5 ps after the start of the annealing, the PDF has a second peak of plateau type. This shape of the second peak is very similar to the second peak of the PDF for a glass at $T = 0.47$, as shown in Fig. 9, which appears in the process of the rapid quenching. Note that this temperature is a little lower than the glass transition temperature $T_g = 0.57$. In fact, not only the second peak but also the whole PDF resembles each other in both cases, thus implying that the atomic configuration in a supercooled liquid is classified between the configuration typical of a liquid and that typical of a glass.

As the annealing proceeds, the peak at $\sqrt{3}a$, a being the interatomic distance, becomes apparent. The appearance of the peak at $\sqrt{3}a$ is regarded as reflecting the growth of the close-packed structures of a layer type because this distance corresponds to the nearest-neighbour positions in a close-packed triangular lattice of two dimension.

The peak at $\sqrt{2}a$ becomes discernible at $t = 5000$ time steps, and gradually develops until it is undoubtedly distinct at $t = 9000$ time steps. This distance is characreristic of two atoms that are placed respectively in the succeeding

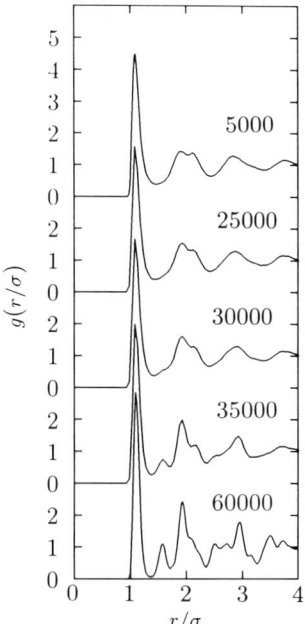

FIG. 36. The PDF $g(r)$ at 40 K in the process of annealing; from top to bottom, $t = 5000$, 25,000, 30,000, 35,000, and 60,000.

two layers of the close-packed triangular lattices, one layer being stacked quite closely on the other.

The peaks at $2a$ and $\sqrt{5}a$ also show up as distinct peaks as the atomic structure in the system relaxes on annealing.

All these results prove that, as a consequence of annealing at 54 K, the atomic distribution achieves a structure with long-range order that is most precisely described by a stacking of the two-dimensional close-packed layers similar to that obtained as a result of the slow cooling described by the small dots in Fig. 22, by the projection of atoms in Fig. 24, and by the atomic configurations in 10 layers as pictured in Fig. 25.

The time evolution of the PDF $g(r)$ on the annealing of a glass at 40 K is shown in Fig. 36. Since the starting structure is a glass, the splitting of the second peak is naturally observed at $t = 10,000$ time steps, which is the first stage of the annealing. From a close examination of the change in the shapes of the second peak, it turns out that the aforementioned two small subpeaks recognized at $t = 10,000$ time steps are located at distances corresponding to the two peaks at $\sqrt{3}a$ and $2a$ characteristic of the crystalline structures, and that these subpeaks eventually become more clearly separated from each other as annealing proceeds.

The intensity of the lower subpeak at $\sqrt{3}a$ increases, indicating the appearance of close-packed configurations of layer type, as is the case for the annealing of a supercooled liquid. The intensity of the higher subpeak at $2a$, on the other hand, decreases as t increases while the peak itself grows to be more distinct. This fact implies that the atoms are distributed more randomly in a glass before the annealing and accordingly there are wider possibilities for the atomic distances around $2a$ when compared with the situation after the annealing.

The time development of the peak at $\sqrt{2}a$ also shows a behaviour similar to that in Fig. 35.

Here again, the mentioned facts serve as the strong evidence for the crystallization of the system on the annealing of a glass at 40 K.

c. *Projections of Atoms*

In the preceding subsection, it has been shown that, by the analysis of the time development of $g(r)$, both a supercooled liquid and a glass finally transform into crystals as a result of structure relaxation on annealing. The relaxation of these systems from the metastable states with disordered structures to the stable states with long-range order is observed more directly by drawing the projections of all atoms in these systems onto some appropriate planes.

In Fig. 37 are shown the projections of all atoms at every 1000 time steps where the volume of the system is normalized in order to make the comparison easier.

The system is undoubtedly disordered at the first stage of the annealing, as seen from the projection at $t = 2000$. The local nucleations are clearly identified at $t = 5000$, and the ordered (or crystalline) regions coexist with the disordered (or still liquidlike) regions between $t = 5000$ and $t = 6000$.

Between $t = 6000$ and $t = 8000$, a rather large-scale adjustment takes place so that the mismatchings of the symmetry axes among different ordered regions are removed, and, after $t = 8000$, the long-range order extends over the whole system, which becomes all the more evident at $t = 10,000$.

The projections of all atoms in the process of annealing a glass at 40 K are shown in Fig. 38 at every 1000 time steps after $t = 32,000$.

The system already contains regions with some kinds of medium-range order even at $t = 32,000$. From the forms of the projected positions, these regions are considered to consist of clusters with the structures of crystalline nature. As time proceeds, these regions grow to be crystalline domains, as seen from the projections between $t = 34,000$ and $t = 37,000$. The mismatchings of the crystalline axes among different domains are gradually lifted, but

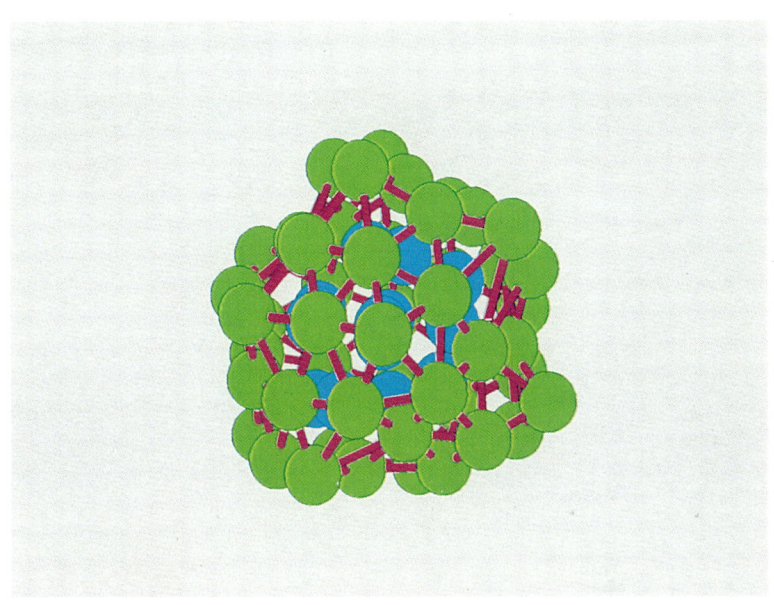

Fig. 48. (a) Icosahedral cluster with double shells of atoms found in the atomic configuration at $t = 10,000$ in the annealing process of glass B. Icosahedral atoms in the center and in the first shell are coloured in blue while icosahedral atoms in the second shell are coloured in green. (b) Only those atoms in the center and in the first shell are presented.

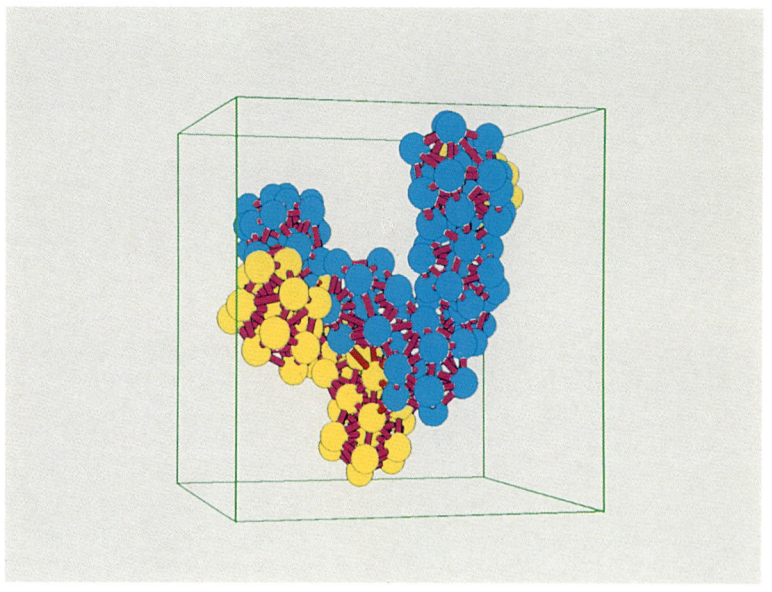

Fig. 49. Persistent clusters, found in the annealing process of glass B, composed of icosahedral atoms coloured in blue and of crystalline atoms coloured in orange. The configurations are observed (a) at $t = 20{,}000$;

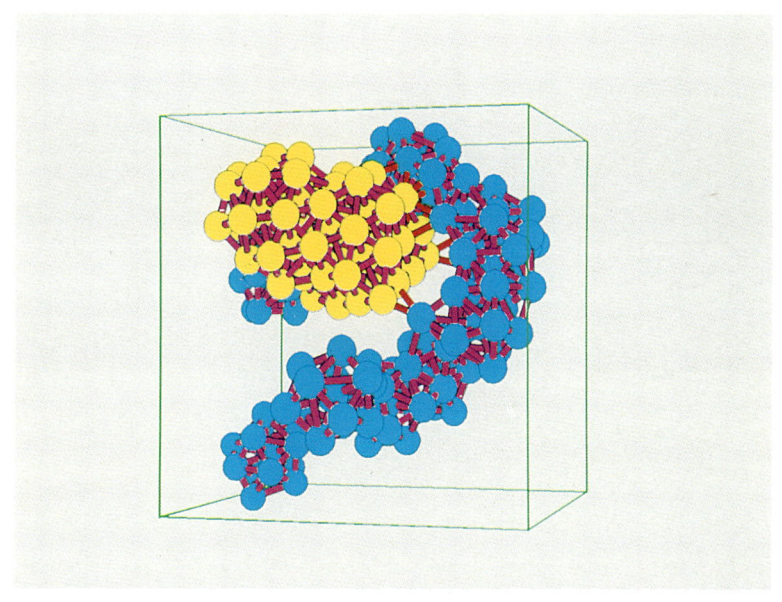

and (b) at $t = 45,000$.

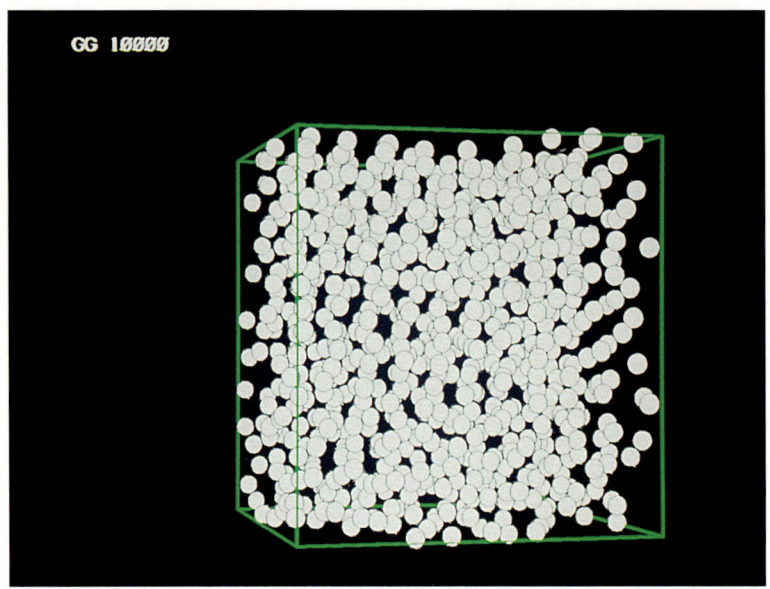

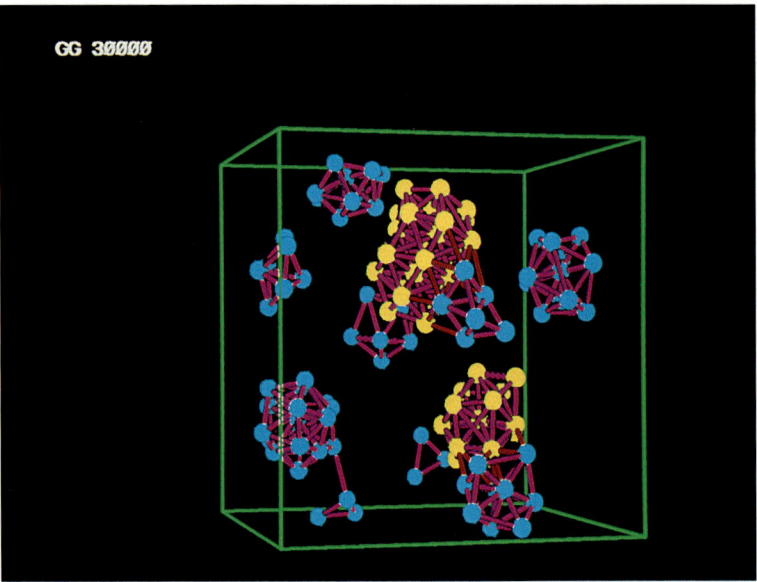

Fig. 50. Configurations of atoms that appear in the annealing process of glass B. (a) The top picture contains all 846 atoms (coloured in white) in the configuration at $t = 10{,}000$. Note that the atomic distribution is completely disordered. (b) The bottom picture, corresponding to the configuration at $t = 30{,}000$, contains only icosahedral atoms (coloured in blue) and crystalline atoms (coloured in orange), while the other atoms are deleted from the picture so that the characteristic features of the clusters are more easily observed.

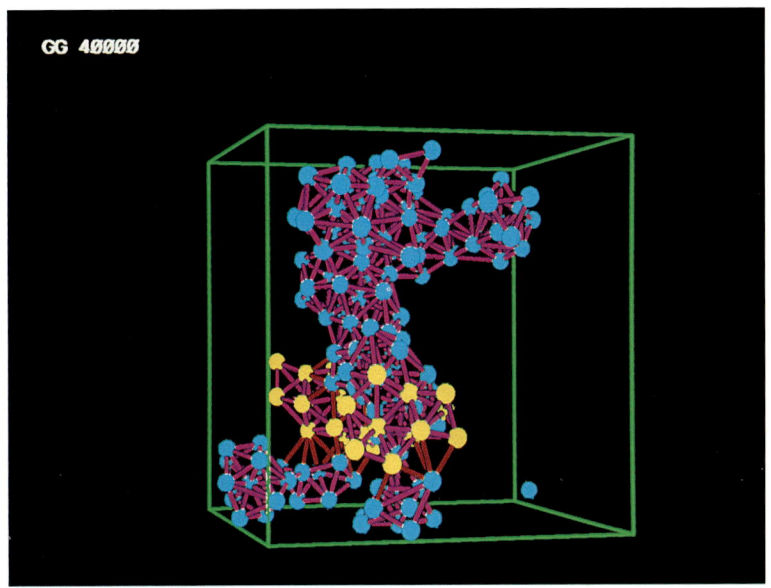

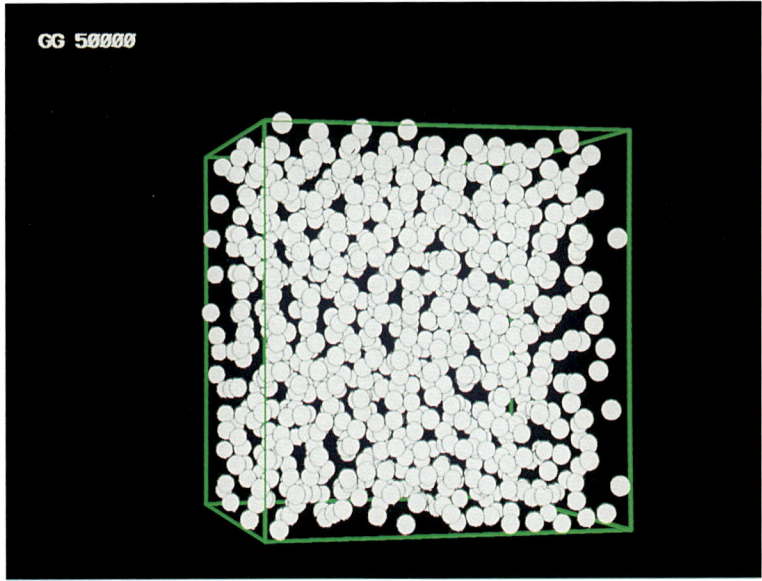

(c) The top picture corresponds to $t = 40,000$. Here again, only icosahedral and crystalline atoms are presented. The coexistence of both kinds of clusters, icosahedral and crystalline, is obvious. Finally, (d) in the bottom picture, all 864 atoms coloured in white are recovered in the configuration at $t = 50,000$; the atomic configuration is still found to be random.

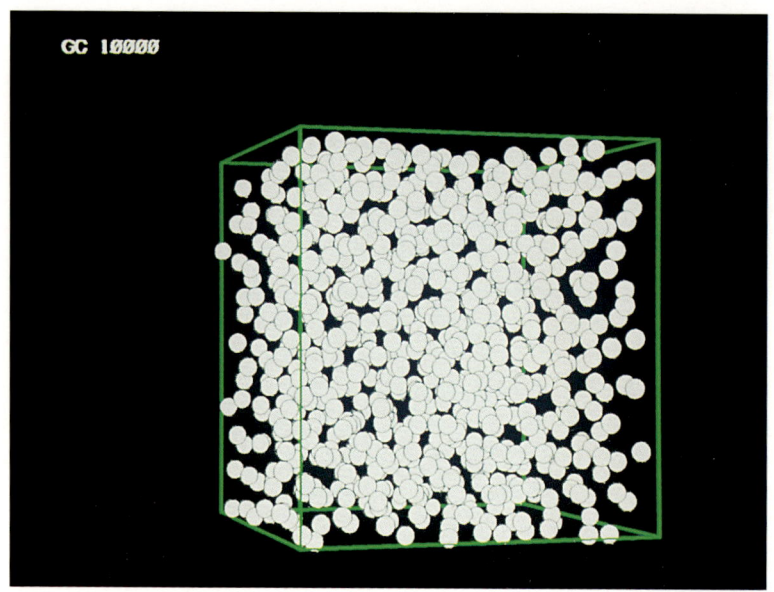

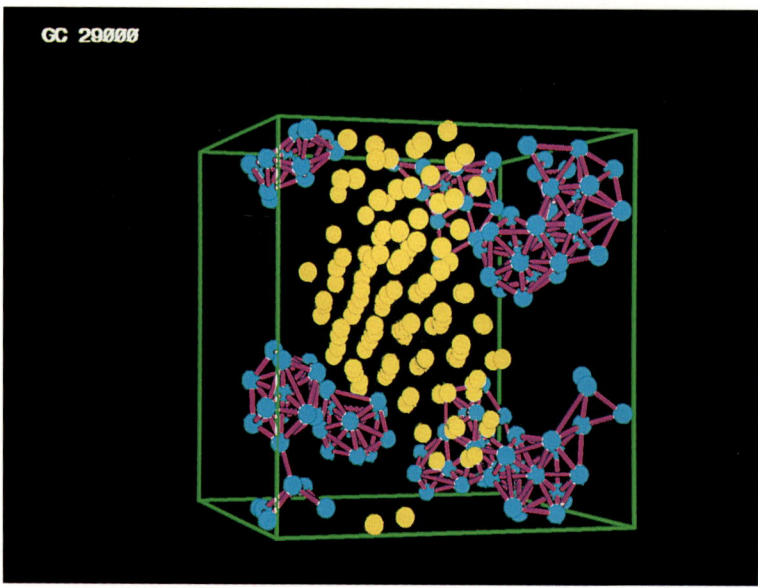

Fig. 51. Configurations of atoms that appear in the annealing process of glass A. (a) The top picture contains all 864 atoms (coloured in white) in the atomic configuration at $t = 10{,}000$. Note that the atomic distribution is completely disordered. (b) The bottom picture corresponds to the configuration at $t = 29{,}000$. Here, only icosahedral and crystalline atoms are presented. Nearest-neighbour bonds in icosahedral clusters composed of blue atoms are shown in red while bonds are omitted in crystalline clusters.

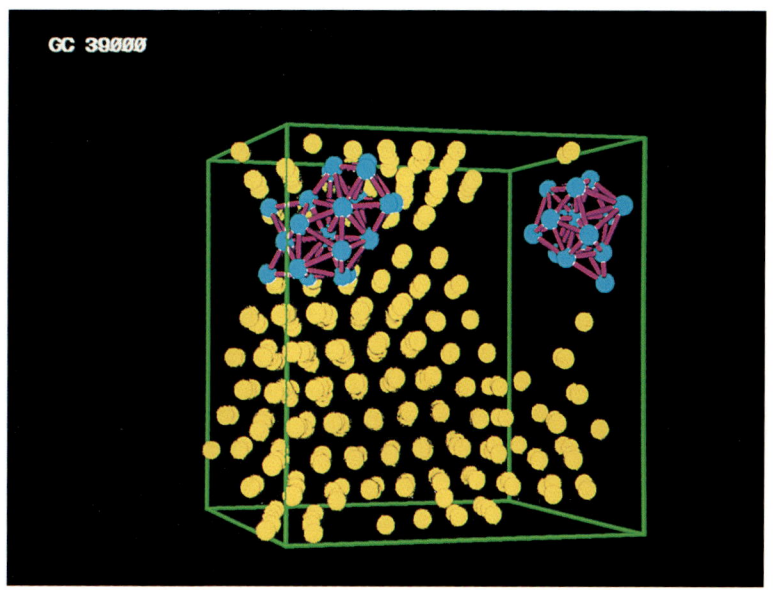

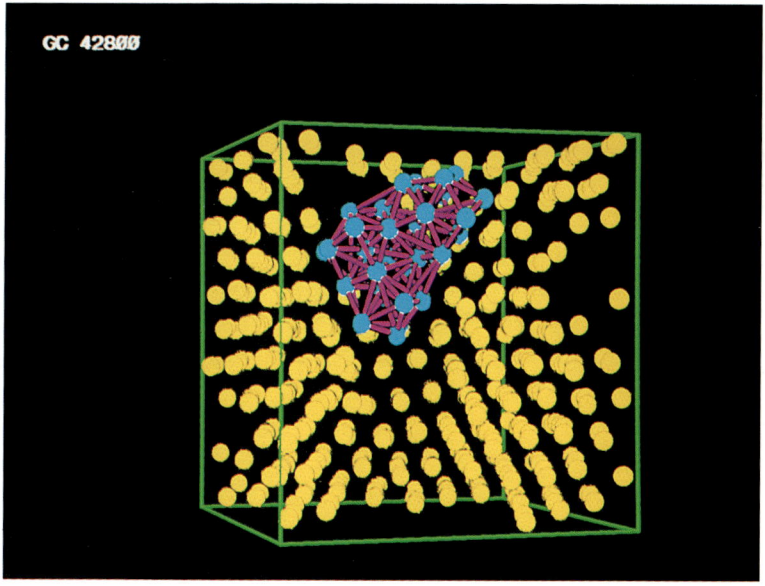

(c) The top picture corresponds to $t = 39,000$, and (d) the bottom picture corresponds to $t = 42,800$. Crystalline layers are growing.

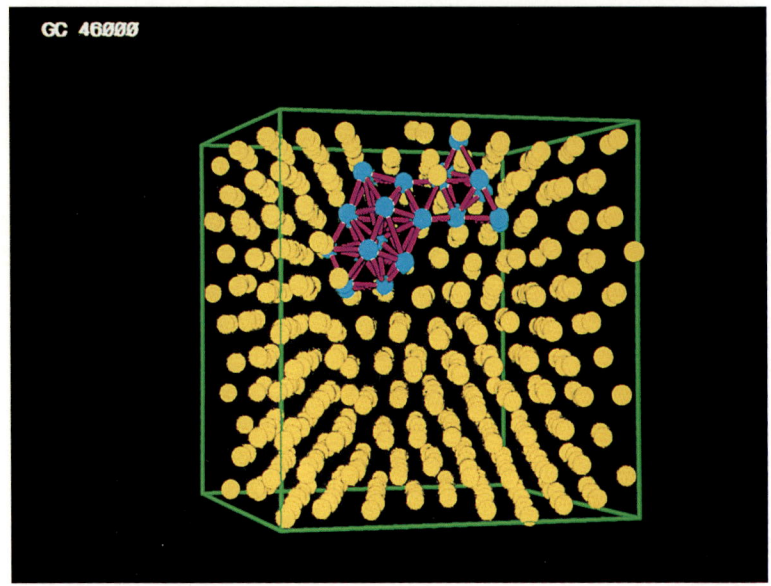

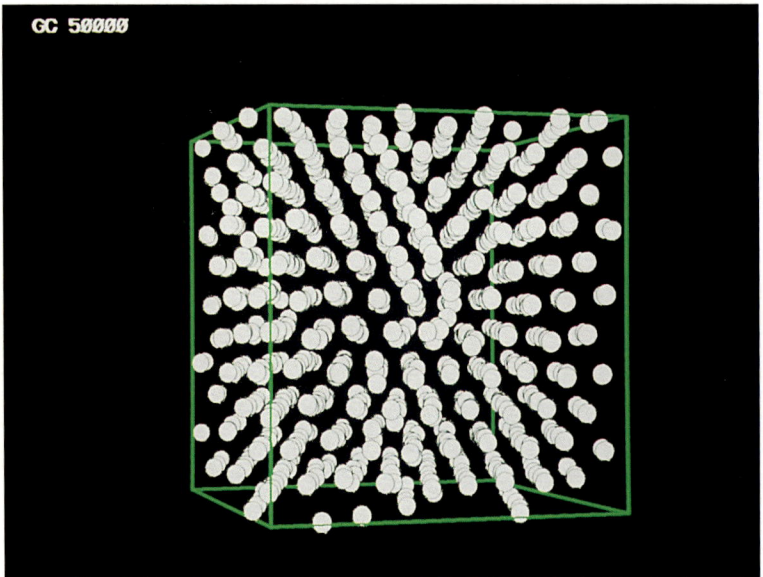

(e) The top picture corresponds to $t = 46,000$, and finally, (f) in the bottom picture, all 864 atoms coloured in white are recovered in the configuration at $t = 50,000$. Obviously, the system has almost crystallized by this timestep.

the pace is very low. Even at $t = 40,000$, there still remains a region in which the deviations from the regular lattice points are fairly large.

This tendency probably occurs because the atoms are packed more closely as compared with the annealing process at 54 K, so a large-scale adjustment of atomic configurations is unlikely to occur. This close packing is also the reason that, in the annealing at 40 K, it takes much longer than annealing at 54 K before complete long-range order prevails in the whole system.

The features demonstrated by the projections in Figs. 37 and 38 will be analyzed systematically in the next subsection.

d. *Microscopic Structure Parameters*

It is widely accepted that the systems in metastable states such as a supercooled liquid and a glass are destined to relax into the most stable states with minimum free energy at the temperature under consideration, which are generally expected to be crystals. The detail of the relaxation in each system is determined by the shape and height of the activation energy that the system has to hurdle.

The microscopic structure parameters as defined in Section 11 give a deeper insight into the world at the atomic level. With a view to understanding the relaxation of atomic configurations in disordered metastable structures, the behaviours of these parameters during the annealing processes are examined in this subsection. Special emphasis is laid on the investigation of the ways in which the relaxation in the mentioned systems is related to the fluctuations of atomic structures. In fact, it will be shown that the relaxation of the structure in a glass (or in other words in an amorphous solid) is fairly distinct from that in a supercooled liquid.

Figure 39 shows several microscopic structure parameters as functions of the time steps in the annealing process at 54 K. Soon after the onset of the annealing, the volume V, the distortion parameter $\langle \xi \rangle_i$, the local symmetry parameter $\langle \hat{w}_6 \rangle_i$, and the Voronoi face parameter n_5 for pentagons begin to decrease at the same time which indicates the appearance of the crystalline nuclei.

The system is considered to be in the first stage of crystallization up to 3000 time steps or so, where the rearrangement of atoms is such that the local symmetry around some atoms changes from icosahedral to crystalline type without causing appreciable volume reduction. Since both δv and $\langle \xi \rangle_i$ remain practically unchanged at this stage, the rearrangement of atoms is still local.

The rearrangement of atoms becomes more extensive at the second stage of crystallization, which corresponds to the time steps between $t = 3000$ and $t = 5000$. All the parameters expressed by solid circles in the figure, except for the fluctuation parameter δv, keep decreasing continuously at this stage,

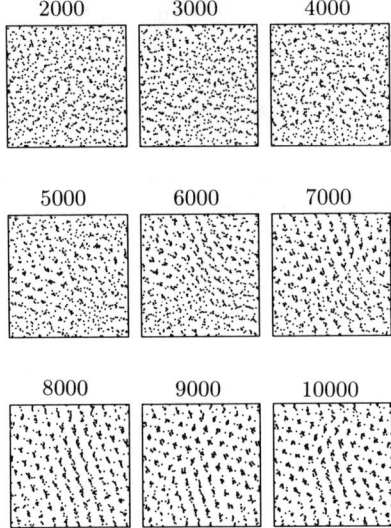

FIG. 37. Projections of atoms onto the xy plane at various time steps at 54 K: every 1000 time steps between $t = 2000$ and $t = 9000$.

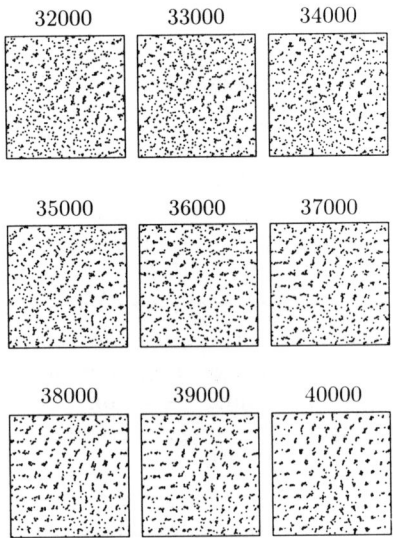

FIG. 38. Projections of atoms onto the xy plane at various time steps at 40 K: every 1000 time steps between $t = 32,000$ and $t = 40,000$.

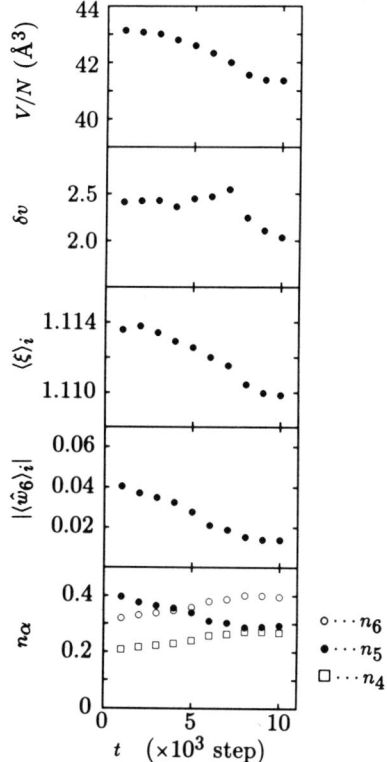

FIG. 39. Microscopic structure parameters versus annealing time at 54 K; V denotes the average of the volumes of the Voronoi polyhedra, δv the fluctuation parameter, ξ the distortion parameter, $\langle \hat{w}_6 \rangle_i$ the local symmetry parameter, and n_α the Voronoi face parameters.

while n_6 for hexagons, denoted by open circles, and n_4 for quadrilaterals, denoted by open squares, are increasing. These behaviours of the parameters suggest that the system is in the coexistence state composed of both crystalline and liquidlike regions.

The transition from this second stage to the final stage of the crystallization seems to take place between $t = 5000$ and $t = 6000$, during which period the parameters $\langle \hat{w}_6 \rangle_i$ and n_5, reflecting the abundance of atoms with the icosahedral symmetry, decrease noticeably, while the fluctuation parameter δv starts increasing.

As implied from the projections at $t = 5000$ in Fig. 37, where the crystalline clusters have grown to considerable sizes, a rather drastic rearrangement of the atomic configuration is required before the system achieves a single-crystal phase. This kind of the drastic rearrangement can be realized if the packing of atoms is not too tight to prohibit local occurrence of a large fluctuation in the atomic distribution.

In the case of annealing a supercooled liquid, there is still enough room for the large fluctuation to take place, as clearly shown in the fluctuation parameter δv at $t = 7000$. This large fluctuation makes possible large-scale rearrangement of atomic positions, which is followed by substantial crystallization. After $t = 8000$, the parameters V, $\langle \hat{w}_6 \rangle_i$, and the Voronoi face parameters (n_4, n_5, and n_6) are all nearly constant, which suggests that the atomic configuration has attained crystalline symmetry after the large fluctuation at $t = 7000$. On the other hand, the fluctuation parameter δv itself decreases further after $t = 8000$, thus indicating that the deviations from the regular lattice points become smaller with time.

It is interesting to study the time development of various microscopic parameters in the process of annealing a glass at 40 K (shown in Fig. 40) in comparison with those in the process of annealing a liquid at 54 K (shown in Fig. 39). In particular, note the differences in time scales as described in the abscissa in Figs. 39 and 40. The relaxation time in the latter case is longer by one order of magnitude than that in the former.

The parameters $\langle \hat{w}_6 \rangle_i$ and n_5 for symmetry start decreasing after $t = 20{,}000$ when the reductions in V and $\langle \xi_i \rangle_i$ are not appreciable. This is different from the situation observed at the first stage of annealing at 54 K, where all four parameters start decreasing at the same time. The difference is considered to come from the fact that, in the glass, the atoms are packed so

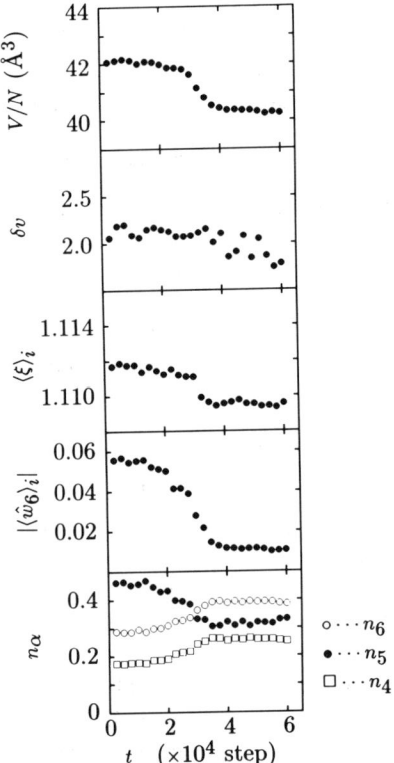

FIG. 40. Microscopic structure parameters versus annealing time at 40 K; V denotes the average of the volumes of the Voronoi polyhedra, δv the fluctuation parameter, ξ the distortion parameter, $\langle \hat{w}_6 \rangle_i$ the local symmetry parameter, and n_α the Voronoi face parameters.

tightly that the clusters with crystalline structures are created as a consequence of local rearrangement of atoms with icosahedral symmetry.

An essential contraction of the system starts after $t = 30,000$, which is accompanied by the rather discontinuous reductions in $\langle \xi_i \rangle_i$, $\langle \hat{w}_6 \rangle_i$, and n_5. These results suggest the progress of the crystallization. A most remarkable point, however, is the behaviour of the fluctuation parameter δv, which increases slightly toward $t = 30,000$ but cannot charge appreciably after that time step. This is again ascribed to the fact that the close packing of the atoms in the glass makes it difficult for an extensive adjustment to develop.

e. *Distribution of the Volumes of the Voronoi Polyhedra*

The assertions in the preceding subsection are reinforced by the analysis concerning the distributions $p(v_i)$ of the volumes of the Voronoi polyhedra during the annealing processes.

Figure 41 is a plot of $p(v_i)$ for the annealing of a supercooled liquid at the same time steps as given in Fig. 37. In each figure, the solid curve represents $p(v_i)$, which is fitted through the least-square method by the sum of two Gaussian curves, described by the dot–dash and broken curves. Curve fitting by two Gaussian distributions gives an overall picture of the coexistence of ordered and disordered regions. The Gaussian curve at lower v_i values is expected to correspond to the ordered regions, while the curve at higher v_i values is regarded as expressing the disordered regions.

According to the figures, there exist the first stage of the crystallization, the coexistence stage, a large fluctuation at 7000 time steps, and substantial crystallization at the final stage. The situation is more clearly demonstrated in Fig. 42 by the time-step dependencies of the mean values and standard deviations of the two Gaussian distributions. The Voronoi volumes of the atoms in the lower-volume group decrease gradually, reflecting the growth of the crystalline clusters as time proceeds. The standard deviation σ_1 of this lower-volume group also decreases monotomically, while the standard deviation σ_2 of the higher-volume group has a distinct maximum at $t = 7000$ at which the fluctuation parameter δv increases markedly. These facts are consistent with each other, lending support to the argument that large-scale rearrangement of atoms is necessary to achieve a single-phase crystallization.

The distributions $p(v_i)$ for glass annealing at 40 K are presented in Fig. 43 for four time steps between $t = 25,000$ and $t = 40,000$. In this case, it is rather difficult to fit the whole curve by using the sum of two Gaussian curves. A detailed analysis indicates that $p(v_i)$ in this case of annealing a glass is decomposed into (1) a Gaussian distribution at the lower Voronoi volumes, denoted by the dot–dash curve in each figure, and (2) the higher-volume region described by solid squares, the latter having a longer tail.

The difference in the characteristic features of the distributions $p(v_i)$ for a supercooled liquid and a glass is probably due to the fact that in a supercooled liquid a clear-cut distinction is possible between the coexisting ordered (crystalline) and disordered (liquidlike) regions, whereas this distinction cannot occur in a glass because the ordered (crystalline) and disordered (amorphous) regions have more or less similar densities.

In this way, microscopic structure analysis has clarified the distinction between the relaxation of a supercooled liquid and of a glass.

13. Relaxation of a Glass Prepared with Lower Quench Rate

The glass treated in the preceding two sections has been prepared with a quench rate of 2×10^{13} K/s, which is hereafter referred to as glass A. Now, it is interesting to study other glasses starting from different initial conditions and/or with different quench rates. Four other glasses are constructed by starting from different initial configurations at $T = 100$ K, but with the same quench rate as the previously described value. It turns out that these four

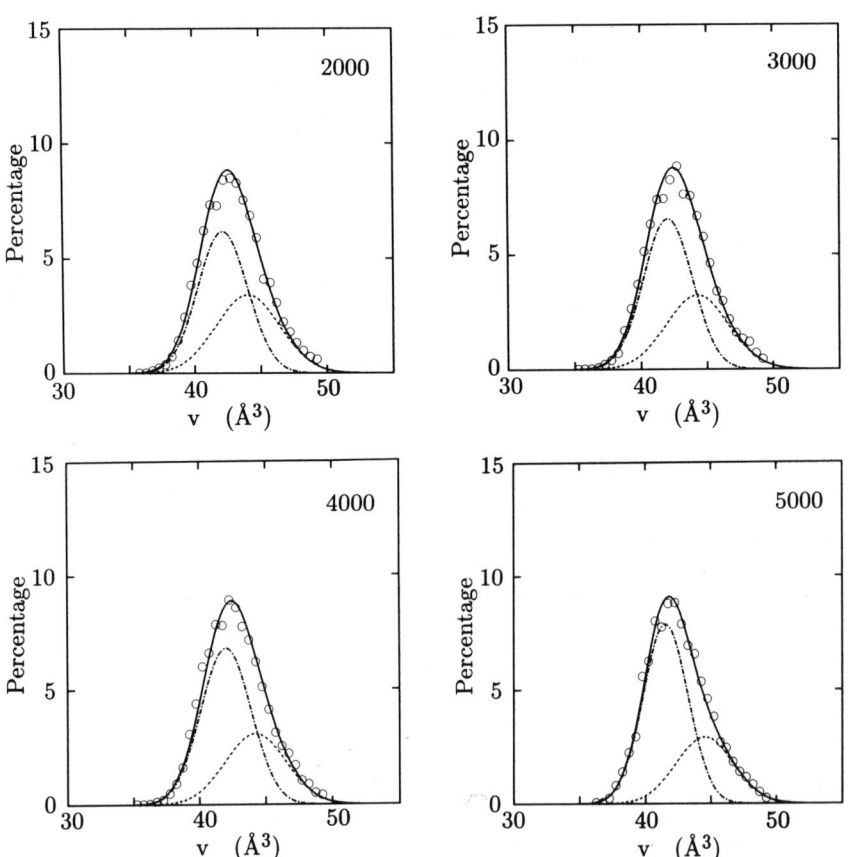

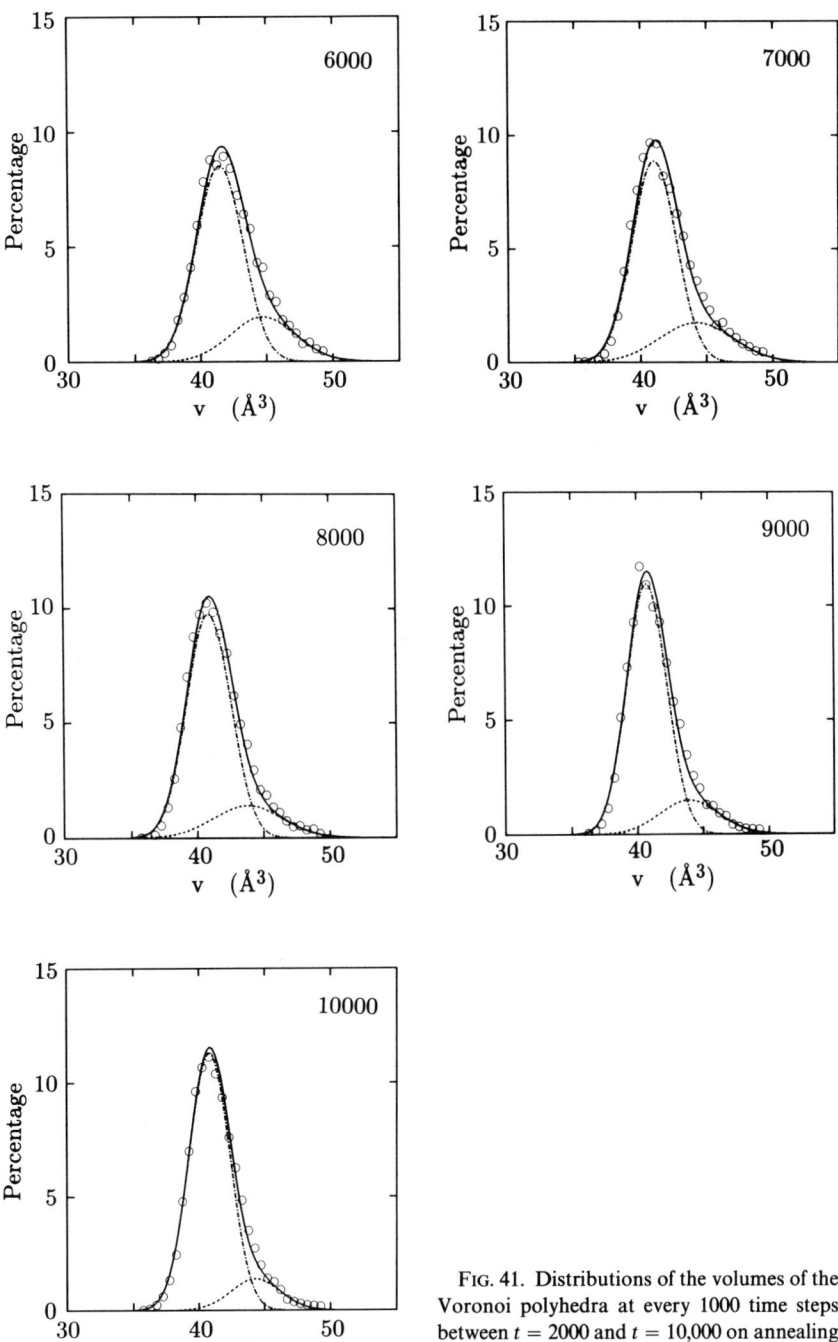

FIG. 41. Distributions of the volumes of the Voronoi polyhedra at every 1000 time steps between $t = 2000$ and $t = 10,000$ on annealing a supercooled liquid at 54 K.

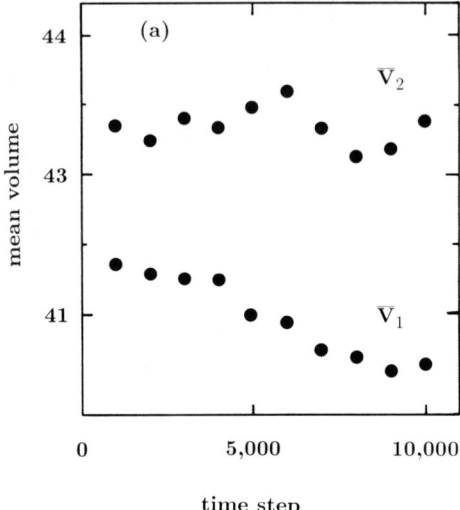

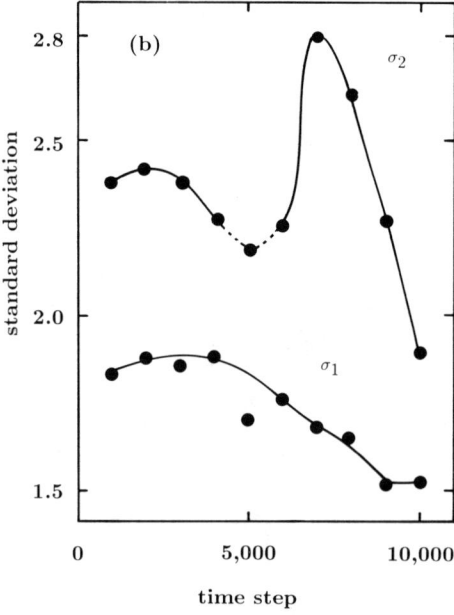

FIG. 42. The means v_1 and v_2 (a) and the standard deviations σ_1 and σ_2 (b) of the two Gaussian curves fitted to $p(v_i)$ during the annealing process at 54 K.

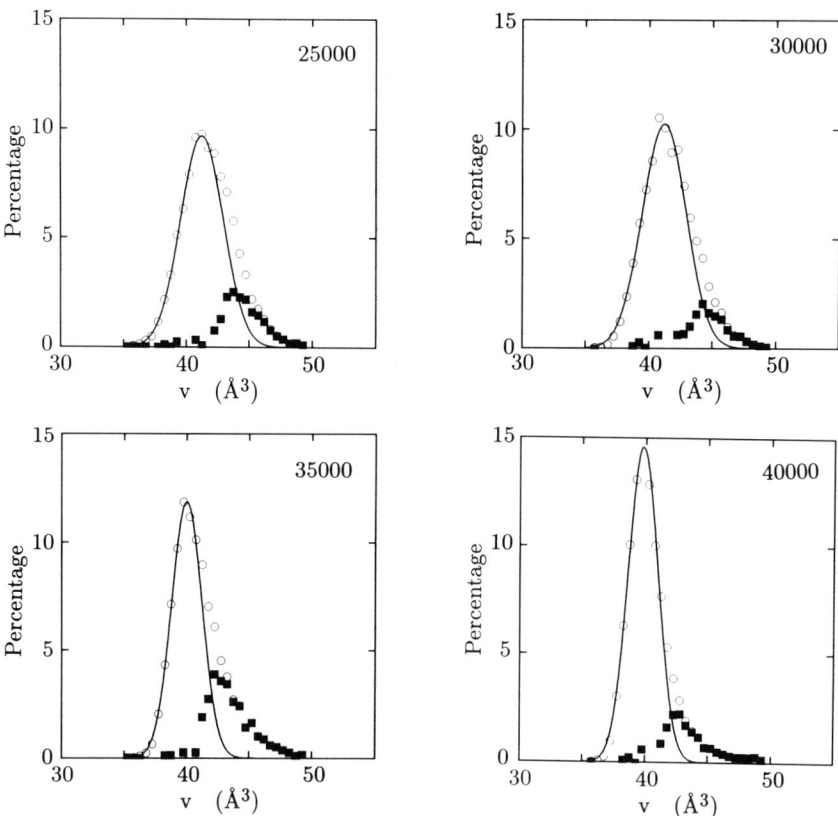

FIG. 43. Distributions of the volumes of the Voronoi polyhedra at every 5000 time steps between $t = 25{,}000$ and $t = 40{,}000$ on annealing a glass at 40 K.

glasses exhibit behaviours essentially similar to those of glass A. In addition to this, five other glasses are produced with a quench rate of 4×10^{11} K/s; and the glass transition temperature T_g is shown to depend on the quench rate as illustrated in Fig. 44—that is, the lower the quench rate, the lower T_g. Among these five glasses, there is one glass that does not relax into a crystalline structure even after the 2×10^5 time steps of annealing; this glass is labeled glass B. The time evolutions of the microscopic structure parameters of glass B are shown up to $t = 60{,}000$ in Fig. 45; no sign of crystallization is evident. Then it is of great interest to find out if there are any specific atomic configurations that stabilize amorphous structures, and, if so, what the characteristic features of such configurations are. In order to cope with this problem, the atomic configurations in glasses or in amorphous

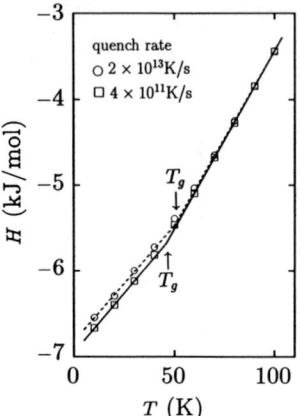

FIG. 44. Enthalpy (H) versus temperature (T) for two quenching processes A and B with quench rates of 2×10^{13} K/s and 4×10^{11} K/s, respectively.

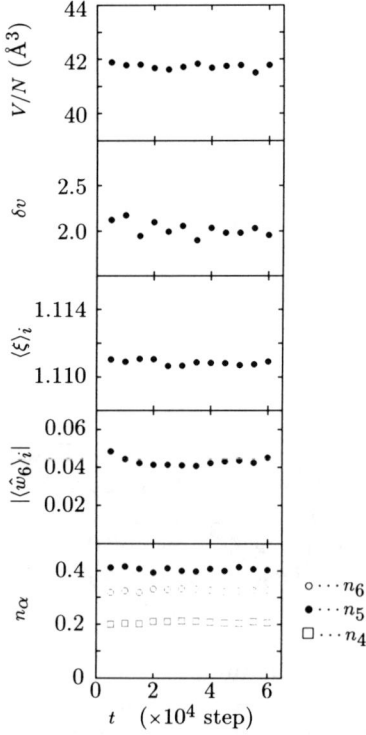

FIG. 45. Microscopic structure parameters versus time for annealing at 40 K of a glass B prepared with quench rate 4×10^{11} K/s; V denotes the average of the volumes of the Voronoi polyhedra, δv the fluctuation parameter, ξ the distortion parameter, $\langle \hat{w}_6 \rangle_i$ the local symmetry parameter, and n_α the Voronoi face parameters.

solids will be visually presented and analyzed in the next section by means of computer graphics.

14. VISUAL PRESENTATION OF ATOMIC CONFIGURATIONS

a. *Local Icosahedral Symmetry*

As stated in Section 11d, the existence of icosahedral clusters in amorphous structures is predicted from energetic considerations. The microscopic structure analyses in the last three sections have proven that this prediction is correct, that there are numerous atoms with local icosahedral symmetry in glasses, and that the abundance of these atoms can be measured by the local symmetry parameter $\langle \hat{w}_6 \rangle_i$ as well as by the Voronoi face parameter n_5.

Therefore, it is now undeniable that local icosahedral configurations are characteristic of amorphous solids consisting of isotropic atoms. The question then arises as to whether these local icosahedral configurations are related to the stability of amorphous structures.

In order to investigate this problem, it is useful to classify atoms in the system according to their local environments. One such way is as follows:[78]

1. Atoms with the environment of icosahedral symmetry.
2. Atoms with the environment of crystalline symmetry.
3. Other atoms.

As the criteria for classification, it is useful to employ $\langle \hat{w}_6 \rangle_i$ and the information concerning the configuration of the neighbouring atoms as described next.

1. An atom is regarded as having icosahedral symmetry when it satisfies the following conditions.

a. The value of $\langle \hat{w}_6 \rangle_i$ is such that

$$0.05 < |\langle \hat{w}_6 \rangle_i| < 0.17. \tag{14.1}$$

b. Among its nearest-neighbour atoms, there is more than one atom with a value of $\langle \hat{w}_6 \rangle_i$ in the region denoted by Eq. (14.1).

2. An atom is regarded as having crystalline symmetry when it satisfies the following conditions.

a. The value of $\langle \hat{w}_6 \rangle_i$ is such that

$$0.005 < |\langle \hat{w}_6 \rangle_i| < 0.02. \tag{14.2}$$

b. Among the bonds connecting the atom under consideration with its nearest-neighbour atoms, there is at least one pair of bonds whose relative angle is between 170° and 190°.

c. Among its nearest-neighbour atoms, there are more than four atoms with a value of $\langle \hat{w}_6 \rangle_i$ in the region denoted by Eq. (14.2).

Here, a nearest-neighbour atom is defined in terms of the Voronoi tessellation; that is, an atom is regarded as the nearest-neighbour atom of the central atom under consideration when these two atoms share one of the faces of their respective Voronoi polyhedra. Also, note that the values of $\langle \hat{w}_6 \rangle_i$ are 0.169 and 0.013, respectively, for an icosahedral cluster and an fcc cluster, each composed of 13 atoms, one at the center and 12 on the nearest-neighbour positions.

On the basis of this classification, the roles of the local icosahedral symmetry will be examined in the following subsections.

b. *Double-Shell Icosahedra*

To see if there is any quench rate dependence in the structural properties of glasses, the quench rates mentioned in the preceding section, 2×10^{13} K/s and 4×10^{11} K/s, are chosen. As far as the number of the atoms with icosahedral symmetry is concerned, glasses prepared with the higher quench rate generally have more atoms with icosahedral symmetry in category 1 than do glasses prepared with the lower quench rate.

It turns out, however, that the number itself of atoms in category 1 does not reflect the stability of the glass. In fact, when the glasses are annealed at some temperature below T_g, say at 40 K, glasses prepared with the higher quench rate tend to relax into crystalline structures more easily than do glasses prepared with the lower quench rate. On the other hand, although the numbers of atoms with icosahedral symmetry are relatively small in glasses with the lower quench rate, the structures in some of these glasses seem to be more stable. Actually, glass B is in this class; it stays in an amorphous state even after the 2×10^5 time steps of annealing.

This result is explained in the following way. When the quench rate is very high, atoms in the system are not allowed to spend enough time to adjust their positions so as to realize a structure with energy as low as possible under the given condition. Accordingly, the redistribution of atoms over a wide range is nearly impossible, and the least energetic configuration is found by adjusting only the immediate neighbours of each atom, thus resulting in a configuration containing many icosahedra.

Icosahedra constructed in this process of high-speed quenching are of the kind shown in Fig. 46a, in which 12 atoms in the first shell surround a core atom and the number of the nearest-neighbour bonds is 42. In the figure, the other clusters composed of 13 atoms are also illustrated for the sake of comparison, the clusters being fcc (face-centered cubic) and hcp (hexagonal close-packed) clusters, either of which has 36 nearest-neighbour bonds. Since each bond gains binding energy comparable with the minimum value of the pairwise interatomic interaction, such as the Lennard–Jones potential, the larger the number of the nearest-neighbour bonds in a cluster, the lower is the energy of the cluster. This is the reason why amorphous solids of isotropic atoms have local icosahedral symmetry.

In high-speed quenching, there exist, as previously described, many icosahedra of this kind, but the extent of each icosahedral symmetry is limited in space, and, consequently, the atoms in the icosahedral clusters easily reconstruct themselves into crystalline configurations.

When the quench rate is not high enough, the atoms are allowed to spend more time in the readjustment of configurations, and therefore the redistribution of atoms beyond the immediate neighbour is not completely excluded. As a result, there may appear clusters whose sizes are larger than a simple icosahedron, as shown in Fig. 46a, and yet whose symmetry is icohedral. One candidate for such a cluster is the double-shell icosahedron, as depicted in Fig. 47a, which consists of 55 atoms and contains 234 nearest-neighbour bonds. An fcc cluster of the same size is shown in Fig. 47b for comparison, which has 216 nearest-neighbour bonds. Here again, the icosahedral cluster has lower energy.

The large icosahedral clusters of this kind are stable once they are constructed because the coherence of the symmetry is longer in space. Another important point about this double-shell icosahedron is that the relative positions of the atoms in the core and in the first and second shells satisfy the fcc crystalline order. In the process of annealing, the double-shell icosahedron can grow by gathering atoms on the surface so that the third shell, the fourth shell, and so on, are formed. These outer shells stack so that they fulfill the fcc order. In this way, coexistence of the icosahedral and

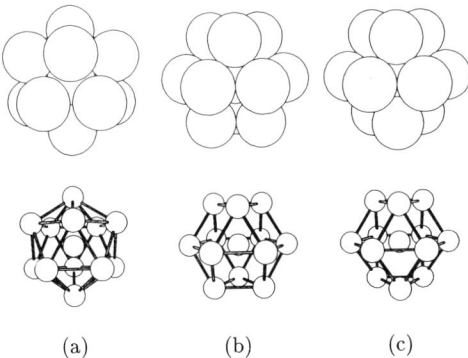

FIG. 46. Clusters in which 12 atoms surround a central atom (the number in the bracket indicates the number of the nearest-neighbour bonds in the cluster); (a) an icosahedral cluster (42); (b) an fcc cluster (36); and (c) an hcp cluster (36).

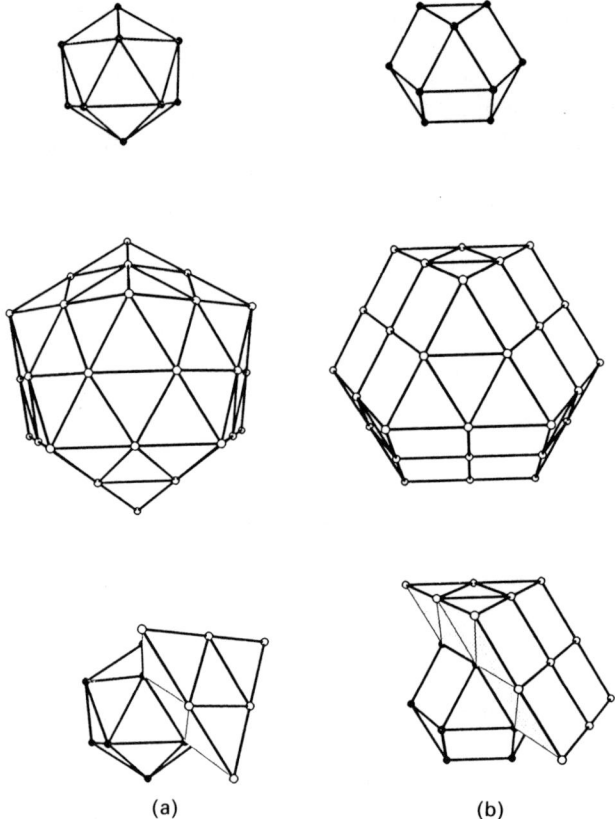

FIG. 47. (a) Illustration of a double-shell icosahedral cluster composed of 55 atoms with 234 nearest-neighbour bonds: (top) an icosahedral cluster in which a central atom is surrounded by 12 atoms belonging to the first shell; (center) 42 atoms in the second shell; (bottom) configurations of atoms in the second shell relative to atoms in the first shell. (b) Illustration of a double-shell fcc cluster composed of 55 atoms with 216 nearest-neighbour bonds: (top) an fcc cluster in which a central atom is surrounded by 12 atoms belonging to the first shell; (center) 42 atoms in the second shell; (bottom) configurations of atoms in the second shell relative to atoms in the first shell.

fcc symmetry becomes possible, which is expected to be stable because quite an amount of energy is required to dissolve such multishell icosahedra.

c. *Annealing of Glass B*

The behaviours of the microscopic structure parameters as shown in Fig. 45 indicate that the amorphous structure in glass B is stable, and the assertion in the preceding subsection suggests that the structure is stabilized by the coexistence of the icosahedral and crystalline clusters. This is found to be so from a close study of the atomic configurations in the annealing process. It is clearly seen in Fig. 48 (see color insert) that the large icosahedron of the double-shell type is actually found in the system.

Figure 49 (see color insert) also demonstrates very clearly the coexistence of the icosahedral and crystalline configurations. The orange atoms connected to the blue atoms by the red bonds cover the surface region of the icosahedral cluster, and, therefore, these orange atoms, though they are classified in category 2 by the criteria stated in Section 14a, construct the outer shells of the multishell icosahedral cluster.

The configuration in Fig. 49 is found at $t = 20{,}000$ in the first stage of annealing. This configuration persists and it is found at $t = 45{,}000$ as well. The thermal vibrations change the details of the configuration, but the underlying structure is the same.

In Fig. 50 (see color insert), the configurations of atoms at four time steps are illustrated. In (a) and (d), all 864 atoms are presented, while in (b) and (c) only those atoms in categories 1 and 2 are shown. It is clearly observed that the coexistence of the icosahedral and crystalline configurations prevent single-phase crystallization from taking place.

d. *Annealing of Glass A*

The time development of the atomic configurations in the annealing of glass A is given in Fig. 51 (see color insert) at various time steps. In (a), all 864 atoms in the system are included and it shows no sign of crystallization. In (b) to (e), only those atoms in categories 1 and 2 are presented. Figure 51b reveals that, at the beginning of the annealing, atoms with icosahedral symmetry dominate, and it is remarkable that there exist no orange atoms with crystalline symmetry at this time step, which means that no double-shell icosahedra are contained in the system. In other words, if they exist, the atoms in the second shell are most probably classified into category 2, and the orange atoms must appear in the figure. Therefore, the absence of the orange atoms in the first stage of the annealing undoubtedly lends support to the assertion in subsection b that the extent of the icosahedral symmetry is limited to the first shell in glasses prepared with very high quench rates, and therefore no such configuration as the double-shell icosahedra can be constructed during the process of quenching.

In (c), (d), and (f) are demonstrated the ways in which the blue atoms with icosahedral symmetry are gradually taken over by the orange atoms with crystalline symmetry. In (f), all 864 atoms are restored, and typical crystalline layers are identified, showing the system has truly undergone crystallization.

VII. Summary and Future Work

Some outcomes from molecular dynamics simulations of the glass transition and of the annealing processes have been discussed with a view to clarifying the nature of amorphous structures. The following properties have been found from the macroscopic and microscopic analyses of the atomic configurations.

1. Melting on heating, crystallization on slow cooling, and a transformation to a glass on rapid cooling are observed in a system composed of Lennard–Jones (LJ) atoms.

2. The critical quench rate q_c which separates crystal-forming quench rates and glass-forming quench rates falls between 4×10^{10} K/s and 4×10^{11} K/s.

3. The thermodynamic, structural, and transport properties of the system, when cooled rapidly, show the behaviours characteristic of the so-called glass transition that have previously been examined in detail by experiments. This fact confirms that the glass transition has really taken place in a simulation box of LJ atoms.

4. Relaxation processes in a supercooled liquid have been compared with those in a glass by laying special emphasis on the relationship between fluctuations and relaxations in disordered systems.

5. The mechanism of the glass transition is essentially of a kinetic nature, which is reflected in most of the physical properties.

6. Atoms in an amorphous system can be classified according to the local symmetry around each atom. For instance, atoms with icosahedral symmetry are typical of amorphous structures.

7. Amorphous structures are stabilized by the existence of double-shell (and possibly multishell) clusters with icosahedral symmetry. This feature is expected to be common to all systems consisting of spherical atoms.

The validity of computer simulations is reinforced by minimizing the effects due to finite sizes and to the periodic boundary condition.

The scope of the molecular dynamics methods is widened by extending the classes of subjects to be studied. Future problems include

- Some spherical pairwise potentials other than the LJ potential
- Nonspherical potentials appropriate for polymers and liquid crystals
- Mixture of different kinds of atoms
- Potentials dependent on the densities of atoms and electrons suitable for metals
- Potentials evaluated from first principles by treating electrons as well as atoms on the same basis.

Author Index

Numbers in parentheses are reference numbers and indicate that an author's work is referred to although his name is not cited in the text.

A

Aaronson, H. I., 83(45)
Aastuen, D. J. W., 18(62)
Abbaschian, G. J., 111(146)
Abraham, F. F., 93(75), 155(291), 156(293, 294, 296), 157, 161(75), 162(313), 201(102, 103), 210(102), 211(102)
Ackerson, B. J., 7(32–34), 14(57), 18(62), 43(117), 68(32–34), 69(33–34), 70(32–34)
Agterof, W. G. M., 6(28)
Ailawadi, N. K., 42(113), 43(113)
Aklonis, J. M., 189(11), 190(11)
Aksay, I. A., 33(101), 34(101), 36(101)
Alder, B. J., 30(85, 86), 198(51–55)
Alexander, S., 13(50), 25(76), 26(76), 33(98, 100, 103), 34(100), 35, 39(103, 109), 40
Alyakov, X., 145(281)
Amini, M., 200(91)
Anderson, H. C., 132(218), 200(74, 84), 201(106), 229(106)
Anderson, I. E., 104(110)
Anderson, P. M., 129(206)
Andres, R. P., 99(94), 100(100)
Angell, C. A., 186(9), 198(50), 200(89, 90)
Antl, L., 6(22)
Armstrong, A. J., 63(147, 150), 64(147, 150), 65(147, 150), 66
Armstrong, P., 138(269), 142(269)
Arora, A. K., 3(13), 5(13), 12(48), 27(80), 44(119), 48(129), 50(132), 51(132), 52(132), 55(13), 58(13), 59(13), 60, 67(156,160)
Arya, K., 219(113)
Asano, S., 219(114)
Asher, S. A., 4(17)
Aur, S., 196(38)
Avrami, M., 127(195)
Avramov, I., 83(42), 145(42), 161(308)
Axe, J. D., 5(18), 15(18), 16, 27(18), 40(18), 43(18), 60(18), 61(18), 67(18). 72(18)

B

Bagchi, B., 70(161)
Ball, M. D., 132(216)
Bangs, L. B., 8(42), 9(42), 11(42)
Bardenhever, P., 108(128), 109(128), 111(142)
Barker, M. F., 137(241), 147(241), 148(241), 156(293–296), 157
Barnes, C. J., 31(92)
Barrat, J. L., 7(38), 54(38)
Barrett, J. C., 170(330), 171
Bartlett, P., 18(61), 56(61), 62(61), 72(61)
Basak, S., 209(108)
Basu, P. K., 159(307), 201(101)
Batschinski, A. J., 195(27)
Battezzati, L., 174(340)
Bayuzick, R. J., 108(131), 109(131), 112(131)
Beall, G. H., 135(236)
Beck, T. L., 132(218)
Becker, R., 82(30), 85(30), 87(30), 98, 156(30)
Belloni, L., 25(76), 26(76)
Bendersky, L. A., 133(222)
Bengtzelius, U., 196(46, 48)
Bennett, C. H., 192(14), 227(14)
Benson, G. C., 161(311)
Benz, K. W., 111(148, 149)
Beresford-Smith, B., 25(75), 26(75)
Berezhnoi, A. I., 135(237)
Berge, A., 8(41)
Bernal, J. D., 6(25), 159(301, 302), 185(5, 6)
Berne, B. J., 10(46), 43(46), 46(123)
Berry, R. S., 132(218)
Biloni, H., 78(5), 91(5)
Binder, K., 155(288), 172(334), 174(336)
Birman, J. L., 219(113)
Bizen, Y., 134(229)
Blank-Bewersdorff, M., 124(188), 125, 131
Blanke, H., 124(189, 190)
Blech, I., 132(215)
Bleckmann, R., 108(128), 109(128), 111(142)
Bockris, J. O., 13(51)

255

Bohn, E., 5(19)
Bonissent, A., 162(313, 314)
Boswell, P. G., 112(161, 162)
Boudart, M., 99(94)
Box, G. E. P., 113(164)
Brenner, S. L., 31(91)
Briels, W. J., 19(65)
Broughton, J. Q., 162(313)
Brown, A. C., 19(64), 199(69), 200(69)
Brown, J. C., 43(114), 46(114)
Buff, F. P., 161(310)
Bugosh, J., 70(163)
Burke, E., 162(313)
Burnett, D. G., 149(285)
Bushnell-Wye, G., 162(314)

C

Cahn, J. W., 78(3), 82(33), 91(60), 132(215), 162(33, 316, 317), 164(322), 168
Callen, H. B., 193(22)
Cannell, D. S., 44(120)
Cantor, B., 91(65)
Cape, J. N., 159(306), 199(62, 63)
Cargill, G. S., III, 133(219), 159(219), 186(8), 190(12, 13), 191(12, 13), 192(12, 13)
Carlier, C. C., 99(95)
Carruthers, J. R., 111(152)
Castillo, C. A., 3(11), 32(11), 36(11)
Cech, R. E., 108(126,133), 109(126, 133), 110(126), 111(126)
Chadwick, G. A., 91(64), 112(160–162), 119(160, 170), 174(337)
Chaikin, P. M., 3(12), 5(18), 7(30, 31), 13(50), 15(18), 16, 25(76), 26(76), 27(18), 33(98–100, 103), 34(100), 35, 39(103), 40(18), 43(18), 55(12), 57(12), 58(99), 60(18), 61(18), 67(12, 18), 68(30, 31), 72(18)
Chakraverty, B. K., 98(93)
Chambey, S., 119(172)
Chan, D. Y. C., 25(75), 26(75), 31(92)
Chanzy, H., 70(164)
Chattopadhyay, K., 112(163)
Cheeseman, P., 200(89, 90)
Chen, L. C., 133(224)
Christian, J. W., 91(62), 127(196)
Christiansen, J. A., 100(98)
Clark, N. A., 7(32–35), 14(57), 18(62), 43(117), 68(32–34), 69(33, 34)
Clark, R., 209(108)
Clarke, J. H. R., 198(50), 199(61, 73), 200(90)
Claus, J. C., 128(204)
Cohen, E. R., 92(72)
Cohen, M. H., 196(33–36), 205(107), 227(107)
Collins, F. C., 98(91)
Coniglio, A., 155(289)
Cooper, A. R., 100(101)
Cormia, R. L., 121(173), 122(173)
Cornel, S. W., 55(139)
Cotter, L. K., 18(62), 43(117)
Courtney, W. G., 100(99)
Craievich, A. F., 135(233)
Crandall, R. S., 7(20), 14(56), 67(29)
Cranmer, D., 136(258), 142(258)
Cruickshank, D. W. J., 136(252)
Curtin, W. A., 162(315)

D

Danilewsky, A., 111(149)
Das, S. P., 196(47)
Day, D. E., 138(267)
de Boisbaudran, L., 79(17)
de Coppet, L. C., 79(18)
Defay, R., 193(23)
Dehoff, R. T., 124(185)
Dekker, M., 78(7)
de Kruif, C. G., 6(23)
Destor, C., 44(120)
Devaud, G., 104(112), 108(130), 109(130)
Dhont, J. K. G., 19(65)
Dickey, J. M., 217(109)
Dickinson, E., 7(36, 37), 52(134), 53(36, 37), 54(37)
DiMarzio, E. A., 194(25, 26)
DeMeglio, J. M., 7(31), 33(99), 58(99), 68(31), 70(31)
Doherty, R. D., 78(6), 91(6, 65)
Donald, W. S., 136(256)
Donovan, P. E., 124(186)
Doolittle, A. K., 195(31)
Döring, W., 82(30), 85(30), 87(30), 98, 156(30)
Douglas, R. W., 149(285), 150(287)
Dozier, W. D., 7(30), 33(99), 58(99), 68(30)
Drehman, A. J., 110(139), 112(155, 156), 133(225), 134
Drifford, M., 25(76), 26(76)
Dubey, K. S., 175(341, 343), 177
Dubois-Violette, E., 14(58)
Duke, D. A., 135(236)

AUTHOR INDEX 257

Dunning, W. J., 78(9)
Duwez, P., 123(179)

E

Eastwood, A. R., 196(37)
Edrei, I., 171(332)
Efremov, I. F., 3(9)
Egami, T., 196(38–40)
Ehrenfest, P., 193(21)
Ellingsen, T., 8(41)
Elliott, R., 112(161)
Elliott, S. R., 185(3), 186(3)
Enderby, J. F., 42(111)
Ermak, D. L., 48(130)
Evans, P. V., 99(96, 97), 124(186), 177(96)
Evans, R., 92(74)
Everett, D. H., 31(92)
Eyring, H., 89(53), 195(29)

F

Fahrenheit, D. B., 78(8)
Fankuchen, I., 6(25)
Farkas, L., 81(29), 85(29)
Feder, J., 78(7), 98(7)
Fehling, J., 108(127), 109(127)
Ferry, J. D., 195(32)
Fijnaut, H. M., 6(28)
Filipovich, V. N., 93(85), 94, 95(88), 102(88), 125(191, 192), 126(191), 138(259), 143(191, 192, 259, 273–275), 146(85), 147(85,88), 148(85, 88, 191), 152, 153(191, 192, 259, 273–275)
Fincham, D., 200(91)
Fink, A., 5(19)
Finney, J. L., 159(304, 306), 162(314), 185(7), 191(7), 192(7), 204(7), 227(7)
Fischer, G. E., 79(10), 82(31), 85(31), 88(31)
Fisher, J. C., 192(15)
Fisher, M. E., 155(288)
Fixman, M., 70(161)
Flaugh, P. L., 4(17)
Fleming, P. D., 198(58)
Flemings, M. C., 104(108, 109), 105(113), 106, 107, 110(140), 111(146)
Flory, P. J., 195(30)
Fokin, V. M., 93(85), 94, 95(88), 102(88), 125(191, 192), 126(191), 143(191, 192, 273), 146(85), 147(85, 88), 148(85, 88, 191), 152, 153(191, 192, 273)

Folda, T., 70(1964)
Fowler, M. E., 100(102)
Fox, J. R., 200(75)
Fox, T. G., 195(30)
Frank, F. C., 132(217), 232(117)
Frankel, N. E., 162(318)
Frenkel, J., 85(50, 51), 88(51), 195(28)
Frisch, H. L., 99(95)
Fuji, Y., 5(18), 15(18), 16, 27(18), 40(18), 43(18), 60(18), 61(18), 67(18), 72(18)

G

Garcia, N. J., 156(297)
Garcia-Escorial, A., 124(186), 132(213)
Garrone, E., 174(340)
Gaskell, P. H., 133(219), 159(219)
Gast, A. P., 10(47), 14(47), 15(47), 40(47), 41, 61(47)
Gaylor, K., 48(130)
Gay Lussac, J. L., 79(12)
Gernez, D., 79(15)
Gerold, V., 81(25)
Gibbons, P. C., 132(216), 134(228)
Gibbs, J. H., 198(25), 199(58)
Gibbs, J. W., 79(21), 161(21)
Gilmer, G. H., 155(290), 162(313)
Gitterman, M., 171(331, 332)
Glasser, F. P., 136(251)
Glasstone, S., 89(53)
Glicksman, M. E., 117(168)
Gomersall, D. W., 108(129), 109(129), 111(145)
Gonzalez-Oliver, C. J. R., 124(182), 136(245), 138(182, 245), 139(270), 140(170), 141(245), 143(270), 147(245), 148(270), 149(270), 151(245)
Goodall, A. R., 2(7), 8(7)
Goodrich, F. C., 170(329)
Goodwin, J. W., 6(22), 43(114)
Gordon, J. M., 198(58)
Gotze, W., 196(148)
Graben, H. W., 201(104, 105)
Grant, M., 168(326)
Grant, P., 25(76), 26(76)
Grantscharova, E., 161(308)
Gratias, D., 132(215)
Graves, J. A., 112(158)
Gray, S. C., 30(88)
Greer, A. L., 83(44), 87(52), 88(52), 89(52), 93(44, 86, 87), 99(52, 86, 87, 99), 100(52),

101, 102(52), 110(139), 111(143), 112(156), 124(181, 186), 125(87), 126(86), 127(193, 197), 128(205), 129(86, 207), 130(156, 210), 132(213), 145(207), 149(52), 153(43, 193), 154
Grest, G. S., 36(105, 106), 37(105, 106), 38–40(106), 72(106), 196(35, 36), 217(110)
Grüner, F., 44(121)
Gulbrand, L., 13(50)
Gunton, J. D., 78(4), 81(4), 168(326)
Gutzow, I., 83(42), 93(82, 83), 135(235), 138(235, 268), 145(42, 279, 281), 147(282), 161(308)

H

Hachisu, S., 13(53, 54), 14(53–55), 30(87), 31(53, 87), 32(53), 55(135, 136, 138), 56(135), 63(151), 70(162), 72(135, 136)
Hafner, J., 227(116)
Haile, J. M., 201(104, 105)
Haller, H. R., 44(120)
Halperin, B. I., 64(155)
Hamaker, H. C., 22(71)
Hamerton, R. G., 99(97)
Handwerker, C. A., 138(261)
Hansen, J. P., 7(38), 37(108), 39(108), 41(110), 42(110), 46(110), 47(110), 48(110), 54(38), 200(74)
Harrowell, P., 70(161), 144(277), 166(277), 167, 168, 169(277, 323, 324)
Härtl, W., 25(77), 26(77), 50(77), 52
Hausen, J. P., 47(126), 48(126)
Haymet, A. D. J., 30(89), 162(313), 166(323, 324)
Hayter, J. B., 43(117), 47(125, 126), 48(126)
Healy, T. W., 21(69)
Hearn, J., 2(7), 8(7)
Heermann, D. W., 155(289)
Heimendahl, M. V., 128(204)
Henderson, D., 24(74), 32(93)
Henderson, D. W., 130(209)
Herman, H., 81(23, 24)
Herold, U., 127(200), 132(214)
Hilderman, G. J., 105(116)
Hill, R. D., 6(22)
Hillert, M., 175(342)
Hilliard, J. E., 81(22), 82(33), 162(33, 316), 164(322), 168
Hillig, W. B., 148(283)

Hiltner, P. A., 2(6), 3(6, 8), 13(6, 8)
Hirano, K.-I., 81(26)
Hirtzel, C. S., 3(11), 32(11), 36(11)
Hishinuma, A., 138(264), 142(264), 143(264)
Hiwatari, Y., 158(300), 199(60, 64, 68), 201(68)
Hockney, R. S., 200(91)
Hoffman, H., 70(164)
Hoffman, J. D., 174(338)
Hofmeister, W. H., 108(131), 109(131), 112(131)
Holz, A., 115(165), 144(165), 175(165)
Holzer, J. C., 134(230)
Hone, D., 13(50), 25(76), 26(76), 33(98, 100, 103), 34(100), 35, 39(103), 40
Honeycutt, J. D., 132(218), 200(84)
Hono, K., 81(26)
Hoover, W. G., 6(24), 17(24), 30(86, 88), 199(59)
Howells, W. S., 42(111)
Hsu, C. S., 159(305), 200(83), 201(98)
Huang, W., 138(267)
Hunter, J. S., 113(164)
Hunter, M. J., 119(170, 171)
Hunter, W. G., 113(164)
Hurd, A. J., 7(35), 14(57)
Hutchinson, J. M., 189(11), 190(11)

I

Ingersoll, L. R., 108(135), 109(135), 110(135)
Inoue, M., 33(97), 132(216), 134(229)
Ise, N., 10(44), 11(44), 14(59), 27(59, 67, 78, 79), 28(66, 67), 29, 43(79, 118)
Ishihara, K. N., 90(58), 91(58)
Ito, K., 10(44), 11(44), 27(78, 79), 43(79, 118)
Ito, M., 136(249)

J

Jackson, K. A., 155(290)
James, P. F., 83(41, 43), 93(84), 124(84, 182), 125(84), 135(232–234), 136(243–245), 137(84, 239–241), 138(84, 243, 245, 263, 269), 139(243, 270), 140(84, 239, 243, 270), 141(243, 263), 142(239, 245, 269), 143(270), 144(40, 239), 146(84), 147(241, 245), 148(84, 241, 243, 270), 150(232, 286), 151(245)
Jansen, J. W., 5(20), 6(23)

AUTHOR INDEX

Jantzen, C. M. F., 81(24)
Johnson, J., 112(154)
Johnson, K. W., 30(88)
Johnson, P. S., 139(270), 140(270), 143(270), 148(270), 149(270)
Johnson, R. A., 67(159)
Johnson, W. A., 127(194)
Johnson, W. C., 83(45)
Jones, D. R. H., 174(337)
Jonssons, B., 13(51)
Jullien, R., 71(166)

K

Kahl, G., 227(116)
Kalinina, A. M., 93(85), 94, 95(88), 102(88), 125(191, 192), 126(191), 138(259, 260), 140(259), 143(191, 192, 259, 273–275), 146(85), 147(85, 88), 148(85, 88, 191), 152, 153(191, 192, 259, 273–275)
Kalos, H., 172(334)
Kantrowitz, A., 93(78)
Kashchiev, K., 83(42), 96(90), 98, 102(103), 144(42)
Kattamis, T. Z., 104(108, 109), 110(140)
Katz, J. L., 92(71, 72)
Kauzmann, W., 194(24)
Keefer, K. D., 71(166)
Kelton, K. F., 83(44), 87(52), 88(52), 89(52), 93(44, 86, 87), 96(89), 99(52, 86, 87), 100(52), 101, 102(52), 125(87), 126(86), 127(193, 198), 129(86, 89), 132(216), 133(198, 221), 134(228, 230), 149(52), 153(43, 193), 154
Kerker, M., 43(116)
Kesavamoorthy, R., 3(13), 5(13), 12(48), 27(80, 81), 44(122), 45, 48(127–129), 49, 50(132), 51(132), 52(132), 55(13), 58(13), 59(13), 60, 63(152), 67 (156, 160)
Khan, A. A., 8(41)
Kikuchi, R., 33(101), 34(101), 36(101), 91(69)
Kiminami, C. S., 112(157), 130(157)
Kimura, M., 200(76), 205(76), 208(76), 209–218(76), 230(76), 231, 233(76)
Kirkwood, J. G., 30(84), 161(310)
Kissinger, H. E., 130(208)
Kitchener, J. A., 23(73)
Klein, L. C., 237(261)
Klein, W., 155(289)
Kleinert, H., 64(154)

Klement, W., 123(179)
Klier, M., 13(52)
Kobayashi, Y., 13(53, 54), 14(53, 54), 31(53), 32(53)
Kose, A., 13(53), 14(53), 31(53), 32(53)
Köster, U., 124(187–190), 125, 127(200), 131, 132(212), 145(280)
Kosterlitz, J. M., 64(155)
Kostorz, G., 81(25)
Kovacs, A. J., 189(11), 190(11)
Kovascheva, T., 145(281)
Kracek, F. C., 136(248)
Kramers, H. A., 169(327)
Kreitlow, G., 136(246)
Kremer, K., 36(105, 106), 37(105, 106), 38–40(106), 72(106)
Krieder, K. G., 133(223)
Krieger, I. M., 2(6), 3(6, 8), 13(6, 8)
Kristensen, W. D., 199(72), 202(72), 203, 204, 205, 205, 232(72)
Kroger, C., 136(246)
Kuhrt, F., 91(67)
Kui, H. W., 111(143), 177(344)
Kuo, K. H., 132(216)

L

Lacy, L. L., 108(132), 109(132), 112(154)
Ladd, A. J. C., 162(313)
Laidler, K. J., 89(53)
Laird, B. B., 162(313)
Lal, K., 112(163)
Lal, M., 7(36), 53(36)
Landau, L. D., 84(46)
Landel, R. F., 195(32)
Langer, J. S., 81(27), 82(27, 34–36), 163(320), 174(34, 35, 320), 165, 166(36), 168
Law, T. J., 108(134), 109(134), 110(134)
Leamy, H. J., 155(290)
Lebeau, S. E., 105(116)
Lebowitz, J. L., 172(334)
Lee, J. K., 83(45), 156(293, 294), 157
Lehmann, W., 44(121)
Lekkerkerker, H. N. W., 19(65), 71(165)
Lele, S., 175(343), 177
Leutheusser, E., 196(41–43)
Levesque, D., 42(112)
Levin, E. M., 136(255)
Levine, L. E., 134(228)
Lifshitz, E. M., 84(46)

Lin, C. J., 128(205)
Lindsay, H. M., 3(12), 5(12), 33(99), 55(12), 57(12), 58(99), 67(12)
Linse, P., 13(51)
Lloyd, D. J., 132(216)
Loper, C. R., Jr., 104(104, 110), 105(122)
Lord, A. E., 112(154), 129(206)
Lothe, J., 78(7), 82(32), 91(32, 66), 92(73), 98(7), 156(32)
Lovett, R., 172(333)
Lowitz, J. T., 79(11)
Lozada-Cassou, M., 24(74)
Luck, W., 13(52)

M

MacMilland, P. W., 135(231), 138(231)
MacRone, R. K., 81(23)
Maeda, K., 196(40)
Maeda, M., 91(58)
Maeda, Y., 70(162)
Maguire, J. F., 199(61)
Mandell, M. J., 158(298, 299), 159(298, 299), 199(66, 67, 71), 200(66, 67), 217(71)
Marcelja, S., 32(96)
March, N. H., 32(95)
Marinov, M., 145(179)
Marro, J., 172(334)
Martin, J. E., 71(166)
Masumoto, T., 132(216), 134(229)
Matsuda, H., 158(300), 198(64, 68), 201(68)
Matsuoka, H., 27(79), 43(79)
Matusita, K., 137(238, 247), 138(247) 140, 142(238), 148(247), 149(247)
Maura, R. D., 138(265)
Mayer, J. E., 6(27)
Mazenko, G. F., 196(47)
McDonald, I. R., 41(110), 42(110), 46(110), 47(110), 48(110)
McGraw, R., 172(334)
McMillan, W. G., 6(27)
McMullen, W. E., 162(315)
McQuarrie, D. A., 32(94)
McTague, J. P., 39(109), 158(298, 299), 159(298, 299), 199(66, 67, 71), 200(66, 67), 217(71)
Mehl, R. F., 127(194)
Mendenhall, C. E., 108(135), 109(135), 110(135)

Merry, G. A., 104(105)
Metropolis, N., 156(292)
Meyer, R. B., 122(177), 143(177)
Mitaku, S. M., 67(157)
Mitchell, D. J., 25(75), 26(75), 32(96)
Mitra, S. K., 200(91, 92)
Miyazaki, J., 156(295, 296)
Miyazawa, Y., 83(39), 118(29), 121
Mockler, R. C., 7(35), 63(147, 150), 64(147, 150), 66
Moir, G. K., 136(251)
Mokrass, B. J., 33(97)
Monovoukas, Y., 10(47), 14(47), 15(47), 40(47), 41, 61(47)
Moore, R. A., 201(96, 97)
Moorish, A. H., 130(211)
Morales, G. J., 25(76), 26(76)
Morey, G. W., 136(250)
Moriya, T., 136(249)
Mork, P. C., 8(41)
Morris, D. G., 128(202), 129
Moss, S. C., 133(223)
Mountain, R. D., 19(64), 61(141), 62(141), 159(307), 199(69), 200(69), 201(100, 101)
Moyal, J. E., 169(328)
Muar, A., 136(257)
Mueller, B. A., 104(111), 105(116, 121), 113(121)
Muller-Krumbhaar, H., 172(334)
Munakata, T., 196(45)
Murray, C. A., 63(146), 64(146), 65(146)
Murray, M. J., 55(137), 56(137)

N

Nagel, G., 111(148, 149)
Nagel, S. R., 209(108), 217(110)
Naumann, R. J., 111(153)
Neilson, G. F., 137(240), 140(271), 142(240), 144
Nelson, D. R., 64(155), 132(218), 200(81), 232(81), 233(81)
Nicholson, R. B., 91(61)
Nieuwenhuis, E. A., 6(28)
Ninham, B. W., 32(96)
Ninomiya, T., 57(140)
Nishioka, K., 91(68), 92(68)
Noack, M. A., 133(226), 134
Nolte, H. J., 27(78)

Nosé, S., 200(77–80, 85, 86), 219(77–80, 85, 86, 111), 220–226(111), 223–225(78, 111), 226(111), 233(77–80, 85, 86, 111, 115), 249(78)
Notheisen, B., 111(149)

O

O'Donnell, S. E., 4(17)
Oehlschlegel, G., 136(242)
Ogawa, T., 158(300), 198(64, 68), 201(68)
Ogita, N., 158(300), 198(64, 68), 201(68)
Ohsaka, K., 105(121), 111(147), 113(121)
Ohtsuki, T., 67(157)
Ok, H. N., 130(211)
Okano, K., 67(157)
Okubo, T., 10(44), 11(44), 27(78)
Okumura, H., 43(118)
Onoda, G. Y., 63(149)
Oriani, R. A., 161(312), 172(334)
Ostlund, S., 72(168)
Ostwald, W., 79(16, 19)
O'Sullivan, W. J., 7(35), 63(147, 150), 64(147, 150), 65(147, 150), 66
Oswatisch, K., 93(77)
Othmer, P., 79(20)
Ottewill, R. H., 6(22), 43(114)
Ou-Yang, H. D., 5(18), 15(18), 16, 27(18), 40(18), 43(18), 60(18), 67(18), 72(18)
Overbeek, J. Th. G., 2(1, 2, 4), 3(1), 4(1), 10(1), 19(1), 21(1), 28(82)
Owens, S. W., 6(22)
Oxtoby, D. W., 92(74), 144(277), 162(315), 166(277, 323, 324), 167, 168, 169(277, 323, 324)

P

Paik, J. S., 105(117, 123), 110(138), 175(123), 176(123)
Pailthorpe, B. A., 23(72)
Pansu, B., 63(145), 64(153)
Papworth, S., 6(22)
Parker, R., 7(36, 37), 53(36, 37), 54(37)
Parrinello, M., 201(99)
Pascova, R., 93(82)
Pask, J. A., 142(272)
Paskin, A., 217(109)
Pecora, R., 10(46), 43(46)
Petton, A. R., 133(226), 134

Penfold, J., 47(125)
Penkov, I., 147(282)
Penn, S. R., 70(161)
Penrose, O., 172(334)
Perepezko, J. H., 104(110, 111), 105(114–117, 121, 123), 106, 107, 110(114, 136, 138), 112(58), 113(121), 150(136), 175(123), 176(123)
Phillips, B., 136(257)
Philpse, A. P., 5(21)
Pieranski, P., 3(10), 4(10), 6(10), 14(10, 58), 63(145, 148), 64(153)
Pincus, P., 12(49), 13(50), 25(76), 26(76), 33(98, 100, 103), 34(100), 35, 39 (103), 40
Pine, D. J., 71(267)
Polk, D. E., 185(2)
Pollock, E. L., 37(108)
Poon, S. J., 133(225, 227), 134
Popov, E., 145(279)
Pound, G. M., 78(7), 82(32), 83(39), 91(33, 66, 68), 92(68, 73), 98(7), 118(39), 121, 156(32, 295, 296)
Powell, G. L. F., 108(24), 109(25)
Price, D. L., 201(94, 95)
Prigogine, I., 193(23)
Punge, Wittler, B., 132(212)
Pusey, P. N., 3(15, 16), 6(15), 7(39), 16(15, 60), 17, 18(61), 25(16), 43(114), 46(124), 56(61), 62(15, 16, 61, 142), 71(167), 72(61)

R

Rabin, Y., 171(331)
Rahman, A., 158(298, 299), 159(298, 299, 303, 305), 198(49), 199(66, 67, 71), 200(66, 67, 83), 201(98, 99), 217(71, 110)
Rajagopalan, R., 3(11), 32(11), 36(11)
Rajalakshmi, M., 27(81)
Ramachandrarao, P., 112(163), 132(216), 175(341, 343), 177
Ramakrishnan, T. V., 33(102), 52(102), 54(102), 64(154), 166(325), 169
Ramaswamy, S., 70(161)
Rangarathan, S., 124(183)
Rao, C. B., 27(81)
Rasmussen, D. H., 104(104, 110), 105(122)
Rathz, T. J., 108(132), 109(132)
Ray, C. S., 138(267)
Reatto, L., 42(112)
Reddy, A. K. N., 13(51)

Ree, F. H., 6(24), 17(24)
Reiss, H., 90(54), 92(70–72), 93(79, 80), 104(105), 143(54)
Rhines, F. N., 124(185)
Rice, O. K., 163(319)
Richmond, J. J., 104(111)
Ridder, J. D., 133(222)
Robbins, M. O., 36(105, 106), 37(105, 106), 38–40(106), 72(106)
Robertson, J. L., 133(223)
Robinson, M. B., 108(131, 132), 109(131, 132), 112(131)
Ronchetti, M., 132(218), 200(81), 231(81), 233(81)
Rosenberg, R. O., 22(70), 23, 33(70), 34(70), 36(70), 37(70), 39(70), 61(141), 62(141)
Rosenbluth, A. W., 156(292)
Rosenbluth, M. N., 156(292)
Ross, M., 199(59)
Roth, R. S., 136(255)
Rowlands, E. G., 137(239, 244), 138(262, 263), 140(239), 141(263), 142(239), 144(239), 150(286)
Russel, W. B., 23(72)
Russell, K. C., 78(7), 83(45), 90(56), 98(7)
Ruthen, F., 14(58)

S

Sabes, P. N., 132(216)
Sadoc, J. F., 57(140)
Sahm, P. R., 112(157), 130(157)
Sahni, P. S., 78(4), 81(4)
Sakaino, T., 136(249)
Sakamoto, S., 200(78–80, 86), 201(93), 219(78–80, 86, 93), 223(78), 224(78), 225(76–80, 86, 93, 115), 226, 233(78–80, 86, 93, 115), 249(78)
Sakuntala, T., 12(48)
Sale, F. R., 112(161)
Salomaa, R., 136(258), 142(258)
Sanders, B. J., 55(137), 56(137)
San Miguel, M., 78(4), 81(4)
Sarkies, K. W., 162(318)
Sastry, G. V. S., 132(216)
Scheil, E., 108(127), 109(127)
Schenkel, J. H., 23(73)
Schneider, T., 162(313)
Schröeder, H., 79(13)
Scott, B., 138(269), 142(269)
Scott, M., 127(201)
Scott, W. D., 142(272)
Servi, I. S., 104(106)
Shaefer, D. W., 10(43), 11(43), 33(103), 39(103), 40, 46(123), 47(43), 71(166)
Shangguan, D. K., 99(97)
Shechtman, D., 132(215)
Shelby, J. E., 136(254)
Shen, Y., 133(227), 134
Shevelev, V. V., 90(55)
Shiflet, G. J., 133(227), 134
Shih, W. Y., 33(101), 34(101), 36(101)
Shingu, P., 91(58)
Shiohara, Y., 105(113), 106, 107
Shirashi, S. Y., 108(129), 109(129), 111(144, 145)
Shizgal, B., 170(330), 171
Shoemaker, C. B., 133(220)
Shoemaker, D. P., 133(220)
Shong, D. S., 112(58)
Shugard, W. J., 93(79)
Shuttleworth, R., 161(311)
Silva, J. M., 33(97)
Singh, H. B., 115(165), 144(165), 175(165)
Singhdeo, A., 112(163)
Sinha, A. K., 133(220)
Sinha, S. K., 5(18), 15(18), 16, 27(18), 40(18), 43(18), 60(18), 61(18), 67(18), 72(18), 133
Sirota, E. B., 5(18), 15(18), 16, 27(18), 40(18), 43(18), 60(18), 61(18), 67(18), 72(18)
Sjögren, L., 196(44, 46)
Sjolander, A., 196(48)
Smalley, M. V., 29(83)
Smart, J., 108(134), 109(134), 110(134)
Smith, C. S., 112(159), 119(159)
Smith, J. S., 105(115)
Smith, K., 6(26)
Smith, P., 70(164)
Smits, C., 19(75)
Snook, I., 36(107), 48(130), 162(314)
Sogami, I. S., 14(59), 18(63), 19(66–68), 27(59, 66, 67), 28(66, 67), 29
Sood, A. K., 3(13, 14), 4(14), 5(14), 6(14), 27(80), 43(115), 44(122), 45, 48(127), 50(132), 51(132), 52(132), 55(13), 58(13), 59, 60
Southin, R. T., 112(160)
Spaepen, F., 90(57), 110(137), 122(176, 177), 123(178), 127(198), 128(205), 129(207), 133(198, 224), 143(176, 177), 144(176, 177), 145(207), 174(339)

Stauffer, D., 155(289), 172(334)
Steinberg, J., 112(154)
Steinhardt, P. J., 72(168), 132(218), 200(81), 232(81), 233(81)
Stillinger, F. H., 200(87, 88), 201(88), 223
Stoll, E., 162(313)
Stookey, S. D., 138(266)
Stowell, M. J., 108(134), 109(134), 110(134)
Strnad, Z., 150(187)
Strober, W., 5(19)
Strzelcki, L., 14(58), 63(145)
Su, Z. B., 219(113)
Sugiura, M., 27(78)
Sundquist, B. E., 161(312), 172(335)
Sur, A., 172(334)
Swalin, R. A., 116(167)
Szilard, L., 81(28), 85(28)

T

Takahashi, T., 104(107), 108(107), 109(107), 143(107)
Takan, K., 63(151)
Takano, K., 14(55), 30(87), 31(87)
Tammann, G., 79(29)
Tanemura, M., 158(300), 199(68), 201(168)
Tang, Y., 63(147), 64(147), 65(147), 66
Tashiro, M., 137(238, 247), 138(247), 139(247), 140, 142(238), 148(247), 149(247)
Tata, B. V. R., 3(13), 5(13), 27(80), 44(122), 45, 48(127, 129), 50(132), 51(132, 133), 52(132), 55(13), 58(13), 59(13), 60, 63(152)
Tausk, R. J., 2(4)
Teller, A. H., 156(292)
Teller, E., 156(292)
Tempkin, D. E., 90(55)
Theodoor, J., 28(82)
Thirumalai, D., 22(70), 23, 33(60), 34(70), 104, 36(70), 37(70), 39(70), 61(141), 70(161)
Thompson, C. V., 87(52), 88(52), 89(52), 90(57), 99(52), 100(52). 101, 102(52), 129(207), 144(207), 149(52), 174(339)
Tiller, W. A., 91(63), 104(107), 108(107), 109(107), 143(107)
Tiwari, R. S., 124(183), 128(203, 204), 129
Toda, M., 31(90)
Tolman, R. C., 161(309)
Torroja, J. M. S., 156(297)
Toschev, S., 138(268), 145(279)
Tough, R. J. A., 46(124), 71(167)

Trinh, E. H., 111(147)
Tsai, A., 132(216), 134(229)
Turnbull, D., 76(1), 82(31, 37), 83(40), 85(31), 88(31), 91(59), 93(81), 99(81), 104(1, 106, 112), 105(118–120), 108–109(126, 130, 133), 110(126, 137, 139), 111(126, 143), 113(1, 37), 115(1), 116, 118(1), 120, 121(1, 173), 122(140, 173, 175), 123(178), 136(40), 142(1, 40), 143(175), 144(1, 175), 145(40), 173, 177(344), 192(15–20), 196(33–34)
Turq, P., 25(76), 26(76)
Turski, L. A., 82(36), 166(36)

U

Ueda, A., 158(300), 199(64, 68), 201(68)
Ueno, Y., 43(118)
Ugelstad, J., 8(41)
Uhlmann, D. R., 137(240, 258), 138(261), 142(240, 258, 264), 143(264)
Ujiie, Y., 112(158)
Underwood, E. E., 124(184)

V

van den Hul, H. J., 2(4, 5), 9(40), 10(5), 12(5), 13(5)
Vanderhoff, J. W., 2(4, 5), 9(40), 10(5), 12(5), 13(5)
van Helden, A. K., 5(20)
van Megen, W., 3(15, 16), 6(15), 16(15, 60), 17, 18(61), 25(16), 36(107), 48(130), 56(61), 62(15, 16, 61, 142), 72(61), 162(314)
van Winkle, D. H., 63(146), 64(146), 65(146)
Verlet, L., 50(131), 199(60), 200(74)
Versmold, H., 25(77), 26(77), 50(77), 52
Verwey, E. J. W., 2(1), 3(1), 4(1), 10(1), 19(1), 21(1)
Violette, C., 79(14)
Vitek, V., 196(40)
Vold, C. L., 117(168)
Vold, M. J., 10(45), 19(45)
Vold, R. J., 10(45), 1945)
Volmer, M., 77(2), 81(2), 85(2, 47, 48), 87(2)
Volterra, V., 100(101)
von Dusch, N., 79(13)
von Heimendahl, M., 124(183)
Vonnegut, B., 82(38), 105(38)
Vrij, A., 5(20, 21), 6(23, 28)

W

Wadati, M., 31(90), 33(97)
Wagner, C. N. J., 57(140)
Wainwright, T. E., 30(85), 198(51–55)
Wakeshima, H., 98(92)
Walker, J. L., 110(141)
Walton, A. G., 117(169), 118(169), 121(174), 123(174)
Wang, C. C., 112(159), 119(159)
Wang, S., 201(96, 97)
Wang, T., 137(241), 147(241), 148(241)
Ward, R. G., 108(129), 109(129), 111(144, 145)
Warten, R. M., 100(102)
Weber, A., 77(2), 81(2), 85(2), 87(2)
Weber, T. A., 200(87, 88), 201(88), 223
Weinberg, M. C., 115(166), 127(199), 135(240), 140(271), 142(240), 144(166), 149(278, 274)
Weiss, J. J., 42(112)
Weitz, D. A., 7(30, 31), 33(99), 58(99), 68(30, 31), 70(31), 71(167)
Wendt, H. R., 201(102), 209(102), 210(102)
Wennerström, H., 13(51)
Wesslau, H., 13(52)
Wiese, G. R., 21(69)
Wilkinson, M. C., 2(7), 8(7)
Willens, R. H., 123(179)
Williams, M., 195(32)
Williams, R., 7(29), 14(56), 67(29)
Williams, R. C., 6(26)
Winter, B., 55(139)
Witten, T. A., Jr., 217(110)
Wolynes, P. G., 199(65)
Wong, J., 186(9)
Wood, G. R., 121(174), 123(174)
Woodcock, L. V., 159(306), 162(313), 198(50, 56, 57), 199(61–63), 200(89, 90)
Wyckoff, R. W. G., 136(250)

Y

Yates, D. E., 31(92)
Ye, H. Q., 132(216)
Yeh, Y., 48(130)
Yerazinus, S., 55(139)
Yinnon, H., 136(258), 142(258)
Yonezawa, F., 57(140), 200(76–80, 85, 86), 205(76), 208(76), 209–218(76), 219(77–80, 85, 86, 111, 112, 114), 220–222(111), 223(78, 111), 224(78, 111), 225(76–80, 85, 86, 111, 112, 115), 230(76), 231, 233(76–80, 85, 86, 111, 112, 115)
Yoshida, H., 43(118)
Yoshimura, S., 55(136), 63(151), 72(135, 136)
Yoshimura, Y., 55(135, 136), 56(135), 72(136)
Yoshiyama, T., 14(59), 18(63), 27(59)
Young, A. P., 64(155)
Young, D. A., 30(86)
Yussouff, M., 33(102), 52(102), 54(102), 166(325), 169

Z

Zachariasen, W. H., 185(1), 186(1)
Zallen, R., 185(4)
Zanotto, E. D., 135(233), 136(243, 256), 138(243), 139(243), 140(243), 141(243), 148(243), 149(278, 284)
Zeldovich, J. B., 85(49), 88(49), 96(49)
Zettlemoyer, A. C., 93(76)
Zhang, Z., 132(216)
Zlateva, E., 145(281)

Subject Index

A

Activation energies, transient nucleation, 147–148
Alder transition, 198–199
Alkali metal potential, glass transition, 201
Alkali metals
  glass transition, pair potentials, 201
  Stilliger–Weber potential, 223–224
  pair potential, 224
  potential dependence, 222–224
Amorphous solid, 181
  atomic configurations, computer graphics, 249–253
  condensed matter, 183–184
  free energies, 181–183
  preparation
    melt-quench method, 181
    quenching, 181
Amorphous structures
  isotropic atoms, 184
  local icosahedral symmetry, 232–233, 251
  tessellation, 227
  Voronoi polyhedra
    distortions, 229–230
    distribution of shape parameters, 229–230
    face parameter, 231–233
    fivefold symmetry, 232–233
    free volume, 228, 230
    local symmetry, 230–233
Anisotropic phases, rod-shaped colloidal particles, 70–71
Annealing
  Gaussian curves, 243–246
  glass A, 253
  glass B, 252–253
  icosahedral clusters, 251–253
  metallic glasses, temperature, 130
  microscopic structure parameters
    crystallization, 240–243
    relaxation, 246–249
  pair distribution function, 235–238
  relaxation
    glass, 234–246
    Lennard–Jones glass, 234–246
    Lennard–Jones liquid, 234–246
    microscopic structure parameters, 240–243, 246–249
    supercooled liquid, 234–246
  structure relaxation, crystallization, 235–240
  treatments, lithium disilicate glass, 152–153
  V-T relation, 234–235
  Voronoi polyhedra
    distortion parameter, 240–243
    distribution of volumes, 243–246
    face parameter, 240–243
    fivefold symmetry, 242–243
    fluctuation parameter, 243
    local symmetry parameter, 240–243
Annealing time, crystal nuclei, lithium disilicate, 153
Aqueous suspensions, charge-stabilized polystyrene spheres, 3–5
Arrhenius rate law, viscosity, glass transition, 187–188
Arrhenius relation, diffusion constant, glass transition, 215
Atomic configuration
  amorphous structures, local icosahedral symmetry, 249–250
  computer graphics
    amorphous solids, 249–253
    glasses, 249–253
  crystalline symmetry, 249–250
  glasses, relaxation processes, molecular dynamics methods, 197–201
  local icosahedral symmetry, amorphous structures, 249–250
  quench rate

SUBJECT INDEX

Atomic configuration (*cont'd.*)
    double-shell icosahedra, 251–252
    icosahedra, 249–253
Attractive potential
    Hamaker constant, 22–23
    London–van der Waals interaction, 22–24

B

Barker–Henderson perturbation theory, effective hard sphere diameter, 32
bcc
    colloidal structures, 4
    Wigner crystals, 32
bcc crystal, Stillinger–Weber potential, quench rate, 224
bcc crystal phase, stability, 38–39
bcc-fcc crystalline phases
    free energies, 33–36
    Lindemann parameter, 33, 36
    self-consistent harmonic approximation, 33–36
bcc-fcc phase boundary
    computer simulation, 37–39
        free energy, 37
        Lindemann criterion, 38
        molecular dynamics computer simulations, 39–40
    density functional theory, 39–40
    harmonic oscillators, 34–36
    Lindemann criterion, 34
    phase diagram, 34–36
    Yukawa potential, 34
bcc-fcc phase diagram, self-consistent phonon theory, 34–36
bcc-fcc transitions
    colloidal polyball suspensions, 30–41
    polyball monodisperse suspensions, 13–16
bcc nucleation, steady-state nucleation, 159
bcc structure, polyball monodisperse suspension, 14
bcc to fcc, monodisperse colloidal suspensions, lattice structure, 13–16
Becker–Doring expression, 156
Becker–Doring formalism, nucleation, condensed systems, 88–89
Bimolecular reactions, cluster size, 86
Binary alloy, phase separation, spinodal mechanism, 80

Binary charged colloidal suspensions, polyballs, MD simulations, 61
Binary mixture
    colloidal alloys, 55–56
    glassy state
        colloids, 56–61
        structure factor, 59–60
    hard sphere colloidal particles, 56
    hard spheres, 56
    order formation, 55
    packing fraction, 56
    packing model, 55–56
Bond-orientational correlation function, 65
Born, Green, and Yuon equation, colloidal liquids, 46
Bragg diffraction, colloidal suspensions, 4
Brownian dynamics, colloidal suspensions, 7
Brownian motion, polyballs, 4
Bulk modulus, colloidal crystals, 66–67
Bulk nucleation, glasses, 124

C

Cage structure
    glass transition, 196, 218
    microscopic structure parameters, 226–227
Capillarity approximation
    cluster size, 160
    free energy, steady-state nucleation, 168
Classical theory
    cluster formation, 81–82
    nucleation
        analysis of, 159–161
        cluster distributions, 160
        kinetic model, 151–154
        nucleation rate, 160
    nucleation rate, 82
    steady-state nucleation, 83–93
    time-dependent nucleation, 93–103
Closure approximation, structure factor, 48
Cluster(s)
    fcc, icosahedra, quench rate, 251–252
    hcp, icosahedra, quench rate, 251–252
Cluster density
    homogeneous nucleation, multicomponent system, 90
    time-dependent, 86
    time-dependent critical, 95

SUBJECT INDEX

Cluster distribution
  classical theory of nucleation, 160
  continuum form, Kramers-Moyal
    equation, 170
  Kashchiev expression, 102-103
  non-steady-state nucleation, 170-171
    diffusion coefficient, 170-171
    drift coefficient, 170-171
    quench rate, 171
  time-dependent nucleation rate, 102
Cluster evolution
  dynamical model, 172
  master equation, coupled differential
    equations, 86
  numerical solutions, 99
Cluster formation
  canonical ensemble, Monte Carlo
    simulation, 156
  classical theory, 81-82
  free energies, Monte Carlo simulation, 156
  thermodynamic fluctuations, 84
  transient nucleation, 96
  undercooling, steady-state nucleation, 168
Cluster growth, steady-state nucleation, 85
Cluster growth rate, silicate glasses,
  nucleation rate, 126
Cluster interface, unbiased molecular, jump
  frequency, 89
Cluster rotation, vapor condensation, 92
Cluster size
  bimolecular reactions, 86
  capillary approximation, 160
  steady-state nucleation, 87-88
  time-dependent flux, 87
Cluster size and shape
  fluctuations, steady-state nucleation, 164
  Langevin force, fluctuations, 164-165
Cluster size distribution
  equilibrium, steady-state nucleation, 84
  nucleation, silicate glasses, 153-154
  silicate glasses, nucleation, 153-154
Cluster translation, vapor condensation, 92
Coarse-grained free-energy density, steady-
  state nucleation, 163, 165
Coated mercury droplets, nucleation rate,
  118
Cold-worked aluminum, recrystallization
  rates, 99
Colloid(s), 2-3
  binary mixture, glassy state, 56-61
Colloidal aggregates, 71
  density-density correlation function, 71

Colloidal alloys, binary mixtures, 55-56
Colloidal crystal-glass surface, two-
  dimensional colloidal crystals, 63
Colloidal crystallite
  growth kinetics, 18
  time evolution, 18-19
Colloidal crystals
  bulk modulus, 66-67
  elastic constants, 66-67
  flow behavior, 68-70
  Kossellines, 14
  nucleation, 18-19
  ordering, 69
  sheared, ordering of, 69
  shear melting, 68-70
  shear-melting transitions, 68-70
  shear modulus, 66-67
  shear stress, 68-70
  shear stress application, 68-70
  two-dimensional, 62-65
  Young modulus, 66-67
Colloidal glass, 56
  dynamic structure factor, 62
  scattering experiments, 58-60
  shear modulus, 57
Colloidal liquids
  Born, Green and Yuon equation, 46
  Coulomb coupling, 47
  Gibbs free energy, 32
  Ornstein-Zernike equation, 47
  perturbation theories, 32
Colloidal particles
  hydrodynamic interactions, 71
  interparticle interactions, 71
  rod-shaped, 70-71
  surface charge, 10-12
    conductivity, 10
  surface potential, 10-12
Colloidal polyball suspensions
  bcc-fcc transition, 30-41
  effective hard sphere diameter, 31-32
  fluid-crystal transition, 30-41
Colloidal sphere(s)
  nearly hard, 5-6
  soft, 3-5
Colloidal sphere systems, disorder-order
  transition, 6
Colloidal structure
  bcc, 4
  polyballs, 4
  DLVO potential approximation, 47-48
  fcc, polyballs, 4

Colloidal structure (*cont'd.*)
  natural, 6
  Sogami potential approximation, 48
Colloidal suspensions
  binary charged polyballs, MD simulations, 61
  Bragg diffraction, 4
  Brownian dynamics, 7
  conventional atomic systems and, 6–7
  elastic constants, 6–7
  energy scale, 7
  glassy state, 4–5
  interaction potential, 42
  length scale, 6
  liquidlike short-range order, 41–42
  particle density, 6
  phase diagram, PMMA particles, 16–17
  polyball, 3–4
  polydispersity, 7
  quasicrystalline phases, 72
  structure factor, 42–52
  supramolecules, 6
Colloidal systems
  model, 3–6
  ordering, 6
Colloid crystallites, ordering
  growth velocity, 18
  time evolution, 18–19
Computer graphics, atomic configurations
  amorphous solids, 249–253
  glasses, 249–253
Computer simulations
  glass transition, 197–201
  microscopic structure parameters
    cage structure, 226–227
    crystal, 224–233
    distortions, 229–230
    fluctuations, 227–228
    glass 224–233
    hidden structures, 225–227
    liquid, 224–233
    local symmetry, 230–233
    Voronoi polyhedra, 227–228
  molecular dynamics, 36–39
    phase diagram, 37–39
    Yukawa potential, 37
  polyballs, 37–39
  physical properties, microscopic structure parameters, 224–233
  steady-state nucleation
    molecular dynamics, 157–159
    Monte Carlo calculations, 155–156
    phase transformation, 155–159
Computer simulations versus analytic theories, bcc-fcc phase boundary, 37–39
Computer studies, disordered structures,
Condensed matter, 183–184
  amorphous solid, 183–184
  crystal, 183–184
  liquid, 183–184
  temperature range, 183–184
Condensed systems
  nucleation, Becker–Doring formalism, 88–89
  phase transitions, Gibbs free energy, 83
Conductivity, surface charge, colloidal particle, 10
Configuration, steady-state nucleation
  metastable and stable, 162
  saddle-point, 162
  time-dependent, 165
Constant-pressure simulations, glass transition
  molecular dynamics simulations, 219
  quench rate dependence, 219
Containerless solidification technique, 111–112
Continuity of volume, glass transition, 193
Continuous random network
  covalent materials, 185
  Zachariasen's model, 185
Continuum approximation, 88, 164
Continuum form, cluster distribution, Kramers–Moyal equation, 170
Conventional atomic systems, colloidal suspensions, 6–7
Cooling rate, metallic glass, 123
Correlation lengths, melting transitions, two-dimensional colloidal crystals, 65
Couette cell, shear experiments, 70
Coulomb coupling, colloidal liquids, 47
Coulomb interaction, interparticle potential, 19
Coupled differential equations, cluster evolution, 86
Coupling constant, one-compound plasma, 34
Covalent materials, continuous random network, 185
Covalent potential, glass transition, 200–201
  pair potential, 200–201

SUBJECT INDEX

Critical cluster, steady-state nucleation
 field-theoretic models, 162
 formation, 164
Critical cluster size, steady-state nucleation, 85
Critical size, time-dependent nucleation rate, 96, 99
 silicate glasses, 126
Crystal
 condensed matter, 183–184
 fcc, molecular dynamics technique, 206
 microscopic structure parameters, 224–233
 number density, 203–204
Crystal-fluid phase transition, polydispersity, 52–55
Crystal-forming quench rates, Lennard–Jones system, 222
Crystal growth, steady-state nucleation, structure factor, 158
Crystalline compound structures, intermetallic compounds, 55
Crystalline phases, bcc-fcc
 free energies, 33–36
 Lindemann parameter, 33, 36
 self-consistent harmonic approximation, 33–36
Crystalline symmetry
 atomic configuration, 249–250
 coexistence with icosahedral symmetry, 251–253
Crystal (bcc)-liquid interface, structure factor, 50
Crystal (ordered) transition, 13–16
Crystal-liquid transition, polyball monodisperse suspensions, 13–16
Crystallites, spherical, 130
Crystallization
 annealing
  microscopic structure parameters, 240–243
  structure relaxation, 235–240
 glasses
  eutectic, 124
  polymorphic, 124
  primary, 124
 hard sphere, 30
 liquid metals, 103–104
 lithium disilicate, 135–137
 quench rate, temperature, 221
 seeding, 79
 water, undercooling, 78
Crystal-melt interface, 162
Crystal nucleation, kinetic theory, 192–193
Crystal nuclei, lithium disilicate, 153
 annealing time, 153
Crystal phase, bcc, stability, 38–39
Crystal size
 nucleation mechanisms and, 124
 silicate glasses, nucleation rates, 125–126
Crystal structure, Voronoi polyhedra
 steady-state nucleation, 158–150
 Wigner–Seitz cell, 158–159
Crystal-vacuum interfacial energy, liquid metals, 115–116
Curvature-dependent interfacial energy, 161
 Tolman approximation, 161

D

Debye–Huckel equation
 effective charge, 26
 repulsive interaction, 20
Degree of undercooling, droplet coating, 112–113
Dense random packing of hard spheres, 204–205
 metallic glasses, 185–186
 quenching, pair distribution functions, 204–205
 quench simulation, pair distribution functions, 204–205
Density–density correlation function, colloidal aggregates, 71
Density functional model, free energy, steady-state nucleation, 168
Density functional theory
 bcc-fcc phase boundary, 39–40
 of freezing, hard spherical colloids, 54
 molecular dynamics computer simulations, 39–40
Derjaguin–Landau–Verway–Overbeek potential, see DLVO
Devitrification, 124
 metallic glasses, 128, 130
 surface crystallization, 130–132
 silicate glasses, 135
DFT, see Density functional theory
Diffusion coefficient, cluster distribution, non-steady state nucleation, 170–171
Diffusion constant
 Arrhenius relation, glass transition, 215

Diffusion constant (*cont'd.*)
  glass transition, 187
    molecular dynamics simulation, 215
    structure relaxation, 215
Dilatometric measurements, mercury emulsions, 75–76
Disorder–order transition, colloidal sphere systems, 6
Distortion parameter, Voronoi polyhedra, annealing, 240–243
Distortions
  microscopic structure parameter, 229–230
  Voronoi polyhedra, amorphous structures, 229–230
Distribution function, Finney's DRPHS structure, 192
Distribution of shape parameters, Voronoi polyhedra, amorphous structures, 229–230
Distribution of volumes, Voronoi polyhedra, annealing, 243–246
DLVO (Derjaguin–Landau–Verwey–Overbeek) potential, 4
  colloidal structures, approximations, 47–48
  dimensionless form, 34
  generalized Madelung lattice sum, 32–33
  lattice sum, 32–33
  molecular dynamics simulation, 39
  phase diagrams, polyball suspensions, 72
  Poisson–Boltzmann equation, 24–25
  polyball surface, 24–27
DLVO (Derjaguin–Landau–Verwey–Overbeek) theory
  effective charge, 24–27
  interparticle interaction potential, 19–27
  repulsive interaction, polyball surface charge, 19–22
Double-shell icosahedra, atomic configuration, quench rate, 251–252
Drift coefficient, cluster distribution, non-steady state nucleation, 170–171
Droplet
  coated mercury, nucleation rate, 118
  coating, degree of undercooling, 112, 113
Droplet morphology, spinodal mechanism, 81
Droplet size, undercooling, degree of, 113
Droplet technique, heterogeneous site removal, 105–110, 112

DRPHS, *see* Dense random packing of hard spheres
Dynamical factor, hydrodynamical treatment, steady-state nucleation rate, 169
Dynamical model, cluster evolution, 172
Dynamical prefactor, saddle-point region steady-state nucleation, 165–166
Dynamical properties, MD simulation, glass transition, 213–218
Dynamic light scattering, macromolecular suspensions, 44–46
Dynamic structure factor
  colloidal glass, 62
  macromolecular suspensions, 44–46

E

Effective charge
  Debye–Huckel equation, 26
  DLVO theory, 24–27
  models, polyball, 26–27
  Poisson–Boltzmann-cell model, 26
  Poisson–Boltzmann equation, 26
  Poisson–Boltzmann-jellium approximation, 26–27
Effective hard sphere diameter
  Barker–Henderson perturbation theory, 32
  colloidal polyball suspensions, 31–32
Effective hard sphere model, fluid-crystal transition, 30–32
Elastic constants
  colloidal crystals, 66–67
  colloidal suspensions, 6–7
Electrostatic interaction
  macroionic solution, 27–29
  polyball suspensions, 27–29
  repulsive potential, polyballs, 72
Electrostatic repulsion
  interaction potential, 19–20
  polyballs, 21
Emulsion polymerization
  micelle, 8
  polyball colloids, 8–10
  swollen, 8
Emulsion polymerization technique, polymeric monodisperse particles, 8
Emulsion technique, heterogeneous site removal, 105–110

SUBJECT INDEX

Energy barrier, steady-state nucleation, 163–164
Energy scale, colloidal suspensions, 7
Enthalpy
 of fusion, free-energy approximations, 173–177
 per particle, MD simulation, glass transition, 211–212
Entrained droplet technique, heterogeneous site removal, 112
Entropy
 continuous, glass transition, 193
 of fusion, free-energy approximations, 173–177
 glass transition, 194
Euler equation, 163
Eutectic glass crystallization, 124
Eutectic transformation, metallic glasses, 130
Excluded volume effect, hard sphere, phase transition, 31

F

Face-centered cubic, see fcc
Face parameter, Voronoi polyhedra
 amorphous structures, 231–233
 annealing, 240–243
fcc clusters, icosahedra, quench rate, 251–252
fcc colloidal structure, polyballs, 4
fcc crystal
 initial system, molecular dynamics technique, 206
 number density, 203–204
fcc structure
 polyball monodisperse suspension, 14
 Voronoi polyhedron indices, 230
$Fe_4^0Ni_{40}P_{14}B_5$, homogeneous nucleation mechanism, 128
Field-theoretic models, steady-state nucleation, 162–169
 critical cluster, 162
 energy barrier to nucleation, 163–164
 free-energy density functional, 162
 nucleation, phase transformations, 162
Finney's DRPHS structure, distribution function, 192
Fivefold symmetry, Voronoi polyhedra
 amorphous structures, 232–233
 annealing, 242–243

Flow behavior
 colloidal crystals, 68–70
 glass transition, 187
Fluctuation
 cluster size and shape
  Langevin force, 164–165
  steady-state nucleation, 164
 microscopic structure parameter, computer simulation, 227–228
Fluctuation parameter, Voronoi polyhedra, 228
 annealing, 243
Fluctuation process, nucleation, 76
Fluid-crystal transition
 colloidal polyball suspensions, 30–41
 effective hard sphere model, 30–32
 repulsive potential, 30
Fluidity, free-volume theory, glass transition, 195
Fluxing technique, heterogeneous site removal, 110–111
Fokker–Planck equation, 88, 164, 170
 non-steady-state nucleation, 171
Forward Kramers–Moyal equation, 169
Free energy
 amorphous solids, 181–183
 bcc-fcc crystalline phases, 33–36
 bcc-fcc phase boundary, computer simulations, 37
 cluster formation, Monte Carlo simulation, 156
 phase space, 181–183
 Singh–Holz approximation, 141
 steady-state nucleation
  capillarity approximation, 168
  density functional model, 168
  undercooling, 168
Free-energy approximations, liquid-crystal, 173–177
 enthalpy of fusion, 173–177
 entropy of fusion, 173–177
Free-energy density functional, steady-state nucleation, field-theoretic theories, 162
Free-energy difference, liquid-crystal, 173
Free-energy functional, steady-state nucleation, 163
 coarse-grained, 165
Free energy surface
 saddle-point configuration, 164

Free energy surface (*cont'd.*)
  saddle transition configuration, 164
Free volume, Voronoi polyhedra,
    amorphous structures, 228, 230
Free-volume theory
  glass transition, 194–196
    cage structure, 196
    fluidity, 195
    Vogel–Fulcher relation, 195–196
  van der Waals volume, 195
Freezing criterion, Hansen–Verlet, 50
Freezing temperature, glass transition, 186
Freezing transition
  hard sphere, phase transition, 30–31
  structure factor, 50–52
Fulcher–Vogel expression, silicate glasses, 140

G

Gallium, solidification rate, 122
Gaussian core potential
  glass transition, 200
  pair potential, 200
Gaussian curves, annealing, 223–246
Generalized Madelung lattice sum, DLVO potential, 32–33
Geometrical factor, repulsive potential, polyball surface, 22
Gibbs–Duhem equation, 29
Gibbs free energy
  colloidal liquids, 32
  condensed systems, phase transition, 83
  mercury emulsion liquid, 76
  undercooled liquids, 76
Gibbs free energy density functional, steady-state nucleation, 162
Glass, 181
  annealing, relaxation, 234–246
  atomic configuration, computer graphics, 249–253
  bulk nucleation, 124
  crystallization
    eutectic, 124
    polymorphic, 124
    primary, 124
  formation
    conditions needed for, 192–193
    from non-glass-formers, quench techniques, 184
  quench rate, 192–193
  microscopic structure parameters, computer simulations, 224–233
  multicomponent systems, 124
  nucleation experiments, 123–154
  preparation, melt-quench method, 181
  time-dependent nucleation, 123
Glass A
  annealing, 253
  higher quench rate, 246
Glass B
  annealing, 252–253
  lower quench rate, 246
Glass-forming quench rates, Lennard–Jones system, 222
Glasslike behavior, polyball suspensions, 57
"Glass" phase, hard sphere model, 17
Glass silicates, *see* Lithium disilicate glass
Glass transition, 186–192
  Alder transition, 198–199
  alkali metal
    pair potentials, 201
    potential, 201
  cage structure
    free-volume theory, 196
    support for, 218
  computer simulations, 197–201
  constant pressure simulations, quench rate dependence, 219
  continuity of volume, 193
  covalent potential, 200–201
    pair potential, 200–201
  diffusion constant, 187
    Arrhenius relation, 215
    molecular dynamics simulation, 215
    structure relaxation, 215
  dynamical properties, 213–218
  entropy
    continuous, 193
    of fusion, 194
  experimental results, 186–187
  flow properties, 187
  fluidity, free-volume theory, 195
  free-volume theory, 194–196
  freezing temperature, 186
  Gaussian core potential, 200
    pair potential, 200
  hard core potential, 198–199
  hidden structures, pair distribution functions, 225–227

SUBJECT INDEX

Kauzmann paradox, 221
kinetic theory, 192–193
Lennard–Jones potential, 199–200
Lennard–Jones system
  quenching, 219–224
  spherical particles, 202
liquid metals, pair distribution function, 190
metallic glasses, pair distribution function, 190
metal-metalloid glasses, 190
  local atomic configurations, 188–192
  pair distribution function, 190
molecular dynamics simulation
  dynamical properties, 213–218
  enthalpy per particle, 211–212
  isothermal compressibility, 213
  mean square displacement function, 213–215
  Newtonian equations of motion, 206
  pair distribution functions, 206–210
  physical properties, 205–218
  pressure, 211
  quench rate, 206
  self-diffusion constant, 213
  shear viscosity, 216–217
  specific heat, 212
  stress autocorrelation function, 216–217
  structural properties, 206–210
  structure factor, 206–210
  temperature, 210–211
  thermodynamical properties, 210–213
  velocity autocorrelation function, 213–218
  Verlet algorithm, 205–206
  V-T curve, 210–211
  Wendt–Abraham parameter, 209–210
non-network liquids, 186–188
pair potential, Lennard–Jones potential, 199–200
percolation theory, free-volume theory, 196
physical properties, 205–218
power spectrum, Lennard–Jones glass, 218
Prigogine–Defay ratio, 193–194
quenching, Lennard–Jones system, 202–205, 219–224
  V-T curve, 219–220
reduced radial distribution function, metallic glass, 190
relaxation, quench rate, 246–249
second order phase transition, 193
simple ionic potential, 200–201
  pair potential, 200–201
soft core potential, 198–199
  pair potential, 199
Stillinger–Weber potential, quench rate, 224
strong network liquids, 186–188
structural properties, 206–210
  glass transition temperature, 209–210
temperature, 186
  quench rate, 202, 220
  structural properties, 209–210
  thermodynamical properties, 210–213
  temperature dependence, quench rate, 188–189
theoretical approaches, 192–196
thermodynamic properties, 193–194, 210–213
  Lennard–Jones fcc crystal, 210–211
  Lennard–Jones glasses, 210–211
velocity autocorrelation function, Lennard–Jones fcc crystal, 217
velocity correlation function, vibrational behaviour, 215, 217–218
viscosity, 187
  Arrhenius rate law, 187–188
  Vogel–Fulcher law, 187–188
  Vogel–Fulcher relation, free-volume theory, 195–196
Glassy phase
  binary mixture, 56–61
  colloids, 56–61
  structure factor, 59–60
colloidal suspensions, 4–5
polyball suspensions, 59
preferred length scale, 61–62
metastable states, 189
monodisperse charged polyball suspensions, 61–62
monodisperse nearly hard spherical particles, 62
monodisperse polyball suspensions, 61–62
Glassy structures, isotropic atoms, 184
Gram-atomic interfacial energy, 116
Gravity, heterogeneous nucleation, 111–112
Growth kinetics, colloid crystallite, 18
Growth mechanism, ordered phase, 18–19

Growth rate, silicate glasses
  nuclei, 148
  polymorphic crystallization, 127
Growth time, silicate glasses, nuclei, 148
Growth velocity, ordering, colloid crystallites, 18

H

Hamaker constant, attractive potential, London–van der Waals interaction, 22–23
Hansen–Verlet freezing criterion, 50
Hard core potential
  glass transition, 198–199
  pair potentials, glass transition, 198
Hard sphere(s)
  binary mixture, 56
  crystallization, 30
  excluded volume effect, 31
  freezing transition, 30–31
  packing model, 56
Hard sphere colloidal particles, binary mixtures of suspensions, 56
Hard sphere diameter, effective
  Barker–Henderson perturbation theory, 32
  colloidal polyball suspensions, 31–32
Hard sphere model
  effective, fluid-crystal transition, 30–32
  "glass" phase, 17
Hard spherical colloids
  density functional theory of freezing, 54
  polydispersity, 54
Harmonic oscillators, bcc-fcc phase boundary, 34–36
hcp clusters, icosahedra, quench rate, 251–252
Heat of fusion, liquid-crystal interfacial energy, 116
Heterogeneous nucleation, 77
  gravity, 111–112
  levitation, 111
  liquid metals, 104
  thermodynamic barrier, 91
Heterogeneous site removal
  containerless solidification technique, 111–112
  droplet technique, 105–110
  emulsion technique, 105–110
  entrained droplet technique, 112
  fluxing technique, 110–111
  mush quenching, 112
  substrate technique, 110
  undercooling, 104–117
    liquid metals, 104–117
Hexatic phase, melting transitions, two-dimensional colloidal crystals, 64–65
Hidden structures
  microscopic structure parameters, computer simulation, 225–227
  pair distribution functions
    glass transitions, 225–227
    Lennard–Jones system, 225–227
Hoffman expression, nucleation rate, metallic glasses, 128–129
Homogeneous nucleation, 76–77
  multicomponent systems, 90–91
  silicate glasses, 135–151
  spherical clusters of crystal, 83–91
  undercooled liquids and gases, 77
  undercooling data, 113–114
Homogeneous nucleation mechanism
  $Fe_{40}Ni_{40}P_{14}B_6$, 128
  metallic glasses, 128, 130
Hydrodynamical treatment, steady-state nucleation rate
  dynamical factor, 169
  statistical factor, 169
Hydrodynamic interactions, colloidal particles, 71
Hydrodynamic model, steady-state nucleation, 161, 166, 168–169
Hypernetted chain approximation, structure factor, 47–50
Hysteresis, melting, Lennard–Jones crystal, 224

I

Icosahedra, quench rate
  atomic configuration, 249–253
  double-shell, atomic configuration, 251–252
  fcc clusters, 251–252
  hcp clusters, 251–252
Icosahedral clusters, annealing, 251–253
Icosahedral order, metallic melts, 132–135
Icosahedral phase transformation, nucleation, 133

Icosahedral symmetry
  coexistence with crystalline symmetry, 251–253
  metallic alloys, 132
Infinite-order differential equation, forward Kramers–Moyal equation, 169
Interaction potential
  colloidal suspensions, 42
  electrostatic repulsion, 19–20
  interparticle, DLVO theory, 19–27
  London–van der Waals attraction, 19, 22–24
  polyball surface, 20–21
  structure factor, 48
Interfacial energy
  crystal-vacuum, liquid metals, 115–116
  curvature-dependent, 161
  gram-atomic, 116
  liquid-crystal, liquid metals, 115
  lithium disilicate, 142–144
  nucleation data, 152
Interfacial structure modeling, liquid metals, 104
Intermetallic compounds, crystalline compound structures, 55
Interparticle interaction potential
  DLVO theory, 19–27
  theory, 19–27
Interparticle interactions, colloidal particles, 71
Interparticle interactive repulsion, polyball suspensions, 12–13
Interparticle potential
  Coulomb interaction, 19
  macroionic solution, 28–29
  sphere systems, 5–6
Interparticle structure factor, macromolecular suspensions, 44
Isothermal compressibility, MD simulation, glass transition, 213
Isotropic atoms
  amorphous structures, 184
  glassy structures, 184

J

Johnson–Mehl–Avrami equation, 127
Jump frequency, unbiased molecular cluster interface, 89

K

Kashchiev expression, 96, 100
  cluster distributions, 102–103
  volume nucleation, 148
  volume nuclei, 146
Kasterlitz–Thouless–Halperin–Nelson–Young theory, melting transition, two-dimensional colloidal crystals, 64, 65
Kauzmann paradox, glass transition, 221
Kinetic model, classical nucleation theory, 151–154
Kinetic theory
  crystal nucleation, 192–193
  glass transition, 192–193
Kirkwood–Alder transition, 30
  charged latex particles suspension, 31
Kirkwood superposition approximation, structure factor, 46–47
Kossellines, colloidal crystal structures, 14
Kramers–Moyal equation, 169
  cluster distribution, continuum form, 170
  forward, master equation, infinite-order differential equation, 169

L

Landau formalism, nucleation, 162
Langevin force, fluctuations, cluster size and shape, 164–165
Latex particles suspension, charged, Kirkwood–Alder transition, 31
Lattice structure
  bcc to fcc, monodisperse colloidal suspensions, 13–16
  stability, 33, 39
  one-compound plasma, 33
Lattice sum, DLVO potential, 32–33
Length scale, colloidal suspensions, 6
Lennard–Jones crystal, hysteresis, melting, 224
Lennard–Jones fcc crystal
  glass transition
    power spectrum, 217
    thermodynamic properties, 210–211
    velocity autocorrelation function, 217
  melting, 224
    hysteresis, 224
Lennard–Jones glass

Lennard–Jones glass (cont'd.)
  annealing, relaxation, 234–246
  glass transition
    power spectrum, 218
    thermodynamic properties, 210–211
Lennard–Jones liquid
  annealing, relaxation, 234–246
  number density, 203–204
  quenching, pair distribution functions, 204–205
  quench simulation, pair distribution functions, 204–205
Lennard–Jones potential
  glass transition, 199–200
  pair potential, glass potential, 199–200
  rare gases, 202
Lennard–Jones system
  crystal-forming quench rates, 222
  glass-forming quench rates, 222
  glass transition, spherical particles, 202
  pair distribution functions, hidden structures, 225–227
  quenching
    glass transition, 202–205, 219–224
    molecular dynamics method, 202–205, 219–224
    pressure, 203
    temperature, 203
    thermodynamics, 203
Levitation, heterogeneous nucleation, 111
Light scattering, dynamic, macromolecular suspensions, 44–46
Lindemann criterion, bcc-fcc phase boundary, 34
  computer simulation, 38
Lindemann parameter, bcc-fcc crystalline phases, 33
Liquid(s)
  condensed matter, 183–184
  microscopic structure parameters, computer simulations, 224–233
  nucleation theories, 82
  transition, 13–16
Liquid-crystal
  free-energy approximations, 173–177
  free-energy difference, 173
Liquid-crystal interfacial energy
  heat of fusion, 116
  liquid metals, 115
Liquid crystallization, MD simulation, 158–159

Liquidlike short-ranged order, 41–52
  colloidal suspensions, 41–42
Liquid metals
  crystallization, 103–104
  crystal-vacuum interfacial energy, 115–116
  heterogeneous nucleation, 104
  interfacial structure modeling, 104
  liquid-crystal interfacial energy, 115
  pair distribution functions, glass transition, 190
  undercooling, 103–104
    heterogeneous site removal, 104–117
  viscosity, 115
Liquid-solid, undercooling, steady-state nucleation, 167–168
Lithium disilicate glass
  annealing time, crystal nuclei, 153
  annealing treatments, 152–153
  crystal nuclei, 153
  crystallization, 135–137
  interfacial energy, 142–144
  temperature-dependent steady-state nucleation rates, 137–140
  time-dependent nucleation, 94–95, 100
  transient nucleation, 95
  time lag, 149
  viscosity, 140
Local icosahedral symmetry, amorphous structures, 232–233, 249–250, 251
Local symmetry,
  microscopic structure parameters, computer simulations, 230–233
  Voronoi polyhedra, amorphous structures, 230–233
Local symmetry parameter, Voronoi polyhedra, annealing, 240–243
London–van der Waals attraction, interaction potential, 19, 22–24
London–van der Waals interaction
  attractive potential, 22–24
  Hamaker constant, 22–23
Lothe–Pound theory, 156

M

Macroionic solution
  electrostatic interaction, 27–29
  interparticle potential, 28–29
  repulsive potential, 28
Macroions, 3

SUBJECT INDEX

Macromolecular suspensions
  dynamic light scattering, 44–46
  dynamic structure factor, 44–46
  interparticle structure factor, 44
  multiple scattering, 44
  particle scattering form factor, 43
  scattered electric field, 43
  scattering wave vector, 43–46
Master equation
  cluster evolution, 86
  forward Kramers–Moyal equation, infinite-order differential equation, 169
MD simulations, see Molecular dynamics simulations
Mean spherical approximation, structure factor, 47–50
Mean square displacement function, 213–215
Melting, Lennard–Jones fcc crystal, 224
  hysteresis, 224
Melting temperature, maximum undercooling and, 113–115
Melting transitions, two-dimensional colloidal crystals, 64–65
  correlation lengths, 65
  hexatic phase, 64–75
  Kasterlitz–Thouless–Halperin–Nelson–Young theory, 65
  power-law exponents, 65
  translational correlation function, 65
Melt-quench method, amorphous solids, preparation, 181
Mercury droplets, coated, nucleation rate measurements, 118
Mercury emulsion liquid, Gibbs free energy, 76
Mercury emulsions
  dilatometric measurements, 75–76
  undercooling, 76
Metallic alloys, icosahedral symmetry, 132
Metallic glasses
  cooling rate, 123
  dense random packing of hard spheres, 185–186
  devitrification, 128, 130
    surface crystallization, 130–132
  eutectic transformation, 130
  homogeneous nucleation mechanism, 128, 130
  nucleation rate, 127–135

Hoffman expression, 128–129
  pair distribution functions, glass transition, 190
  quasicrystals, 132–135
  quenched-in nuclei, 130, 134
  reduced radial distribution function, glass transitions, 190
  spherical crystallites, 130
  steady-state nucleation rate, 128, 130
  transient nucleation, 128–132, 134
  viscosity
    nucleation rate, 128
    Stokes–Einstein relation, 128
Metallic melts, icosahedral order, 132–135
Metal-metalloid glasses, glass transitions
  local atomic configurations, 188–192
  pair distribution functions, 190
Metastability limit, super saturated solutions, 79
Metastable and stable configuration, steady-state nucleation, 162
Metastable regions, phase transformations, 79–80
Metastable states, glassy states, 189
Micelle
  emulsion polymerization, 8
  monodispersity, 8
Microscopic structure parameters
  annealing, crystallization, 240–243
  computer simulations
    cage structure, 226–227
    crystal, 224–233
    distortion, 229–230
    fluctuations, 227–228
    glass, 224–233
    hidden structures, 225–227
    liquid, 224–233
    local symmetry, 230–233
    physical properties, 224–233
    Voronoi polyhedra, 227–228
  relaxation
    annealing, 246–249
    annealing process, 240–243
  R-tau curve, 227
Model colloidal systems, 3–6
  Monte-Carlo simulation, 36–39
  steady-state nucleation, 157–159
Molecular dynamics simulations, 36–39
  bcc-fcc phase boundary, 39–40
  crystal nucleation from liquid, 157–159
  density functional theory, 39–40

Molecular dynamics simulations (cont'd.)
  DLVO potential, 39
  fcc crystal, initial system, 206
  glass transition, 197–201
    constant-pressure simulations, 219
    diffusion constant, 215
    dynamical properties, 213–218
    enthalpy per particle, 211–212
    isothermal compressibility, 213
    mean square displacement function, 213–215
    Newtonian equations of motion, 206
    pair distribution function, 206–210
    physical properties, 205–218
    pressure, 211
    quench rate, 206
    self-diffusion constant, 213
    shear viscosity, 216–217
    specific heat, 212
    stress autocorrelation function, 216–217
    structural properties, 206–210
    structure factor, 206–210
    temperature, 210–211
    thermodynamical properties, 210–213
    velocity autocorrelation function, 213–218
    Verlet algorithm, 206
    V-T curve, 210–211
    Wendt–Abraham parameter, 209–210
  liquid crystallization, 158–160
  Monte Carlo simulation, 36–39
  phase diagram, 37–39
  polyballs, 37–39
    binary charged colloidal suspensions, 61
  polydispersity, 53–54
  quenching, Lennard–Jones system, 202–205, 219–224
  relaxation processes, atomic configurations in glasses, 197–201
  steady-state nucleation, 157–159
Monodisperse charged polyball suspensions, glassy phase, 61–62
Monodisperse colloidal suspensions, lattice structures, bcc to fcc, 13–16
Monodisperse nearly hard spherical particles, glassy phase, 62
Monodisperse polyball suspensions, glassy phase, 61–62
Monodispersity, micelle, 8
Monte Carlo simulation
  cluster formation
    canonical ensemble, 156
    free energies, 156
    molecular dynamics, 36–39
    steady-state nucleation, 155–156
Multicomponent system
  glass, 124
  homogeneous nucleation, 90–91
    cluster density, 90
Multiple scattering, macromolecular suspensions, 44
Mush quenching, heterogeneous site removal, 112

N

n-alkane liquids, nucleation rate, 122–123
Nearest-neighbor atom, Voronoi tessellation, 250
Near-freezing measurements
  polydispersity, 51
  structure factor, 50–52
Nearly hard colloidal spheres, 5–6
Nearly hard sphere systems, phase diagram, 16–18
Negative entropy (negentropy), temperature dependence, nucleation rate, 122
Newtonian equations of motion, molecular dynamic technique, glass transitions, 206
Non-glass-formers, glass formation from, 184
Non-metallic systems, undercooling data, 117
Non-network liquids, glass transition, 186–188
Non-network systems, spherical atoms, 190
Non-steady-state nucleation
  cluster distribution, 170–171
  diffusion coefficient, 170–171
  drift coefficient, 170–171
  Fokker–Planck equations, 171
Nucleation
  bcc, steady-state nucleation, 159
  colloid crystals, 18–19
  condensed systems, Becker–Doring formalism, 88–89
  crystal, kinetic theory, 192–193
  fluctuation process, 76
  glasses, 123–154
    experimental studies, 123–154

SUBJECT INDEX

icosahedral phase transformation, 133
interfacial energy, 152
Landau formalism, 162
mechanisms, crystal sizes, 124
of a solid
  solid density expansion, 166–167
  steady-state nucleation, 167–168
  undercooling
phase transformations, field-theoretic theories, 162
postcritical clusters, 85
pseudobinary systems, 150
silicate glasses, cluster size distribution, 153–154
spinodal transformation, 80–81, 160, 164
steady-state nucleation rate, 120
undercooled liquid metals, 142–143
vapor condensation, 92
Nucleation-initiated phase transformation, 81
Nucleation prefactor, 89
Nucleation rate
  classical theory, 82
  classical theory of nucleation, 160
  coated mercury droplets, 118
  definition, 76
  measurement, 117–123
    mercury droplets, coated, 118
  metallic glasses, 127–135
    Hoffman expression, 128–129
  n-alkane liquids, 122–123
  per mole, steady-state nucleation, 87
  saddle-point, steady-state nucleation, 165–166
  silicate glasses, 124–127, 160
    cluster growth rate, 126
    crystal size, 125–126
    two-stage annealing treatment, 124–125
  temperature dependence, negative entropy (negentropy), 122
  viscosity, metallic glasses, 128
  water, 123
  Zeldovich–Frenkel equation, 88
Nucleation theories, liquids, 82
Nucleation theory, classical
  analysis of, 159–161
  kinetic model, 151–154
Nuclei, silicate glasses
  growth rate, 148
  growth time, 148

Nuclei structure, steady-state nucleation, 155
Number density
  fcc crystals, 203–204
  Lennard–Jones liquids, 203–204
Numerical solutions
  cluster evolution, 99
  time-dependent nucleation rate, 99–102

O

One-compound plasma
  coupling constant, 34
  lattice stability, 33
Ordered phase, growth mechanisms, 18–19
Order formation, binary mixtures, 55
Ordering
  colloidal systems, 6
  colloid crystallites
    growth velocity, 18
    time evolution, 18–19
  rod-shaped colloidal particles, 70–71
  sheared colloidal crystals, 69
  synthetic model colloidal systems, 6
Organic liquids, undercooling data, 117
Ornstein–Zernike equation, colloidal liquids, 47
Outer Helmholtz layer, polyball surface, 12

P

Packing fraction, binary mixtures, 56
Packing model, binary mixtures, 55–56
  hard spheres, 56
Pair distribution function
  annealing process, 235–238
  glass transition
    hidden structures, 225–227
    liquid metals, 190
    metal-metalloid glasses, 190
    metallic glasses, 190
    molecular dynamics simulation, 206–210
  hidden structures, Lennard–Jones system, 225–227
  quenching
    dense random packing of hard spheres, 204–205
    Lennard–Jones liquids, 204–205
Pair potential
  alkali metals

Pair potential (cont'd.)
  glass transition, 201
  Stillinger–Weber potential, 224
 glass transition
  covalent potential, 200–201
  Gaussian core potential, 200
  hard core potential, 198
  Lennard–Jones potential, 199–200
  simple ionic potential, 200–201
  soft core potential, 199
Particle density, colloidal suspensions, 6
Particle scattering form factor, macromolecular syspensions, 43
$Pd_{40}Ni_{40}P_{20}$, spherical crystallites, 130
$Pd_{77.5}Cu_6Si_{16}$, spherical crystallites, 130
Percolation theory, free-volume theory, glass transition, 196
Percus–Yevick approximation, structure factor, 47–50
Perturbation theory
 Barker–Henderson, effective hard sphere diameter, 32
 colloidal liquids, 32
Phase boundary, bcc-fcc
 computer simulation versus analytic theories, 37–39
 density functional theory, 39–40
 free energy, 37
 harmonic oscillators, 34–36
 Lindemann criterion, 34, 38
 molecular dynamics computer simulations, 39–40
 phase diagram, 34–36
 Yukawa potential, 34
Phase diagram
 bcc-fcc
  phase boundary, 34–36
  self-consistent phonon theory, 34–36
 experimental, charged polyball suspensions, 40–41
 molecular dynamics, computer simulations, 37–39
 nearly hard sphere systems, 16–18
 PMMA particles, colloidal suspension, 16–17
 polyball suspensions, DLVO potential, 72
 soft sphere systems, polyball suspensions, 12–16
 Yukawa potential
  computer simulation, 37
  polyball suspensions, 72

Phase equilibrium, metastable regions, phase transformations, 79
Phase separation, spinodal mechanism, binary alloy, 80
Phase space, free energies, 181–183
Phase transformation
 computer simulation, steady-state nucleation, 155–159
 field-theoretic theories, nucleation, 162
 metastable regions, 79–80
 nucleation-initiated, 81
 system trajectory, 158
Phase transition
 condensed systems, Gibbs free energy, 83
 crystal-fluid, polydispersity, 52–55
 second order, glass transition, 193
Physical properties
 microscopic structure parameters, computer simulations, 224–233
 molecular dynamics simulation, glass transition, 205–218
PMMA particles, phase diagram, colloidal suspension, 16–17
Poisson–Boltzmann-cell model, effective charge, 26
Poisson–Boltzmann equation
 DLVO potential, 24–25
 effective charge, 26
 repulsive interaction, 20
Poisson–Boltzmann-jellium approximation, effective charge, 26–27
Polyball(s), 3–4
 bcc colloidal structure, 4
 binary charged colloidal suspensions, MD simulations, 61
 binary charged polyballs, MD simulations, 61
 Brownian motion, 4
 colloidal suspensions, 3–4
 effective charge, models, 26–27
 electrostatic repulsion, 21
 fcc colloidal structure, 4
 hard sphere packing model, 56
 molecular dynamics, computer simulations, 37–39
 monodisperse suspension
  bcc-fcc transitions, 13–16
  crystal-liquid transitions, 13–16
  fcc structure, 14
 repulsive interaction, 3–5
 repulsive potential, electrostatic

SUBJECT INDEX

interaction, 72
screened Coulombic repulsion, 3
screening length, 3–4
surface
  DLVO potential, 24–27
  geometrical factor, 22
  interaction potential, 20–21
  outer Helmholtz layer, 12
  repulsive potential, 21–22
  Stern layer, 10
  Yukawa potential, 22
surface charge, 10–12
DLVO theory, repulsive interaction, 19–22
suspensions, see Polyball suspensions
two-dimensional colloidal crystal, 63
Polyball colloids, emulsion polymerization, 8–10
Polyball suspensions
  charged, experimental phase diagram, 40–41
  colloidal
    bcc-fcc transitions, 30–41
    fluid-crystal transitions, 30–41
    electrostatic interaction, 26–29
    glasslike behavior, 57
    glassy state, 59
    interparticle interactive repulsion, 12–13
    molecular dynamics simulations, 61
    monodisperse, glassy phase, 61–62
    phase diagram
      DLVO potential, 72
      Yukawa potential, 72
    polydispersity, 72
    soft sphere systems, phase diagram, 12–16
    viscoelastic behavior, 67
Polydispersity
  colloidal suspensions, 7
  crystal-fluid phase transition, 52–55
  hard spherical colloids, 54
  molecular dynamics simulations, 53–54
  near-freezing measurements, 51
  polyball suspensions, 72
Polymeric monodisperse particles
  diameter, 8
  emulsion polymerization technique, 8
  synthesis, 8
Polymerization, see Emulsion polymerization
Polymorphically crystallizing liquids, undercooling, 103–117
Polymorphic crystallization, silicate glasses, growth rate, 127
Polymorphic glass, crystallization, 124
Polystyrene spheres, charge-stabilized, aqueous suspensions, 3–5
Postcritical clusters, nucleation, 85
Potential dependence, alkali metals, Stilliger–Weber potential, 222–224
Power-law exponents, melting transitions, two-dimensional colloidal crystals, 65
Power spectrum
  glass transition
    Lennard–Jones fcc crystal, 217
    Lennard–Jones glass, 218
Preferred length scale, glassy state, 61–62
Pressure
  molecular dynamics simulation, glass transition, 211
  quenching, Lennard–Jones system, 203
Prigogine–Defay ratio, glass transition, 193–194
Primary glass, crystallization, 124
Probability flow, saddle-point, steady-state nucleation, 165–166
Pseudobinary systems, nucleation, 150

Q

Quasicrystalline phases, colloidal suspensions, 72
Quasicrystals, metallic glasses, 132–135
Quenched-in distribution, transient nucleation, 171
Quenched-in nuclei, metallic glasses, 130, 134
Quenching
  amorphous solid, preparation, 181
  Lennard–Jones system, 203
    glass transition, 202–205, 219–224
    V-T curve, 219–220
  molecular dynamics method, 202–205, 219–224
  pressure, 203
  temperature, 203
  pair distribution functions
    dense random packing of hard spheres, 204–205
    Lennard–Jones liquids, 204–205
Quench rate, 130
  atomic configuration
    double-shell icosahedra, 251–252

icosahedra, 249–253
cluster distribution, 171
crystal-forming, Lennard–Jones system, 222
fcc clusters, icosahedra, 251–252
glass A, 246
glass formation, 192–193
glass transition temperature, 202
hcp clusters, icosahedra, 251–252
molecular dynamics simulation, glass transition, 206
relaxation, glass transition, 246–249
Stillinger–Weber potential
  bcc crystal, 224
  glass transition, 224
temperature
  crystallization, 221
  glass transition, 220
temperature dependence, glass transition, 188–189
Quench rate dependence, constant-pressure simulations, glass transition, 219
Quench techniques, glass formation from non-glass-formers, 184

R

Rare gases, Lennard–Jones potential, 202
Recrystallization rates, cold-worked aluminum, 99
Reduced radial distribution function, metallic glass, glass transition, 190
Relaxation
  annealing
    glass, 234–246
    Lennard–Jones glass, 234–246
    microscopic structure parameters, 240–243, 246–249
    supercooled liquid, 234–246
  atomic configurations in glasses, molecular dynamics methods, 197–201
  glass transition, quench rate, 246–249
  Lennard–Jones liquid, 234–246
Repulsive interaction
  Debye–Huckel equation, 20
  DLVO theory, polyball surface charge, 19–22
  Poisson–Boltzmann equation, 20
  polyballs, 3–5
Repulsive potential

electrostatic interaction, polyballs, 72
fluid-crystal transition, 30
macroionic solution, 28
polyball surface, 21–22
  geometrical factor, 22
  Yukawa potential, 22
Rescaled mean spherical approximation, structure factor, 47–50
Rod-shaped colloidal particles, 70–71
  anisotropic phases, 70–71
  ordering, 70–71
Rotational partition function, vapor condensation, 92
R-tau curve
  microscopic structure parameter, 227
  Wendt–Abraham parameter, 227

S

Saddle-point
  steady-state nucleation
    nucleation rate, 165–166
    probability flow, 165–166
Saddle-point configuration
  free energy surface, 164
  steady-state nucleation, 162, 164
Saddle-point region, steady-state nucleation
  dynamical prefactor, 165–166
  statistical prefactor, 165–166
Saddle transition configuration, free energy surface, 164
Scattered electric field, macromolecular suspensions, 43
Scattering experiments
  colloidal glass, 58–60
  structure factor, 43–52
Scattering wave vector, macromolecular suspensions, 43–46
Screened Coulombic repulsion, polyballs, 3
Screening length, polyballs, 3–4
Seeding, crystallization, 79
Self-consistent harmonic approximation, bcc-fcc crystalline phases, 33–36
Self-consistent phonon theory, bcc-fcc phase diagram, 34–36
Self-diffusion constant, MD simulation, glass transition 213
Shape parameter, distribution of, Voronoi polyhedra, 229–230
Sheared colloidal crystals, ordering, 69

SUBJECT INDEX

Shear experiments, Couette cell, 70
Shear melting, colloidal crystal, 68–70
Shear-melting transitions, colloidal crystals, 68–70
Shear modulus
  colloidal crystals, 66–67
  colloid glass, 57
Shear stress, colloidal crystals, 68–70
Shear viscosity, MD simulation, glass transition, 216–217
Silicate glasses
  cluster size distribution, nucleation, 153–154
  critical sizes, time-dependent nucleation, 126
  crystal size, nucleation rates, 125–126
  devitrification, 135
  Fulcher–Vogel expression, 140
  growth rate, polymorphic crystallization, 127
  growth time, nuclei, 148
  homogeneous nucleation, 135–151
  nucleation, cluster size distributions, 153–154
  nucleation rate, 124–127, 160
    cluster growth rate, 126
  nuclei, growth rate, 148
  steady-state nucleation rate, 135–145, 146
  surface nucleation, 150–151
  transient nucleation, 141, 145–149
    activation energies, 147–148
    time lag, 146–149
  transient times, 149
  two-state annealing treatment, nucleation rates, 124–125
  volume nucleation, 150–151
    rates, 149–150
Simple ionic potential, glass transition, 200–201
  pair potential, 200–201
Singh–Holz approximation, free-energy, 141
Soft colloidal spheres, 3–5
Soft core potential, glass transition, 198–199
  pair potential, 199
Soft sphere systems, phase diagram, polyball suspensions, 12–16
Sogami potential, 28
  colloidal structures, approximations, 48
Solid density expansion, nucleation of a solid, 166–167

Solidification rate, Gallium data, 122
Specific heat, MD simulation, glass transition, 212
Sphere systems, interparticle potentials, 5–6
Spherical atoms, non-network systems, 190
Spherical clusters of crystal, homogeneous nucleation, 83–91
Spherical crystallites
  metallic glasses, 130
  $Pd_{40}Ni_{40}P_{20}$, 130
  $Pd_{77.5}Cu_6Si_{16}$, 130
Spherical particles, Lennard–Jones system, glass transition, 202
Spinodal curve, 80
Spinodal mechanism
  binary alloy, 80
  droplet morphology, 81
  phase separation, binary alloy, 80
Spinodal transformation, nucleation, 80–81, 160, 164
Stability
  bcc crystal phase, 38–39
  lattice structure, 33, 39
Statistical factor, hydrodynamical treatment, steady-state nucleation rate, 169
Statistical prefactor, saddle-point region, steady-state nucleation, 165–166
Steady-state nucleation
  bcc nucleation, 159
  classical theory, 83–93
  cluster growth, 85
  cluster size and shape, fluctuation, 164
  cluster sizes, 87–88
  coarse-grained free-energy density, 163, 165
  computer simulations, 155–159
    Monte Carlo calculations, 155–156
    molecular dynamics, 157–159
  critical cluster formation, 164
  critical cluster size, 85
  crystal growth, structure factor, 158
  crystal structure, Voronoi polyhedra, 158–160
  energy barrier, 163–164
  equilibrium cluster size distribution, 84
  field-theoretic models, 162–169
    critical cluster, 162
    energy barrier to nucleation, 163–164
    free-energy density functional, 162
  free energy

Steady-state nucleation (*cont'd.*)
 capillarity approximation, 168
 density functional model, 168
 undercooling, 168
 free-energy functional, 163
 Gibbs free energy density functional, 162
 hydrodynamic model, 161, 166, 168–169
 liquid-solid, undercooling, 167–168
 metastable and stable configuration, 162
 nucleation of a solid, undercooling, 167–168
 nucleation rate per mole, 87
 nuclei structure, 155
 saddle-point
  configuration, 162, 164
  nucleation rate, 165–166
  probability flow, 165–166
 saddle-point region
  dynamical prefactor, 165–166
  statistical prefactor, 165–166
 time-dependent cluster density, 86
 undercooling, cluster formation, 168
Steady-state nucleation rate, 87–90
 hydrodynamical treatment
  dynamical factor, 169
  statistical factor, 169
 metallic glasses, 128, 130
 nucleation data and, 120
 silicate glasses, 135–145, 146
 temperature-dependent, lithium disilicate, 137–140
 undercooling, 103
 Zeldovich factor, 88
Stern layer, polyball surface, 10
Stilliger–Weber potential
 alkali metals, 222–224
  pair potential, 224
  potential dependence, 222–224
 bcc crystal, quench rate, 224
 glass transition, quench rate, 224
Stokes–Einstein relation, viscosity, metallic glasses, 128
Stress autocorrelation function, MD simulation, glass transition, 216–217
Strong network liquids, glass transition, 186–188
Structural properties
 glass transition, 206–210
 glass transition temperature, 209–210
 molecular dynamics simulation, 206–210
 two-dimensional colloidal crystals, 63–64
Structure factor
 binary mixtures, glassy state, 59–60
 closure approximation, 48
 colloidal suspensions, 42–52
 crystal(bcc)-liquid interface, 50
 dynamic
  colloidal glass, 62
  macromolecular suspensions, 44–46
 freezing transitions, 50–52
 hypernetted chain approximation, 47–50
 interaction potential, 48
 Kirkwood superposition approximation, 46–47
 mean spherical approximation, 47–50
 molecular dynamics simulation, glass transition, 206–210
 near-freezing measurements, 50–52
 Percus–Yevick approximation, 47–50
 rescaled mean spherical approximation, 47–50
 scattering experiments, 43–52
 steady-state nucleation, crystal growth, 158
Structure relaxation
 annealing, crystallization, 235–240
 diffusion constant, glass transition, 215
Substrate technique, heterogeneous site removal, 110
Supercooled liquid, annealing, relaxation, 234–246
Supersaturated solutions, metastability limit, 79
Supramolecules, colloidal suspensions, 6
Surface charge
 colloidal particle, 10–12
 conductivity, 10
 polyball, 10–12
Surface crystallization, devitrification, metallic glasses, 130–132
Surface nucleation
 metallic glasses, devitrification, 130–132
 silicate glasses, 150–151
Surface potential, colloidal particle, 10–12
Suspensions
 binary mixtures, hard sphere colloidal particles, 56
 charged latex particles, Kirkwood–Alder transition, 31
System trajectory, phase transformation, 158

T

temperature
  annealing, metallic glasses, 130
  crystallization, quench rate, 221
  glass transition, quench rate, 220
  molecular dynamics simulation, glass transition, 210–211
  quenching, Lennard–Jones system, 203
Temperature dependence
  glass transition, quench rate, 188–189
  nucleation rate, negative entropy (negentropy), 122
  steady-state nucleation rates, lithium disilicate, 137–140
Temperature range, condensed matter, 183–184
Tessellation, amorphous structures, 227
Thermodynamic barrier, heterogeneous nucleation, 91
Thermodynamic fluctuations
  cluster formation, 84
  glass transition, 193–194, 210–213
    Lennard–Jones fcc crystal, 210–211
    Lennard–Jones glasses, 210–211
    molecular dynamics simulation, 210–213
  glass transition temperature, 210–213
Thermodynamics, quenching, Lennard–Jones system, 203
Time-dependent cluster density, 86
  steady-state nucleation, 86
Time-dependent critical cluster density, 95
Time-dependent flux, cluster sizes, 87
Time-dependent nucleation, 78
  classical theory, 93–103
  glasses, 123
    lithium disilicate glass, 94–95, 100
  silicate glasses, 93, 96
    critical sizes, 127
  transient nucleation, 93
Time-dependent nucleation rate
  analytical solutions, 96–99
  cluster distributions, 102
  critical size, 96, 99
  numerical solutions, 99–102
Time-dependent steady-state nucleation, configuration, 165
Time evolution, ordering, colloid crystallites, 18–19

Time lag, transient nucleation
  lithium disilicate, 149
  silicate glasses, 146–149
Tolman approximation, curvature-dependent interfacial energy, 161
Transient nucleation, *see also* Time-dependent nucleation
  cluster formation, 96
  lithium disilicate, time lag, 149
  lithium disilicate glass, 95
  metallic glasses, 128–132, 134
  quenched-in distribution, 171
  silicate glasses, 141, 145–149
    activation energies, 147–148
    time lag, 146–149
  time-dependent nucleation, 93
Transient time, 98, 148
  silicate glasses, 149
Transitions
  bcc-fcc
    colloidal polyball suspensions, 30–41
    polyball monodisperse suspensions, 13–16
  crystal (ordered), 13–16
  crystal–liquid, polyball monodisperse suspensions, 13–16
  liquid (disordered), 13–16
Translational correlation function
  melting transitions, two-dimensional colloidal crystals, 65
  vapor condensation, 92
Two-dimensional colloidal crystals, 62–65
  colloidal crystal-glass surface, 63
  melting transitions, 64–65
    bond-orientational correlation function, 65
    correlation lengths, 65
    hexatic phase, 64–65
    Kasterlitz–Thouless–Halperin–Nelson–Young theory, 64
  polyballs, 63
  structural transitions, 63–64
  wedge geometry, 63–63
Two-stage annealing treatment, silicate glasses, nucleation rates, 124–125

U

Undercooled liquid metals, nucleation, 142–143

Undercooled liquids, Gibbs free energy, 76
Undercooled liquids and gases,
  homogeneous nucleation, 77
Undercooling
  cluster formation, steady-state nucleation, 168
  degree of, droplet size and, 113
  free energy, steady-state nucleation, 168
  heterogeneous site removal, 104–117
    liquid metals, 104–117
  homogeneous nucleation, 113–114
  liquid metals, 103–104
    heterogeneous site removal, 104–117
    viscosity, 115
  liquid–solid, steady-state nucleation, 167–168
  maximum
    melting temperature and, 113–115
    values, 113–117
  mercury emulsions, 76
  nonmetallic systems, 117
  nucleation of a solid, steady-state nucleation, 167–168
  organic liquids, 117
  polymorphically crystallizing liquids, 103–117
  steady-state nucleation rate, 103
  water crystallization, 78

V

van der Waals volume, free-volume theory, 195
Vapor condensation
  cluster rotations, 92
  cluster translations, 92
  nucleation, 92
  rotational partition function, 92
  translational partition function, 92
Velocity autocorrelation function, glass transition
  Lennard–Jones fcc crystal, 217
  molecular dynamics simulation, 213–218
  vibrational behaviour, 215, 217, 218
Verlet algorithm, molecular dynamic method, glass transitions, 206
Vibrational behaviour, velocity autocorrelation function, glass transition, 215, 217–218
Virial theorem, 203

Viscoelastic behavior, polyball suspensions, 67
Viscosity
  glass transition, 187
    Arrhenius rate law, 187–188
    Vogel–Fulcher law, 187–188
  lithium disilicate, 140
  metallic glasses
    nucleation rate, 128
    Stokes–Einstein relation, 128
  undercooling, liquid metals, 115
Vogel–Fulcher law, viscosity, glass transition, 187–188
Vogel–Fulcher relation, free-volume theory, glass transition, 195–196
Volume nucleation
  Kashchiev expression, 148
  silicate glasses, 150–151
Volume nucleation rates, silicate glasses, 149–150
Volume nuclei, Kashchiev expression, 146
Voronoi polyhedra
  amorphous structures
    distortions, 229–230
    distribution of shape parameters, 229–230
    face parameter, 231–233
    fivefold symmetry, 232–233
    free volume, 228, 230
    local symmetry, 230–233
    shape parameter, 229
  annealing
    distortion parameter, 240–243
    distribution of volumes, 243–246
    face parameter, 240–243
    fivefold symmetry, 242–243
    fluctuation parameter, 243
    local symmetry parameter, 240–243
  crystal structure
    steady-state nucleation, 158–160
    Wigner–Seitz cell, 158–159
  indices, fcc structure, 230
  microscopic structure parameter, computer simulations, 227–228
Voronoi tessellation, nearest-neighbor atom, 250
V-T curve, glass transition
  molecular dynamics simulation, 210, 211
  rapid quenching of Lennard–Jones system, 219–220
V-T relation, annealing, 234–235

W

Water, nucleation rate, 123
Water crystallization, undercooling, 78
Wedge geometry, two-dimensional colloidal crystals, 63–64
Wendt–Abraham parameter
  molecular dynamics simulation, glass transition, 209–210
  R-tau curve, 227
Wigner crystals, bcc, 32
Wigner glass, 57
Wigner–Seitz cell, Voronoi polyhedra, crystal structure, 158–159, 227

Y

Young modulus, colloidal crystals, 66–67

Yukawa potential
  bcc-fcc phase boundary, 34
  phase diagram
    computer simulation, 37
    polyball suspensions, 72
  repulsive potential, polyball surface, 22

Z

Zachariasen's model, continuous random network, 185
Zeldovich–Frenkel equation, 170–171
  nucleation rates, 88
Zeldovich expression, 100
Zeldovich factor, 88, 166
  steady-state nucleation rate, 88

Cumulative Author Index, Volumes 1–45

A

Abrikosov, A. A.: Supplement 12—Introduction to the Theory of Normal Metals

Adler, David: Insulating and Metallic States in Transition Metal Oxides, **21**, 1

Adrian, Frank J.: *see* Gouary, B. S.

Akamatu, Hideo: *see* Inokuchi, H.

Alexander, H., and Haasen, P.: Dislocations and Plastic Flow in the Diamond Structure, **22**, 28

Allen, Philip B., and Mitrović, Božidar: Theory of Superconducting T_c, **37**, 1

Amelinckx, S., and Dekeyser, W.: The Structure and Properties of Grain Boundaries, **8**, 327

Amelinckx, S.: Supplement 6—The Direct Observation of Dislocations

Anderson, Philip W.: Theory of Magnetic Exchange Interactions: Exchange in Insulators and Semiconductors, **14**, 99

Appel, J.: Polarons, **21**, 193

Ashcroft, N. W., and Stroud, D.: Theory of the Thermodynamics of Simple Liquid Metals, **33**, 1

B

Bastard, G., Brum, J. A., and Ferreira, R.: Electronic States in Semiconductor Heterostructures, **44**, 229

Becker, J. A.: Study of Surfaces by Using New Tools, **7**, 379

Beenakker, C. W. J., and van Houten, H.: Quantum Transport in Semiconductor Nanostructures, **44**, 1

Beer, Albert C.: Supplement 4—Galvanomagnetic Effects in Semiconductors

Bendow, Bernard: Multiphonon Infrared Absorption in the Highly Transparent Frequency Regime of Solids, **33**, 249

Beyers, R., and Shaw, T. M.: The Structure of $Y_1Ba_2Cu_3O_{7-\delta}$ and Its Derivatives, **42**, 135

Blatt, Frank J.: Theory of Mobility of Electrons in Solids, **4**, 200

Blount, E. I.: Formalisms of Band Theory, **13**, 306

Borelius, G.: Changes of State of Simple Solid and Liquid Metals, **6**, 65

Borelius, G.: The Changes in Energy Content, Volume, and Resistivity with Temperature in Simple Solids and Liquids, **15**, 1

Bouligand, Y.: Liquid Crystals and Their Analogs in Biological Systems, *in* Supplement 14—Liquid Crystals, 259

Boyce, J. B.: *see* Hayes, T. M.

Brill, R.: Determination of Electron Distribution in Crystals by Means of X-Rays, **20**, 1

Brown, E.: Aspects of Group Theory in Electron Dynamics, **22**, 313

Brown, Frederick C.: Ultraviolet Spectroscopy of Solids with the Use of Synchrotron Radiation, **29**, 1

Brum, J. A.: *see* Bastard, G.

Bube, Richard H.: Imperfection Ionization Energies in CdS-Type Materials by Photoelectronic Techniques, **11**, 223

Bullett, D. W.: The Renaissance and Quantitative Development of the Tight-Binding Method, **35**, 129

Bundy, F. P., and Strong, H. M.: Behavior of Metals at High Temperatures and Pressures, **13**, 81

Busch, G., and Güntherodt, H.-J.: Electronic Properties of Liquid Metals and Alloys, **29**, 235

Busch, G. A., and Kern, R.: Semiconducting Properties of Gray Tin, **11**, 1

C

Callaway, Joseph: Electron Energy Bands in Solids, 7, 100

Callaway, J., and March, N. H.: Density Functional Methods: Theory and Applications, 38, 136.

Cardona, Manuel: Supplement 11—Optical Modulation Spectroscopy of Solids

Cargill, G. S., III: Structure of Metallic Alloy Glasses, 30, 227

Carlsson, A. E.: Beyond Pair Potentials in Elemental Transition Metals and Semiconductors, 43, 1

Charvolin, Jean, and Tardieu, Annette: Lyotropic Liquid Crystals: Structures and Molecular Motions, in Supplement 14—Liquid Crystals, 209

Chou, M. Y.: see de Heer, Walt A.

Clendenen, R. L.: see Drickamer, H. G.

Cohen, Jerome B.: The Internal Structure of Guinier–Preston Zones in Alloys, 39, 131

Cohen, M. H., and Reif, F.: Quadrupole Effects in Nuclear Magnetic Resonance Studies of Solids, 5, 322

Cohen, Marvin L., and Heine, Volker: The Fitting of Pseudopotentials to Experimental Data and Their Subsequent Application, 24, 38

Cohen, Marvin L.: see de Heer, Walt A.

Cohen, Marvin L.: see Joannopoulos, J. D.

Compton, W. Dale, and Rabin, Herbert: F-Aggregate Centers in Alkali Halide Crystals, 16, 121

Conwell, Esther M.: Supplement 9—High Field Transport in Semiconductors

Cooper, Bernard R.: Magnetic Properties of Rare Earth Metals, 21, 393

Corbett, J. W.: Supplement 7—Electron Radiation Damage in Semiconductors and Metals

Corciovei, A., Costache, G., and Vamanu, D.: Ferromagnetic Thin Films, 27, 237

Costache, G.: see Corciovei, A.

Currat, R., and Janssen, T.: Excitations in Incommensurate Crystal Phases, 41, 202

D

Dalven, Richard: Electronic Structure of PbS, PbSe, and PbTe, 28, 179

Das, T. P., and Hahn, E. L.: Supplement 1—Nuclear Quadrupole Resonance Spectroscopy

Davison, S. G., and Levine, J. D.: Surface States, 25, 1

Dederichs, P. H.: Dynamical Diffraction Theory by Optical Potential Methods, 27, 136

de Fontaine, D.: Configurational Thermodynamics of Solid Solutions, 34, 74

de Gennes, P. G.: Macromolecules and Liquid Crystals: Reflections on Certain Lines of Research, in Supplement 14—Liquid Crystals, 1

de Heer, Walt A., Knight, W. D., Chou, M. Y., and Cohen, Marvin L.: Electronic Shell Structure and Metal Clusters, 40, 94

de Jeu, W. H.: The Dielectric Permittivity of Liquid Crystals, in Supplement 14—Liquid Crystals, 109

Dekeyser, W.: see Amelinckx, S.

Dekker, A. J.: Secondary Electron Emission, 6, 251

de Launay, Jules: The Theory of Specific Heats and Lattice Vibrations, 2, 220

Deuling, H. J.: Elasticity of Nematic Liquid Crystals, in Supplement 14—Liquid Crystals, 77

Devreese, J. T.: see Peeters, F. M.

de Wit, Roland: The Continuum Theory of Stationary Dislocations, 10, 249

Dexter, D. L.: Theory of the Optical Properties of Imperfections in Nonmetals, 6, 355

Dimmock, J. O.: The Calculation of Electronic Energy Bands by the Augmented Plane Wave Method, 26, 104

Doran, Donald G., and Linde, Ronald K.: Shock Effects in Solids, 19, 230

Drickamer, H. G.: The Effects of High Pressure on the Electronic Structure of Solids, 17, 1

Drickamer, H. G., Lynch, R. W., Clendenen, R. L., and Perez-Albuerne, E. A.: X-Ray Diffraction Studies of the Lattice Parameters of Solids under Very High Pressure, 19, 135

Dubois-Violette, E., Durand, G., Guyon, E., Manneville, P., and Pieranski, P.: Instabilities in Nematic Liquid Crystals, in Supplement 14—Liquid Crystals, 147

Duke, C. B.: Supplement 10—Tunneling in Solids

Durand, G.: see Dubois-Violette, E.

E

Echenique, P. M., Flores, F., and Ritchie, R. H.: Dynamic Screening of Ions in Condensed Matter, **43**, 230

Ehrenreich, H., and Schwartz, L. M.: The Electronic Structure of Alloys, **31**, 150

Einspruch, Norman G.: Ultrasonic Effects in Semiconductors, **17**, 217

Eshelby, J. D.: The Continuum Theory of Lattice Defects, **3**, 79

F

Fan, H. Y.: Valence Semiconductors, Germanium and Silicon, **1**, 283

Ferreira, R.: see Bastard, G.

Flores, F.: see Echenique, P. M.

Frederikse, H. P. R.: see Kahn, A. H.

Fulde, Peter, Keller, Joachim, and Zwicknagl, Gertrud: Theory of Heavy Fermion Systems, **41**, 1

G

Galt, J. K.: see Kittel, C.

Geballe, Theodore H.: see White, Robert M.

Gilman, J. J., and Johnston, W. G.: Dislocations in Lithium Fluoride Crystals, **13**, 148

Givens, M. Parker: Optical Properties of Metals, **6**, 313

Glicksman, Maurice: Plasmas in Solids, **26**, 275

Goldberg, I. B.: see Weger, M.

Gomer, Robert: Chemisorption on Metals, **30**, 94

Gouray, Barry S., and Adrian, Frank J.: Wave Functions for Electron-Excess Color Centers in Alkali Halide Crystals, **10**, 128

Gschneidner, Karl A., Jr.: Physical Properties and Interrelationships of Metallic and Semimetallic Elements, **16**, 275

Guinier, André: Heterogeneities in Solid Solutions, **9**, 294

Güntherodt, H.-J.: see Busch, G.

Guttman, Lester: Order-Disorder Phenomena in Metals, **3**, 146

Guyer, R. A.: The Physics of Quantum Crystals, **23**, 413

Guyon, E.: see Dubois-Violette, E.

H

Haasen, P.: see Alexander, H.

Hahn, E. L.: see Das, T. P.

Halperin, B. I., and Rice, T. M.: The Excitonic State at the Semiconductor-Semimetal Transition, **21**, 116

Ham, Frank S.: The Quantum Defect Method, **1**, 127

Hashitsume, Natsuki: see Kubo, R.

Hass, K. C.: Electronics Structure of Copper-Oxide Superconductors, **42**, 213

Haydock, Roger: The Recursive Solution of the Schrödinger Equation, **35**, 216

Hayes, T. M., and Boyce, J. B.: Extended X-Ray Absorption Fine Structure Spectroscopy, **37**, 173

Hebel, L. C., Jr.: Spin Temperature and Nuclear Relaxation in Solids, **15**, 409

Hedin, Lars, and Lundqvist, Stig: Effects of Electron-Electron and Electron-Phonon Interactions on the One-Electron States of Solids, **23**, 2

Heeger, A. J.: Localized Moments and Nonmoments in Metals: The Kondo Effect, **23**, 284

Heer, Ernst, and Novey, Theodore B.: The Interdependence of Solid State Physics and Angular Distribution of Nuclear Radiations, **9**, 200

Heiland, G., Mollwo, E., and Stöckmann, F.: Electronic Processes in Zinc Oxide, **8**, 193

Heine, Volker: see Cohen, M. L.

Heine, Volker: The Pseudopotential Concept, **24**, 1

Heine, Volker, and Weaire, D.: Pseudopotential Theory of Cohesion and Structure, **24**, 250

Heine, Volker: Electronic Structure from the Point of View of the Local Atomic Environment, **35**, 1

Hensel, J. C., Phillips, T. G., and Thomas, G. A.: The Electron-Hole Liquid in Semiconductors: Experimental Aspects, **32**, 88

Herzfeld, Charles M., and Meijer, Paul H. E.: Group Theory and Crystal Field Theory, **12**, 2

Hideshima, T.: see Saito, N.

Huebener, R. P.: Thermoelectricity in Metals and Alloys, **27**, 64

Huntington, H. B.: The Elastic Constants of Crystals, **7**, 214

Hutchings, M. T.: Point-Charge Calculations of Energy Levels of Magnetic Ions in Crystalline Electric Fields, **16**, 227

I

Inokuchi, Hiroo, and Akamatu, Hideo: Electrical Conductivity of Organic Semiconductors, **12**, 93
Ipatova, I. P.: *see* Maradudin, A. A.
Isihara, A.: Electron Correlations in Two Dimensions, **42**, 271
Iwayanagi, S.: *see* Saito, N.

J

James, R. W.: The Dynamical Theory of X-Ray Diffraction, **15**, 55
Jan, J.-P.: Galvanomagnetic and Thermomagnetic Effects in Metals, **5**, 3
Janssen, T.: *see* Currat, R.
Jarrett, H. S.: Electron Spin Resonance Spectroscopy in Molecular Solids, **14**, 215
Joannopoulos, J. D., and Cohen, Marvin L.: Theory of Short-Range Order and Disorder in Tetrahedrally Bonded Semiconductors, **31**, 71
Johnston, W. G.: *see* Gilman, J. J.
Jørgensen, Kluxbüll, Chr.: Chemical Bonding Inferred from Visible and Ultraviolet Absorption Spectra, **13**, 376
Joshi, S. K., and Rajagopal, A. K.: Lattice Dynamics of Metals, **22**, 160

K

Känzig, Werner: Ferroelectrics and Antiferroelectrics, **4**, 5
Kahn, A. H., and Frederikse, H. P. R.: Oscillatory Behavior of Magnetic Susceptibility and Electronic Conductivity, **9**, 257
Keller, Joachim: *see* Fulde, P.
Keller, P. and Liebert, L.: Liquid-Crystal Synthesis for Physicists, *in* Supplement 14—Liquid Crystals, 19
Kelly, M. J.: Applications of the Recursion Method to the Electronic Structure from an Atomic Point of View, **35**, 296

Kelton, K. F.: Crystal Nucleation in Liquids and Glasses, **45**, 75
Kern, R.: *see* Busch, G. A.
Keyes, Robert W.: The Effects of Elastic Deformation on the Electrical Conductivity of Semiconductors, **11**, 149
Keyes, Robert W.: Electronic Effects in the Elastic Properties of Semiconductors, **20**, 37
Kittel, C., and Galt, J. K.: Ferromagnetic Domain Theory, **3**, 439
Kittel, C.: Indirect Exchange Interactions in Metals, **22**, 1
Klemens, P. G.: Thermal Conductivity and Lattice Vibrational Modes, **7**, 1
Klick, Clifford C., and Schulman, James H.: Luminescence in Solids, **5**, 97
Knight, W. D.: Electron Paramagnetism and Nuclear Magnetic Resonance in Metals, **2**, 93
Knight, W. D.: *see* de Heer, Walt A.
Knox, Robert S.: Bibliography of Atomic Wave Functions, **4**, 413
Knox, R. S.: Supplement 5–Theory of Excitons
Koehler, J. S.: *see* Seitz, F.
Kohn, W.: Shallow Impurity States in Silicon and Germanium, **5**, 258
Kondo, J.: Theory of Dilute Magnetic Alloys, **23**, 184
Koster, G. F.: Space Groups and Their Representations, **5**, 174
Kothari, L. S., and Singwi, K. S.: Interaction of Thermal Neutrons with Solids, **8**, 110
Kröger, F. A., and Vink, H. J.: Relations between the Concentrations of Imperfections in Crystalline Solids, **3**, 310
Kubo, Ryogo, Miyake, Satoru J., and Hashitsume, Natsuki: Quantum Theory of Galvanomagnetic Effect at Extremely Strong Magnetic Fields, **17**, 270
Kwok, Philip C. K.: Green's Function Method in Lattice Dynamics, **20**, 214

L

Lagally, M. G.: *see* Webb, M. B.
Lang, Norton D.: The Density-Functional Formalism and the Electronic Structure of Metal Surfaces, **28**, 225
Laudise, R. A., and Nielsen, J. W.: Hydrothermal Crystal Growth, **12**, 149

Lax, Benjamin, and Mavroides, John G.: Cyclotron Resonance, **11**, 261

Lazarus, David: Diffusion in Metals, **10**, 71

Leibfried, G., and Ludwig, W.: Theory of Anharmonic Effects in Crystals, **12**, 276

Levine, J. D.: *see* Davison, S. G.

Lewis, H. W.: Wave Packets and Transport of Electrons in Metals, **7**, 353

Liebert, L.: *see* Keller, P.

Linde, Ronald K.: *see* Doran, D. G.

Liu, S. H.: Fractals and Their Applications in Condensed Matter Physics, **39**, 207

Lobb, C. J.: *see* Tinkham, M.

Low, William: Supplement 2—Paramagnetic Resonance in Solids

Low, W., and Offenbacher, E. L.: Electron Spin Resonance of Magnetic Ions in Complex Oxides. Review of ESR Results in Rutile, Perovskites, Spinel, and Garnet Structures, **17**, 136

Ludwig, G. W., and Woodbury, H. H.: Electron Spin Resonance in Semiconductors, **13**, 223

Ludwig, W.: *see* Leibfried, G.

Lundqvist, Stig: *see* Hedin, L.

Lynch, R. W.: *see* Drickamer, H. G.

M

McClure, Donald S.: Electronic Spectra of Molecules and Ions in Crystals. Part I. Molecular Crystals, **8**, 1

McClure, Donald S.: Electronic Spectra of Molecules and Ions in Crystals. Part II. Spectra of Ions in Crystals, **9**, 400

McGreevy, Robert L.: Experimental Studies of the Structure and Dynamics of Molten Alkali and Alkaline Earth Halides, **40**, 247

MacKinnon, A.: *see* Miller, A.

MacLaughlin, Douglas E.: Magnetic Resonance in the Superconducting State, **31**, 1

McQueen, R. G.: *see* Rice, M. H.

Mahan, G. D.: Many-Body Effects on X-Ray Spectra of Metals, **29**, 75

Manneville, P.: *see* Dubois-Violette, E.

Maradudin, A. A., Montroll, E. W., Weiss, G. H., and Ipatova, I. P.: Supplement 3—Theory of Lattice Dynamics in the Harmonic Approximation

Maradudin, A. A.: Theoretical and Experimental Aspects of the Effects of Point Defects and Disorder on the Vibrations of Crystals—1, **18**, 274

Maradudin, A. A.: Theoretical and Experimental Aspects of the Effects of Point Defects and Disorder on the Vibrations of Crystals—2, **19**, 1

March, N. H.: *see* Callaway, J.

Markham, Jordan J.: Supplement 8—F-Centers in Alkali Halides

Mavroides, John G.: *see* Lax, B.

Meijer, Paul H. E.: *see* Herzfeld, C. M.

Mendelssohn, K., and Rosenberg, H. M.: The Thermal Conductivity of Metals at Low Temperatures, **12**, 223

Miller, A., MacKinnon, A., and Weaire, D.: Beyond the Binaries—The Chalcopyrite and Related Semiconducting Compounds, **36**, 119

Mitra, Shashanka S.: Vibration Spectra of Solids, **13**, 1

Mitrović, Božidar: *see* Allen, P. B.

Miyake, Satoru J.: *see* Kubo, R.

Mollwo, E.: *see* Heiland, G.

Montgomery, D. J.: Static Electrification of Solids, **9**, 139

Montroll, E. W.: *see* Maradudin, A. A.

Muto, Toshinosuke, and Takagi, Yutaka: The Theory of Order–Disorder Transitions in Alloys, **1**, 194

N

Nagamiya, Takeo: Helical Spin Ordering—1 Theory of Helical Spin Configurations, **20**, 306

Nelson, D. R., and Spaepen, Frans: Polytetrahedral Order in Condensed Matter, **42**, 1

Newman, R., and Tyler, W. W.: Photoconductivity in Germanium, **8**, 50

Nichols, D. K., and van Lint, V. A. J.: Energy Loss and Range of Energetic Neutral Atoms in Solids, **18**, 1

Nielsen, J. W.: *see* Laudise, R. A.

Nilsson, P. O.: Optical Properties of Metals and Alloys, **29**, 139

Novey, Theodore B.: *see* Heer, E.

Nussbaum, Allen: Crystal Symmetry, Group Theory, and Band Structure Calculations, **18**, 165

O

Offenbacher, E. L.: *see* Low, W.
Okano, K.: *see* Saito, N.

P

Pake, G. E.: Nuclear Magnetic Resonance, **2**, 1
Parker, R. L.: Crystal Growth Mechanisms: Energetics, Kinetics, and Transport, **25**, 152
Peercy, P. S.: *see* Samara, G. A.
Peeters, F. M., and Devreese, J. T.: Theory of Polaron Mobility, **38**, 82
Perez-Albuerne, E. A.: *see* Drickamer, H. G.
Peterson, N. L.: Diffusion in Metals, **22**, 409
Pettifor, D. G.: A Quantum-Mechanical Critique of the Miedema Rules for Alloy Formation, **40**, 43
Pfann, W. G.: Techniques of Zone Melting and Crystal Growing, **4**, 424
Phillips, J. C.: The Fundamental Optical Spectra of Solids, **18**, 55
Phillips, J. C.: Spectroscopic and Morphological Structure of Tetrahedral Oxide Glasses, **37**, 93
Phillips, T. G.: *see* Hensel, J. C.
Pieranski, P.: *see* Dubois-Violette, E.
Pines, David: Electron Interaction in Metals, **1**, 368
Piper, W. W., and Williams, F. E.: Electroluminescence, **6**, 96
Platzman, P. M., and Wolff, P. A.: Supplement 13—Waves and Interactions in Solid State Plasmas

R

Rabin, Herbert: *see* Compton, W. D.
Rajagopal, A. K.: *see* Joshi, S. K.
Rasolt, M.: Continuous Symmetries and Broken Symmetries in Multivalley Semiconductors and Semimetals, **43**, 94
Reif, F.: *see* Cohen, M. H.
Reitz, John R.: Methods of the One-Electron Theory of Solids, **1**, 1
Rice, M. H., McQueen, R. G., and Walsh, J. M.: Compression of Solids by Strong Shock Waves, **6**, 1

Rice, T. M.: *see* Halperin, B. I.
Rice, T. M.: The Electron-Hole Liquid in Semiconductors: Theoretical Aspects, **32**, 1
Ritchie, R. H.: *see* Echenique, P. M.
Roitburd, A. L.: Martensitic Transformation as a Typical Phase Transformation in Solids, **33**, 317
Rosenberg, H. M.: *see* Mendelssohn, K.

S

Safran, S. A.: Stage Ordering in Intercalation Compounds, **40**, 183
Saito, N., Okano, K., Iwayanagi, S., and Hideshima, T.: Molecular Motion in Solid State Polymers, **14**, 344
Samara, G. A.: High-Pressure Studies of Ionic Conductivity in Solids, **38**, 1
Samara, G. A., and Peercy, P. S.: The Study of Soft-Mode Transitions at High Pressure, **36**, 1
Scanlon, W. W.: Polar Semiconductors, **9**, 83
Schafroth, M. R.: Theoretical Aspects of Superconductivity, **10**, 295
Schnatterley, S. E.: Inelastic Electron Scattering Spectroscopy, **34**, 275
Schulman, James H.: *see* Klick, C. C.
Schwartz, L. M.: *see* Ehrenreich, H.
Seitz, Frederick: *see* Wigner, E. P.
Seitz, Frederick, and Koehler, J. S.: Displacement of Atoms during Irradiation, **2**, 307
Sellmyer, D. J.: Electronic Structure of Metallic Compounds and Alloys: Experimental Aspects, **33**, 83
Sham, L. J., and Ziman, J. M.: The Electron-Phonon Interaction, **15**, 223
Shaw, T. M.: *see* Beyers, R.
Shull, C. G., and Wollan, E. O.: Applications of Neutron Diffraction to Solid State Problems, **2**, 138
Singh, Jai: The Dynamics of Excitons, **38**, 295
Singwi, K. S.: *see* Kothari, L. S.
Singwi, K. S., and Tosi, M. P.: Correlations in Electron Liquids, **36**, 177
Slack, Glen A.: The Thermal Conductivity of Nonmetallic Crystals, **34**, 1
Smith, Charles S.: Macroscopic Symmetry and Properties of Crystals, **6**, 175
Sood, Ajay K.: Structural Ordering in Colloidal Suspensions, **45**, 1

Spaepen, Frans: see Nelson, D. R.
Spector, Harold N.: Interaction of Acoustic Waves and Conduction Electrons, **19**, 291
Stern, Frank: Elementary Theory of the Optical Properties of Solids, **15**, 300
Stöckmann, F.: see Heiland, G.
Strong, H. M.: see Bundy, F. P.
Stroud, D.: see Ashcroft, N. W.
Sturge, M. D.: The Jahn–Teller Effect in Solids, **20**, 92
Swenson, C. A.: Physics at High Pressure, **11**, 41

T

Takagi, Yutaka: see Muto, T.
Tardieu, Annette: see Charvolin, Jean
Thomas, G. A.: see Hensel, J. C.
Thomson, Robb: Physics of Fracture, **39**, 1
Tinkham, M., and Lobb, C. J.: Physical Properties of the New Superconductors, **42**, 91
Tosi, Mario P.: Cohesion of Ionic Solids in the Born Model, **16**, 1
Tosi, M. P.: see Singwi, K. S.
Turnbull, David: Phase Changes, **3**, 226
Tyler, W. W.: see Newman, R.

V

Vamanu, D.: see Corciovei, A.
van Houten, H.: see Beenakker, C. W. J.
van Lint, V. A. J.: see Nichols, D. K.
Vink, H. J.: see Kröger, F. A.

W

Wallace, Duane C.: Thermoelastic Theory of Stressed Crystals and Higher-Order Elastic Constants, **25**, 302
Wallace, Philip R.: Positron Annihilation in Solids and Liquids, **10**, 1
Walsh, J. M.: see Rice, M. H.
Weaire, D.: see Heine, V.
Weaire, D.: see Miller, A.
Weaire, D.: see Wooten, F.
Webb, M. B., and Lagally, M. G.: Elastic Scattering of Low-Energy Electrons from Surfaces, **28**, 302
Weger, M., and Goldberg, I. B.: Some Lattice and Electronic Properties of the β-Tungstens, **28**, 1

Weiss, G. H.: see Maradudin, A. A.
Weiss, H.: see Welker, H.
Welker, H., and Weiss, H.: Group III–Group V Compounds, **3**, 1
Wells, A. F.: The Structures of Crystals, **7**, 426
White, Robert M., and Geballe, Theodore H.: Supplement 15—Long Range Order in Solids
Wigner, Eugene P., and Seitz, Frederick: Qualitative Analysis of the Cohesion in Metals, **1**, 97
Williams, F. E.: see Piper, W. W.
Wokaun, Alexander: Surface-Enhanced Electromagnetic Processes, **38**, 224
Wolf, E. L.: Nonsuperconducting Electron Tunneling Spectroscopy, **30**, 1
Wolf, H. C.: The Electronic Spectra of Aromatic Molecular Crystals, **9**, 1
Wolff, P. A.: see Platzman, P. M.
Wollan, E. O.: see Shull, C. G.
Woodbury, H. H.: see Ludwig, G. W.
Woodruff, Truman O.: The Orthogonalized Plane-Wave Method, **4**, 367
Wooten, F., and Weaire, D.: Modeling Tetrahedrally Bonded Random Networks by Computer, **40**, 1

Y

Yafet, Y.: g Factors and Spin-Lattice Relaxation of Conduction Electrons, **14**, 1
Yeomans, Julia: The Theory and Application of Axial Ising Models, **41**, 151
Yonezawa, Fumiko: Glass Transition and Relaxation of Disordered Structures, **45**, 179

Z

Zak, J.: The kq-Representation in the Dynamics of Electrons in Solids, **27**, 1
Zheludev, I. S.: Ferroelectricity and Symmetry, **26**, 429
Zheludev, I. S.: Piezoelectricity in Textured Media, **29**, 315
Ziman, J. M.: see Sham, L. J.
Ziman, J. M.: The Calculation of Bloch Functions, **26**, 1
Zunger, Alex: Electronic Structure of $3d$ Transition-Atom Impurities in Semiconductors, **39**, 276
Zwicknagl, Gertrud: see Fulde, P.